Lecture Notes in Computer Science 2210

Edited by G. Goos, J. Hartmanis, and J. van Leeuwen

Lecture Notes in Computer Science 2210
Edited by G. Goos, J. Hartmanis, and J. van Leeuwen

Springer
Berlin
Heidelberg
New York
Barcelona
Hong Kong
London
Milan
Paris
Tokyo

Yong Liu   Kiyoshi Tanaka   Masaya Iwata
Tetsuya Higuchi   Moritoshi Yasunaga (Eds.)

Evolvable Systems: From Biology to Hardware

4th International Conference, ICES 2001
Tokyo, Japan, October 3-5, 2001
Proceedings

 Springer

Volume Editors

Yong Liu
University of Aizu
Tsuruga, Ikki-machi, Aizu-Wakamatsu City, Fukushima 965-8580, Japan
E-mail: yliu@u-aizu.ac.jp

Kiyoshi Tanaka
Shinshu University, Faculty of Engineering
4-17-1 Wakasato, Nagano 380-8553, Japan
E-mail: ktanaka@gipwc.shinshu-u.ac.jp

Masaya Iwata, Tetsuya Higuchi
Nat. Inst. of Adv. Ind. Science and Technology, Adv. Semiconductor Research Center
AIST Tsukuba Central 2, 1-1-1 Umezono, Tsukuba, Ibaraki 305-8568, Japan
E-mail: {m.iwata/t-higuchi}@aist.go.jp

Moritoshi Yasunaga
University of Tsukuba, Institute of Information Sciences and Electronics
Tsukuba, Ibaraki 305-8573, Japan
E-mail: yasunaga@is.tsukuba.ac.jp

Cataloging-in-Publication Data applied for

Die Deutsche Bibliothek - CIP-Einheitsaufnahme

Evolvable systems : from biology to hardware ; 4th international conference ;
proceedings / ICES 2001, Tokyo, Japan, October 3 - 5, 2001. Yong Liu ...
(ed.). - Berlin ; Heidelberg ; New York ; Barcelona ; Hong Kong ; London ;
Milan ; Paris ; Tokyo : Springer, 2001
   (Lecture notes in computer science ; Vol. 2210)
   ISBN 3-540-42671-X

CR Subject Classification (1998): B.6, B.7, F.1, I.6, I.2, J.2, J.3

ISSN 0302-9743
ISBN 3-540-42671-X Springer-Verlag Berlin Heidelberg New York

Springer-Verlag Berlin Heidelberg New York
a member of BertelsmannSpringer Science+Business Media GmbH

http://www.springer.de

© Springer-Verlag Berlin Heidelberg 2001
Printed in Germany

Typesetting: Camera-ready by author, data conversion by Steingräber Satztechnik GmbH, Heidelberg
Printed on acid-free paper   SPIN: 10840795   06/3142   5 4 3 2 1 0

Preface

On behalf of the ICES 2001 Conference Committee, it is our pleasure to present to you the proceedings of the fourth International Conference on Evolvable Systems: From Biology to Hardware, ICES 2001, held in Tokyo, Japan, on 3-5 October 2001, addressing the latest developments and discussing challenges facing the field of evolvable systems.

The idea of evolving machines, whose origins can be traced back to the cybernetics movement of the 1940s and the 1950s, has recently re-emerged in the form of the nascent field of bio-inspired systems and evolvable hardware. Following the workshop, Towards Evolvable Hardware, which took place in Lausanne, Switzerland, in October 1995, the First International Conference on Evolvable Systems: *From Biology to Hardware* (ICES96), was held at the Electrotechnical Laboratory (MITI), Tsukuba, Japan, in October 1996. The second and the third International Conferences on Evolvable Systems: *From Biology to Hardware* (ICES98 and ICES 2000) were respectively held in Lausanne in September 1998, and in Edinburgh in April 2000.

Following the success of these past events, ICES 2001 was dedicated to the promotion and advancement of all aspects of evolvable systems, including hardware, software, algorithms, and applications. By bringing together researchers who use biologically inspired concepts to implement real systems in artificial intelligence, artificial life, robotics, VLSI design, and related domains, ICES 2001 reunited this burgeoning community.

High quality papers were selected and presented at the conference. These contributions cover a wide variety of aspects pertaining to evolvable systems and applications: evolutionary design of electronic circuits, embryonic electronics, biological-based systems, evolutionary robotics, evolutionary optimization, evolutionary learning, and the various applications of evolvable hardware.

We wish to thank all the authors for submitting their work, as well as the Program Committee members and reviewers for their enthusiasm, time, and expertise. Finally, we would like to thank the sponsors, who helped in one way or another to achieve our goals for the conference. These sponsors are Real World Computing Partnership, the Japanese Society for Artificial Intelligence, SIG on AI Challenges, Logic Systems Laboratory (EPFL), and the National Institute of Advanced Industrial Science and Technology (AIST).

August 2001

Yong Liu
Kiyoshi Tanaka
Masaya Iwata
Tetsuya Higuchi
Moritoshi Yasunaga

Organization

ICES 2001 Conference Committee

General chair: Tetsuya Higuchi, National Institute of Advanced Industrial
Science and Technology (AIST), Japan
Program co-chair: Daniel Mange, Swiss Federal Institute of Technology,
Switzerland
Program co-chair: Moritoshi Yasunaga, University of Tsukuba , Japan
Program co-chair: Yong Liu, University of Aizu, Japan
Publicity chair: Kiyoshi Tanaka, Shinshu University, Japan
Finance chair: Hiroshi Okuno, Kyoto University, Japan
Local chair: Masaya Iwata, National Institute of Advanced Industrial Science
and Technology (AIST), Japan
Conference secretary: Ayako Suzuki, National Institute of Advanced Industrial
Science and Technology (AIST), Japan

ICES 2001 Program Committee

Peter Dittrich, University of Dortmund, Germany
Rolf Drechsler, Albert-Ludwigs-University, Germany
Tim Goredon, University College London, U.K.
Pauline Haddow, Norwegian University of Science and Technology (NTNU),
Norway
Alister Hamilton, Edinburgh University, U.K.
Masaya Iwata, National Institute of Advanced Industrial Science and
Technology (AIST), Japan
Isamu Kajitani, National Institute of Advanced Industrial Science and
Technology (AIST), Japan
Tatiana Kalganova, Brunel University, U.K.
Lishan Kang, Wuhan University, P. R. China
Michael Korkin, Genobyte, Inc., U.S.A.
Pierre Marchal, Centre Suisse d'Electronique et de Microtechnique, Switzerland
Karlheinz Meier, University of Heidelberg, Germany
David Montana, BBN Technologies Robert Popp, U.S.A.
Masahiro Murakawa, National Institute of Advanced Industrial Science and
Technology (AIST), Japan
Andres Perez-Uribe, University of Fribourg, Switzerland
Hidenori Sakanashi, National Institute of Advanced Industrial Science and
Technology (AIST), Japan
Eduardo Sanchez, Swiss Federal Institute of Technology, Switzerland
Moshe Sipper, Swiss Federal Institute of Technology, Switzerland
Kiyoshi Tanaka, Shinshu University, Japan
Gianluca Tempesti, Swiss Federal Institute of Technology, Switzerland

Table of Contents

Evolutionary Robotics

Evolutionary Optimization

Evolutionary Learning

Applications

Two-Step Incremental Evolution of a Prosthetic Hand Controller Based on Digital Logic Gates

Jim Torresen

Department of Informatics, University of Oslo,
P.O. Box 1080 Blindern, N-0316 Oslo, Norway,
jimtoer@ifi.uio.no, http://www.ifi.uio.no/~jimtoer

Abstract. Evolvable Hardware (EHW) has been proposed as a new method for designing systems for real-world applications. In this paper it is applied for evolving a prosthetic hand controller. It is shown that better generalization performance than neural networks can be obtained. The proposed architecture is based on digital logic gates and its configuration is determined by two separate steps of evolution.

1   Introduction

To enhance the lives of people who has lost a hand, prosthetic hands have existed for a long time. These are operated by the signals generated by contracting muscles – named electromyography (EMG) signals, in the remaining part of the arm [1]. Presently available systems normally provide only two motions: Open and close hand grip. The systems are based on the user adapting *himself* to a fixed controller. That is, he must train himself to issue muscular motions trigging the wanted action in the prosthetic hand. Long time is often required for rehabilitation.

By using Evolvable Hardware (EHW) it is possible to make the *controller* itself adapt to each disabled person. The controller is constructed as a pattern classification hardware which maps input patterns to desired actions of the prosthetic hand. Adaptable controllers have been proposed based on neural networks [2]. These require a floating point CPU or a neural network chip. However, by using gate level EHW, a much more compact implementation can be provided making it more feasible to be installed inside a prosthetic hand.

Experiments based the EHW approach have already been undertaken by Kajitani et al [3]. The research on adaptable controllers is based on designing a controller providing six different motions in three different degrees of freedom. Such a complex controller could probably only be designed by *adapting* the controller to each dedicated user. It consists of AND gates succeeded by OR gates (Programmable Logic Array). The latter gates are the outputs of the controller, and the controller is evolved as one complete circuit. The simulation indicates a similar performance as artificial neural network but since the EHW controller requires a much smaller hardware it is to be preferred.

One of the main problems in evolving hardware systems seems to be the limitation in the chromosome string length [4,5]. A long string is normally required

Y. Liu et al. (Eds.): ICES 2001, LNCS 2210, pp. 1–13, 2001.

for representing a complex system. However, a larger number of generations is required by genetic algorithms (GA) as the string increases. This often makes the search space too large. Thus, work has been undertaken to try to diminish this limitation. Various experiments on speeding up the GA computation have been undertaken [6]. The schemes involve fitness computation in parallel or a partitioned population evolved in parallel – by parallel computation. Other approaches to the problem have been by using variable length chromosome [7] and reduced genotype representation [8]. Another option, called function level evolution, is to evolve at a higher level than gate level [9]. Most work is based on fixed functions. However, there has been work in Genetic Programming for *evolving* the functions [10]. The method is called Automatically Defined Functions (ADF) and is used in software evolution.

Another improvement to artificial evolution – called co-evolution, has been proposed [11]. In co-evolution, a part of the data, which defines the problem, co-evolves simultaneously with a population of individuals solving the problem. This could lead to a solution with a better generalization than a solution evolved based on the initial data. Further overview of related works can be found in [12].

Incremental evolution for EHW was first introduced in [13] for a character recognition system. The approach is a divide-and-conquer on the evolution of the EHW system, and thus, named *increased complexity evolution*. The goal is to develop a scheme that could evolve systems for complex real-world applications. In this paper, it is applied to the application of a prosthetic hand controller circuit. Several improvements in the EHW architecture as well as how incremental evolution is applied are to be introduced. These should improve the generalization performance of gate level EHW and make it a strong alternative to artificial neural networks.

The next two sections introduce the concepts of the evolvable hardware based prosthetic hand controller. Results are given in Section 4 with conclusions in Section 5.

2   Prosthetic Hand Control

The research on adaptable controllers presented in this paper is based on designing controllers providing six different motions in three different degrees of freedom: Open and Close hand, Extension and Flection of wrist, Pronation and Supination of wrist. The data set consists of the same motions as used in earlier work [3], and it is collected by Dr. Kajitani at Electrotechnical Laboratory in Japan.

The published results on adaptive controllers are usually based on data for non-disabled persons. Since you may observe the hand motions, a good training set can be generated. For the disabled person this is not possible since there is no hand observe. The person would have to by himself distinguish the different motions. Thus, it would be a harder task to get a high performance for such a training set but it will indicate the expected response to be obtainable by the prosthesis user. This kind of training set is applied in this paper. No other publication is yet available – to make a comparison of the results, where this data set is used.

2.1   Data Set

The absolute value of the EMG signal is integrated for 1 s and the resulting value is coded by *four* bits. To improve the performance of the controller it is beneficial to be using several channels. In these experiments *four* channels were used in total, giving an input vector of 4 x 4 = 16 bits.

The *output* vector consists of one binary output for each hand motion, and therefore, the output vector is coded by *six* bits. For each vector only *one* bit is "1". Thus, the data set is collected from a disabled person by considering one motion at a time. For each of the six possible motions, a total of 50 data vectors are collected, resulting in a total of: 6 x 50 = 300 vectors. Further, *two* such sets were made, one to be used for evolution (training) and the others to be used as a separate test set for evaluating the best circuit *after* evolution is finished.

3   A Gate Architecture for Incremental Evolution

The evolution scheme – introduced as *increased complexity evolution,* has been proposed to overcome the problem of a long chromosome string. The idea is to evolve a system gradually. Evolution is first undertaken individually on a set of basic units. Each of these could contain gates or higher level functions as building blocks. The evolved functions are the basic blocks used in further evolution (or assembly) of a larger and more complex system. This may continue until a final system is at a sufficient level of complexity. In this paper, a novel method applying two separate and succeeding evolution steps are proposed.

3.1   Approaches to Increased Complexity Evolution

The main advantage of the method is that evolution is not performed in one operation on the complete evolvable hardware unit but rather in a bottom-up way. It may be looked at as a division of the *problem* domain. The challenge of the approach would be how to define the fitness functions for the lower level subsystems. Two alternatives seem possible:

- **Partitioned training vectors.** A first approach to incremental evolution is by partitioning the training vectors. For evolving a truth table - i.e. like those used in digital design, each separate output could be evolved separately. In this method, the fitness function is given explicitly as a subset of the complete fitness function.
- **Partitioned training set.** A second approach is to divide the training set into several subsets. This corresponds to the way humans learns: Learning to walk and learning to talk are two different learning tasks. The fitness function would have to be designed for each task individually and used together with a global fitness function, when the tasks are to be assembled. This may not be a trivial problem.

The benefits of applying the *increased complexity evolution* are several:

- Making the search space *simpler* by having the complexity of the problem to be evolved reduced for each subsystem.
- Making the search space *smaller* by having a shorter chromosome string. This is because the circuit is smaller.

Both of these two items concern the evolutionary search space as illustrated in Fig. 1. A simple search space is shown in Fig. 1a), while a more complex one is given in Fig. 1b). In the former figure, several instances would give a near maximum fitness while it becomes more important hitting the exact maximum in the latter. Thus, when both are having the same deviation $\triangle$ to reach maxiumum fitness, there is a large difference in their corresponding fitness deviation $\triangle f$. In the proposed method, the complexity of the problem is reduced by reducing the *amount* of information to be represented in a digital circuit.

The indices along the x-axis indicate the step size in the evolutionary search space. This is given by the chromosome length. A short string would imply a small number of indices compared to a longer string. The former would be more beneficial as this would reduce the search space. Thus, the goal of the *increased complexity evolution* approach is to *both* reduce the number of indices (by having a short chromosome string) as well as making the search space more smooth (by reducing the complexity of the problem to be evolved).

3.2   The Architecture of the Prosthetic Hand Controller

In this section, the proposed architecture for the controller is described. This includes the algorithm for undertaking the incremental evolution.

The architecture is illustrated in Fig. 2. It consists of one subsystem for *each* of the six prosthetic motions. In each subsystem, the binary inputs $x_0 \ldots x_{15}$ are processed by a number of different units, starting by the AND-OR unit. This is a layer of AND gates followed by a layer of OR gates. Each gate has the same number of inputs, and the number can be selected to be two, three or four. The outputs of the OR gates are routed to the Selector. This unit selects which of these outputs that are to be counted by the succeeding counter. That is, for each new input, the Counter is counting the number of *selected* outputs being

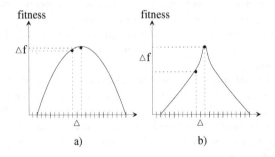

Fig. 1. Illustration of evolutionary search spaces.

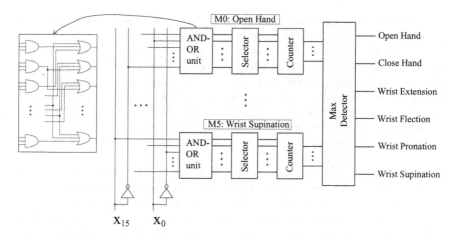

Fig. 2. The digital gate based architecture of the prosthetic hand controller.

"1" from the corresponding AND-OR unit. Finally, the Max Detector outputs which counter – corresponding to *one* specific motion, is having the largest value. Each output from the Max Detector is connected to the corresponding motor in the prosthesis. If the Counter having the *largest* value corresponds to the correct hand motion, the input has been correctly classified.

A scheme, based on using multi-input AND gates together with counters, has been proposed earlier [14]. However, the architecture proposed in this paper is distinguished by including OR-gates, together with the selector units involving incremental evolution.

The incremental evolution of this system can be described by the following steps:

1. **Step 1 evolution.** Evolve the AND-OR unit for each subsystem *separately* one at a time. Apply *all* vectors in the training set for the evolution of each subsystem. There are no interaction among the subsystems at this step, and the fitness is measured on the output of the AND-OR units. A largest possible number of OR gates should be "1" for the 50 patterns corresponding to the motion the subsystem is set to respond to. For all other patterns, the number of gates outputting "1' should be as small as possible. That is, each subsystem should ideally respond only to the patterns for one specific prosthesis motion.
2. **Step 2 evolution.** Assemble the six AND-OR units into one system as seen in Fig. 2. The AND-OR units are now fixed and the *Selectors* are to be evolved in the assembled system. Which outputs, from each of the AND-OR units, to select for making the *total* performance highest possible are now determined. Thus, the fitness is measured using the same training set as in step 1 but the evaluation is now on the output of the Max Selector.
3. The system is now ready to be applied in the prosthesis.

In the first step, subsystems are evolved separately, while in the second step these are evolved together. The motivation for evolving separate subsystems – instead of a single system in one operation, is that earlier work has shown that the evolution time can be substantially reduced by this approach [12,13]. In this paper, a less flexible AND-OR unit is used rather than the more general multi-layer gate array used in the earlier works. This is due to the initial experiments indicated that this restriction of flexibility was beneficial to make GA easier find better performing circuits. This is reasonable since only the connections are represented in the chromosome. In the architecture used earlier, the *function* of each gate was included as well.

This first step of evolution corresponds to the partitioned training vector approach presented in Section 3.1. In the following experiments, evolving sub-systems, as described above, are compared to evolving a system directly. In the latter case, the system is evolved in one operation to classify all the six motions in the training set. The system consists of only one AND-OR unit with one gate output for each motion.

The layers of AND and OR gates in one AND-OR unit consist of 32 gates each. This number has been selected to give a chromosome string of about 1000 bits which has been shown earlier to be appropriate for GA. A larger num-ber would have been beneficial for expressing more complex Boolean functions. However, the search space for GA could easily become too large. For the step 1 evolution, each gate's *inputs* are determined by evolution. The encoding of each gate in the binary chromosome string is as follows:

Input 1 (5 bit)	Input 2 (5 bit)	(Input 3 (5 bit))	(Input 4 (5 bit))

As described in the previous section, the EMG signal input consists of 16 bits. Inverted versions of these are made available on the inputs as well, making up a total of 32 input lines to the gate array. The evolution will be based on gate level. However, since several output bits are used to represent one motion, the signal resolution becomes increased from the two binary levels.

For the step 2 evolution, each line in each selector is represented by *one* bit in the chromosome. If a bit is "0", the corresponding line should *not* be input to the counter, whereas if the bit "1", the line *should* be input.

3.3 Fitness Function

The fitness function is important for the performance of GA in evolving circuits. For the step 1 evolution, the fitness function – applied for each AND-OR unit separately, is as follows for the motion m ($m \in [0,5]$) unit:

$$F_1(m) = \frac{1}{s} \sum_{j=0}^{50m-1} \sum_{i=1}^{O} x + \sum_{j=50m}^{50m+49} \sum_{i=1}^{O} x + \frac{1}{s} \sum_{j=50m+50}^{P-1} \sum_{i=1}^{O} x$$

$$\text{where } x = \begin{cases} 0 \text{ if } y_{i,j} \neq d_{m,j} \\ 1 \text{ if } y_{i,j} = d_{m,j} \end{cases}$$

where $y_{i,j}$ in the computed output of OR gate i and $d_{m,j}$ is the corresponding target value of the training vector j. P is the total number of vectors in the training set ($P = 300$). As mentioned earlier, each subsystem is trained for one motion. This includes outputting "0" for input vectors for other motions.

The s is a scaling factor to implicit emphasize on the vectors for the motion the given subsystem is assigned to detect. An appropriate value ($s = 4$) was found after some initial experiments. The O is the number of outputs included in the fitness function and is either 16 or 32 in the following experiments (referred to as "fitness measure" in the result section).

The fitness function for the step 2 evolution is applied on the complete system and is given as follows:

$$F_2 = \sum_{j=0}^{P-1} x \ \text{ where } x = \begin{cases} 1 \text{ if } d_{m,j} = 1 \text{ and } m = i \text{ for which } \max_{i=0}^{5}(Counter_i) \\ 0 \text{ else} \end{cases}$$

This fitness function counts the number of training vectors for which the target *output*[1] being "1" *equals* the *id* of the counter having the maximum output.

3.4   The GA Simulation

Various experiments were undertaken to find appropriate GA parameters. The ones that seemed to give the best results were selected and fixed for all the experiments. This was necessary due to the large number of experiments that would have been required if GA parameters should be able vary through all the experiments. The preliminary experiments indicated that the parameter setting was not a major critical issue.

The simple GA style – given by Goldberg [15], was applied for the evolution with a population size of 50. For each new generation an entirely new population of individuals is generated. Elitism is used, thus, the best individuals from each generation are carried over to the next generation. The (single point) crossover rate is 0.8, thus the cloning rate is 0.2. Roulette wheel selection scheme is applied. The mutation rate – the probability of bit inversion for each bit in the binary chromosome string, is 0.01.

The proposed architecture fits into most FPGAs. The evolution is undertaken off-line using software simulation. However, since no feed-back connections are used and the number of gates between the input and output is limited to n, the real performance should equal the simulation. Any spikes could be removed using registers in the circuit.

For each experiment presented in the next section, four different runs of GA were performed. Thus, *each* of the four resulting circuits from step 1 evolution is taken to step 2 evolution and evolved for four runs.

---

[1] As mentioned earlier only *one* output bit is "1" for each training vector.

4   Results

This section reports the experiments undertaken to search for an optimal configuration of the prosthetic hand controller. They will be targeted at obtaining the best possible performance for the test set.

Table 1. The results of evolving the prosthetic hand controller in several different ways.

Type of system	# inp/gate	Step 1 evolution			Step 1+2 evolution		
		Min	Max	Avr	Min	Max	Avr
A: Fitness measure 16 (train)	3	63.7	69.7	65.5	71.33	76.33	73.1
A: Fitness measure 16 (test)	3	50.3	60.7	55.7	44	67	55.1
B: Fitness measure 32 (train)	3	51	57.7	53.4	70	76	72.9
B: Fitness measure 32 (test)	3	40	46.7	44.4	45	54.3	50.1
C: Fitness measure 16 (train)	2	51.3	60.7	54.8	64.3	71.3	67.5
C: Fitness measure 16 (test)	2	46	51.7	49	44.3	54.7	50
D: Fitness measure 16 (train)	4	59.3	71.3	65.5	70	76	73.4
D: Fitness measure 16 (test)	4	52.7	59.7	55.3	48.3	56.3	52.7
E: Direct evolution (train)	4	56.7	63.3	59.3	-	-	-
E: Direct evolution (test)	4	32.7	43.7	36.6	-	-	-

Table 1 shows the main results – in percentage correct classification. Several different ways of evolving the controller are included. The training set and test set performances are listed on separate lines in the table. The *fitness measure* – introduced in Section 3.3, will be discussed later in this section. The "# inp/gate" column includes the number of inputs for each gate in the AND-OR unit. The columns beneath "Step 1 evolution" report the performance after only the *first* step of evolution. That is, each subsystem is evolved separately, and afterwards they become assembled to compute their total performance. The "Step 1+2 evolution" columns show the performance when the *selector units* have been evolved too (step 2 of evolution). In average, there is an improvement in the performance for the latter. Thus, the proposed *increased complexity evolution* give rise to improved performances.

In total, the best way of evolving the controller is the one listed first in the table. The circuit evolved with the best *test set* performance obtained 67% correct classification. Fig. 3 shows the step 2 evolution of this circuit. The training set performance is monotone increasing which is demanded by the fitness function. The test set performance is increasing with a couple of valleys. The circuit had a 60.7% test set performance after step 1 evolution[2]. Thus, the step 2 evolution provides a substantial increase up to 67%. Other circuits didn't perform that well, but the important issue is that it has been shown that the proposed architecture provides the *potential* for achieving high degree of generalization.

---

[2] Evaluated with all 32 outputs of the subsystems.

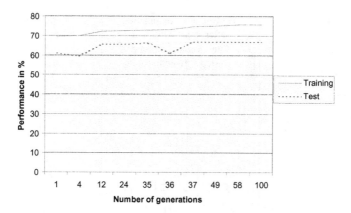

Fig. 3. Plot of the step 2 evolution of the best performing circuit.

The results presented in this paper are from the first experiments undertaken. A lot more should be conducted to optimize parameters as well determining how to measure the fitness during evolution to secure a high test set performance in general.

A feed-forward neural network was trained and tested with the same data sets. The network consisted of (two weight layers with) 16 inputs, 40 hidden units and 6 outputs. In the best case, a test set performance of 58.8% correct classification was obtained. The training set performance was 88%. Thus, a higher training set performance but a lower test set performance than for the best EHW circuit. This shows that the EHW architecture holds good generalisation properties.

The experiment B is the same as A except that in B all 32 outputs of each AND-OR unit are used to compute the fitness function in the step 1 evolution. In A, each AND-OR unit also has 32 outputs but only 16 are included in the computation of the fitness function, see Fig. 4. The 16 outputs not used are included in the chromosome and have random values. That is, their value do not affect the fitness of the circuit. What is amazing about this is that the performance of A in the table for the step 1 evolution is computed by using *all* the 32 outputs. Thus, over 10% better training set as well as the test set performance (in average) are obtained by having 16 outputs "floating" rather than measuring their fitness during the evolution!

What is also interesting is that if the performance of the circuits in A is measured (after step 1 evolution) with only 16 outputs, the performance is not very impressive. Thus, "floating" outputs in the evolution substantially improve the performance – including the test set performance.

This may seem strange but has a reasonable explanation. Only the OR gates in the AND-OR unit are "floating" during the evolution since all AND gates may be inputs to the 16 OR gates used by the fitness function. The 16 "floating" OR-gates then provide additional combination of these *trained* AND gates. To

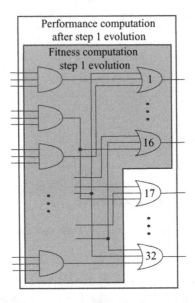

Fig. 4. A "fitness measure" equal to 16.

explain why not B shows better performance, the description in Section 3.1 is appropriate. In this experiment, the chromosome is longer since in A the bits assigned to the 16 "floating" OR-gates are not used. Other numbers of "floating" OR gates (8 and 24) were tested but the results were best for 16.

A further improvement of A could be by introducing one *more* step of evolution. After finishing step 1 evolution, the AND-gates and the OR-gates covered by the fitness function F_1 could be fixed, and a new step of evolution could evolve the 16 "floating" OR gates. This could increase the performance, but not necessarily, since the step 2 evolution already removes the bad performing OR gates.

The C and D rows in the table contain the results when the gates in the AND-OR units each consists of two and four inputs, respectively. The lowest figures are for two input gates indicating that the architecture is too small to represent the given problem. Four inputs, on the other hand, could be too complex since having *three* input gates give a slightly better results.

Each subsystem is evolved for 10,000 generations each, whereas the step 2 evolution was applied for 100 generation. These numbers were selected after a number of experiments. One comparison of the step 1 evolution (each gate having three inputs) is included in Figure 5 and shows that the best average performance is achieved when evolving for 10,000 generations.

The circuits evolved with direct evolution (E) were undertaken for 100,000 generations[3]. The training set performance is impressive when thinking of the

---

[3] This is more than six times 10,000 which were used in the other experiments.

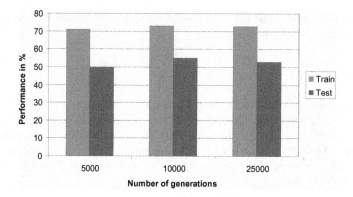

Fig. 5. Performance of three different numbers of generations in step 1 evolution (average of four runs).

simple circuit used. Each motion is controlled by a *single* four input OR gate. However, the test set performance is very much lower than what is achieved by the other approaches. This is explained by having an architecture that is too small to provide good generalization. A larger one, on the other hand, would make the chromosome string longer with the problems introduced in Section 3.1. This once again emphasizes the importance of applying the *increased complexity evolution* scheme.

An interesting observation in the experiments is that the number of inputs to a gate is not always equal to the preselected value. Several inputs to *one* gate can be connected to the same source and thus, reducing the number of active inputs. Further, both inverted and non-inverted version of the same signal can be input to an AND gate. This makes the output becoming fixed to "0" and the OR gates connecting to it would then "reduce" its number of inputs. Such behavior illustrates the good adaptivity features within the AND-OR units.

According to these results there are several interesting topics for future work. The step 2 evolution could include a selection of *subsystem* for each motion *in addition* to the selector parameters. That is, for each motion a *set* of separately evolved subsystems could be available for the step 2 evolution to select among. In this way you would be able to assemble the best combination of many of the subsystems evolved in the step 1 evolution, rather than only six. Further, the architecture is a general one and should be applied to other problems within pattern recognition/classification.

5   Conclusions

In this paper, a new EHW architecture for pattern classification including incremental evolution has been proposed. The best circuit evolved shows a better generalization performance that what was obtained by artificial neural networks.

The results illustrate that this is a promising approach to evolving systems for complex real-world applications.

Acknowledgments

The author would like to thank the group leader Dr. Higuchi and the researchers in the Evolvable Systems Laboratory, Electrotechnical Laboratory, Japan for inspiring discussions and fruitful comments on my work, during my visit there in January-April 2000. Further, I will express my gratefulness to the Japan Science and Technology Corporation (JST) for awarding me the STA fellowship making the visit possible.

References

1. R.N. Scott and P.A. Parker. Myoelectric prostheses: State of the art. *J. Med. Eng. Technol.*, 12:143–151, 1988.
2. S. Fuji. Development of prosthetic hand using adaptable control method for human characteristics. In *Proc. of Fifth International Conference on Intelligent Autonomous Systems.*, pages 360–367, 1998.
3. I. Kajitani and other. An evolvable hardware chip and its application as a multi-function prosthetic hand controller. In *Proc. of 16th National Conference on Artificial Intelligence (AAAI-99)*, 1999.
4. W-P. Lee et al. Learning complex robot behaviours by evolutionary computing with task decomposition. In Andreas Brink and John Demiris, editors, *Learning Robots: Proc. of 6th European Workshop, EWLR-6 Brighton*. Springer, 1997.
5. X. Yao and T. Higuchi. Promises and challenges of evolvable hardware. In T. Higuchi et al., editors, *Evolvable Systems: From Biology to Hardware. First Int. Conf., ICES 96*. Springer-Verlag, 1997. Lecture Notes in Computer Science, vol. 1259.
6. E. Cantu-Paz. A survey of parallel genetic algorithms. *Calculateurs Parallels*, 10(2), 1998. Paris: Hermes.
7. M. Iwata et al. A pattern recognition system using evolvable hardware. In *Proc. of Parallel Problem Solving from Nature IV (PPSN IV)*. Springer Verlag, September 1996. Lecture Notes in Computer Science, vol. 1141.
8. P. Haddow and G. Tufte. An evolvable hardware FPGA for adaptive hardware. In *Proc. of Congress on Evolutionary Computation*, 2000.
9. M. Murakawa et al. Hardware evolution at function level. In *Proc. of Parallel Problem Solving from Nature IV (PPSNIV)*. Springer Verlag, September 1996. Lecture Notes in Computer Science, vol. 1141.
10. J. R. Koza. *Genetic Programming II: Automatic Discovery of Reusable Programs*. The MIT Press, 1994.
11. W.D. Hillis. Co-evolving parasites improve simulated evolution as an optimization procedure. In *Physica D*, volume 42, pages 228–234. 1990.
12. J. Torresen. Scalable evolvable hardware applied to road image recognition. In *Proc. of the 2nd NASA/DoD Workshop on Evolvable Hardware*. Silicon Valley, USA, July 2000.

13. J. Torresen. A divide-and-conquer approach to evolvable hardware. In M. Sipper et al., editors, *Evolvable Systems: From Biology to Hardware. Second Int. Conf., ICES 98*, pages 57–65. Springer-Verlag, 1998. Lecture Notes in Computer Science, vol. 1478.

14. M. Yasunaga et al. Genetic algorithm-based design methodology for pattern recognition hardware. In J. Miller et al., editors, *Evolvable Systems: From Biology to Hardware. Third Int. Conf., ICES 98*. Springer-Verlag, 2000. Lecture Notes in Computer Science, vol. 1801.

15. D. Goldberg. *Genetic Algorithms in search, optimization, and machine learning.* Addison Wesley, 1989.

Untidy Evolution:
Evolving *Messy* Gates for Fault Tolerance

Julian F. Miller[1] and Morten Hartmann[2,3]

[1] School of Computer Science, University of Birmingham,
Birmingham B15 2TT, UK,
j.miller@cs.bham.ac.uk
[2] Department of Computer and Information Science,
The Norwegian University of Science and Technology,
7491 Trondheim, Norway,
Morten.Hartmann@idi.ntnu.no

Abstract. The exploitation of the physical characteristics has already been demonstrated in the intrinsic evolution of electronic circuits. This paper is an initial attempt at creating a world in which "physics" can be exploited in simulation. As a starting point we investigate a model of gate-like components with added noise. We refer to this as a kind of *messiness*. The principal idea behind these *messy* gates is that artificial evolution makes a *virtue* of the untidiness. We are ultimately trying to study the question: What kind of components should we use in artificial evolution? Several experiments are described that show that the messy circuits have a natural robustness to noise, as well as an implicit fault-tolerance. In addition, it was relatively easy for evolution to generate novel circuits that were surprisingly efficient.

1 Introduction

Natural evolution has produced the most subtle and complex bio-chemical information processing machines known (i.e. living creatures). In addition to this complexity living systems possess a remarkable degree of fault tolerance and robustness. At this point it is necessary to clarify the exact meanings of the terms: fault-tolerance and robustness. Robustness deals primarily with problems that are expected to occur and must be protected against. By contrast, fault tolerance primarily deals with problems that are unexpected. Humans can design systems to be robust but true fault tolerance is a much more difficult problem[1, 19]. This is particularly acute in digital electronics. Digital gates are robust from the point of view of minor changes in input voltages but systems built from them are fragile to stuck-at faults.

Another aspect of human designed systems is that they are usually built from production line components (especially if they are electronic). Living systems are built from components (i.e. cells) that vary considerably in their properties. It is now recog-

---

[3] The work was carried out while in the School of Computer Science, University of Birmingham.

Y. Liu et al. (Eds.): ICES 2001, LNCS 2210, pp. 14–25, 2001.

nised that many of the advantageous properties of living systems emerge from the way in which the individual properties of the components are exploited.

Artificial neural networks (ANN) were one of the first bio-inspired circuits to be developed. It has been suggested that ANNs exhibit graceful degradation in the presence of faults [2]. However more recent work has indicated that ANNs are not intrinsically fault tolerant [9,15].

In recent decades the use of design algorithms that employ the principles of Darwinian evolution have been applied to the design of electronic systems [3,7,16,17]. Such work has become known as Evolvable Hardware. Already researchers are using bio-inspired ideas to design electronic hardware with more fault-tolerance [4,10]. One of the most intriguing findings in this field is that of Adrian Thompson [11]. He showed that it was possible for artificial evolution to create FGPA designs that exploited the physical characteristics of the silicon substrate to efficiently carry out a particular task. Thompson found that unconstrained artificial evolution explored very unusual ways of solving problems *precisely* because it was able to exploit the subtle and incidental physical characteristics. It can be argued that evolution has produced such complex systems because it can make use of the full, *unmodellable* richness of the physical world.

Another question raised from Thompson's work is the following: *What basic components should we be using in artificial evolution?* It does not seem very likely that the electronic components that have been created for human design should happen to be particularly useful in artificial evolution. Indeed it could be seen as a testament to the power of evolution that using them, artificial evolution could produce anything useful at all. If one is going to build practical systems using artificial evolution it appears that one has to go back to basics and try to design radically new forms of electronic components, or use existing components in novel configurations [e.g. 18] that might assist the artificial evolutionary process. This could be done in two ways. First, one could search for special materials and then subject them to *intrinsic* artificial evolution, and second, one could try to define new kinds of components in simulation and subject them to *extrinsic* evolution. The former is obviously a good approach, however it is potentially *very expensive* and likely to be very difficult. The latter though, perhaps, is not as exciting, however it is *feasible* and could actually assist the former goal by helping us to identify what kinds of properties are important. These thoughts were the starting point for the work reported in this paper.

The paper is divided as follows: Section 2 is concerned with the definition of the new component model (which is described as *messy*). Section 3 describes the genotype representation and evolutionary algorithm that was used. Section 4 describes the experiments performed. In section 5 the experimental results are presented. In section 6 the concluding remarks are made. The paper finishes with a description of possible future work in section 7.

2 Messy gates

Since the aim is not only to create new types of components for use in artificial evolution but also to *understand* exactly how they work, it was necessary to choose a new model that was adjustable. It was desirable to be able to change a parameter continuously so that the new component model could become the same as a familiar traditional component. A natural choice of traditional component was a digital logic gate. Accordingly, a model that was that of a digital multiplexer with additional randomness on the output was created. The new model took real valued input combined the inputs to give a real valued output. Random noise was then superimposed on the output (figure 1). These gates are being referred to as *messy* and as having a degree of *messiness*.

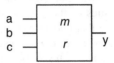

Fig. 1. Model of messy MUX

The equation below describes the messy MUX:

$$y = mr + a\bar{c} + bc \ . \tag{1}$$

where m represents the constant value of messiness chosen for the entire circuit and r represents a real random number uniformly sampled from the interval [-1.0, 1.0]. The inputs to the messy MUX are a and b, and c is the control input. The bar over c refers to *1-c*. All variables are real-valued. Clearly when $m=0$ and a, b, and c are only allowed to be 0 or 1, the digital MUX is recovered and all the rules of Boolean logic apply. In the experiments a messy MUX with one input inverted (input b would then become 1-b) was used as well. Once real values for variables are allowed it is not clear how to decide real-valued equivalence of the Boolean operators. Arithmetic add and multiply are the closest analogues of the inclusive-OR and the AND operators. The model we have used was not designed to be a feasible electronic model, since the primary aim was to investigate the ability of artificial evolution to use *any* chosen model. No claims are made about the "correctness" of the model we used other than it possessed the property of being a superset of the digital case. Now that the system is designed, more complex and physically realistic models can easily be added later (even though it may be hard to design the model itself).

3 Evolutionary algorithm and genotype

The genotype representation is the same as that used for evolving digital designs [5,6]. It is best explained with a small example. In figure 2 are shown four messy MUX (mMUX) gates. The numbers (0-3) refer to the four primary inputs of the target function. The numbers on the inputs to the mMUX refer to the connections to the primary inputs or the outputs of other mMUX. The outputs of the mMUX are labeled (sequentially following on from the inputs). Thus the second mMUX on the left has the "a" input connected to the output of the first mMUX, the other two mMUX inputs are connected to the two primary inputs 3 and 0. In this example it is assumed that the target function has four outputs. Thus the genotype has four output connections (4 5 7 3). The numbers in bold refer to which of the two mMUX were being used (0 refers to mMUX with no inputs inverted, 1 refers to the "b" input being inverted). The numbers printed in grey refer to inactive genes or mMUX (i.e the third mMUX does not have its output connected). This paper only considers feed-forward circuits. The representation allows any mMUX to have its input connected to any other mMUX on its left.

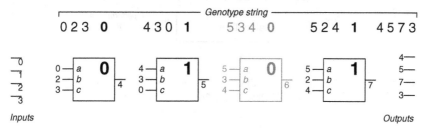

Fig. 2. Example genotype and resulting circuit with four inputs and four outputs

The evolutionary algorithm employed was a simple form of (1+4) evolutionary strategy (ES). In this case a population of 5 random chromosomes is randomly generated and the fittest chromosome selected. The new population is then filled with mutated versions of this. Random mutation is defined as a percentage of genes in the population that were mutated. The mutation operator respects the feed-forward nature of the circuits and also the different alphabets associated with connections and functions.

Each *circuit* has a fixed value of messiness m. However each gate has its own random value r (equation 1). This is illustrated in figure 3.

$$r_1 \quad r_2 \quad r_3 \quad r_4$$

Fig. 3. Circuit schematic with individual random values

The fitness of an individual is measured by testing the chromosome with all possible combinations of inputs and comparing the output values with the target Boolean truth table. For all experiments described in this paper, the target is a 2-bit multiplier. Thus, there are 4 inputs, 4 outputs and 2^4 possible input test vectors yielding a total of 64 output bits in the truth table.

Fitness is equal to the number of output bits of a circuit being equal to the corresponding bit in the target truth table. This is referred to as bitwise correctness. A perfectly functional circuit would have all its output bits equal to the output bits of the truth table, and thus the bitwise correctness would equal 64 in the case of the 2-bit multiplier. In the case of messy circuits, the real valued internal signals were rounded at the output before being compared to the target truth table.

4 Experiments

Several experiments were performed to investigate the nature of the mMUX and its influence on the evolved circuits and the evolutionary algorithm. These experiments are described below.

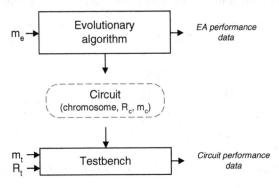

Fig. 4. Experimental setup

The setup for the experiments is shown in figure 4. The evolutionary algorithm produces a circuit consisting of a chromosome, a set of random values R_c and the messiness used ($m_c = m_e$). A test bench allows measuring the performance of a circuit by using different values of messiness (m_t), different sets of random values (R_t) and introducing stuck-at faults.

The first two experiments investigated whether it was feasible to evolve circuits that exhibited a natural robustness to internal random noise within some specified range.

A set of about 500 chromosomes was evolved for six different values of m. The fitness measure was based on taking an average of the bitwise correctness over 15 sets of random values r (for each gate). Evolution was halted each time the fitness was equal to 64 (all 15 chromsomes had bitwise correctness equal to 64). This set of chromosomes was used to generate the results of the first three experiments. Thus, these experiments investigate different configurations of the testbench, with the fitness of each circuit measured as an average over 50 trials. This was done both to introduce new random values R_t for every test performed, as well as testing different randomly chosen gates for stuck-at faults in the case of the third experiment.

4.1 First experiment

The first experiment sought to investigate how computationally demanding it is to evolve circuits with high values of m, as well as the general performance of the circuits in an environment equally noisy to the one in which it was evolved ($m_t=m_c$). In all other respects the experiment was the same as that described previously.

4.2 Second experiment

Thompson has demonstrated that it is possible to evolve robust circuits intrinsically by exposing them to various environments [12, 13]. In the second experiment, the same set of chromosomes was tested with the messiness of the test bench m_t, being set to increasingly higher values. This simulates the circuits running in increasingly more noisy environments, and disregards whatever value m_e used when the chromosomes were evolved. This was done to investigate the robustness of the chromosomes with regards to the amplification m_t of the internal noise, and its relation to the value of m_e used in the evolution of the different chromosomes.

4.3 Third experiment

An interesting property of a digital circuit is its fault tolerance. An experiment was carried out to measure the tolerance in the evolved circuits to stuck-at faults.

The test bench was set up to subject the circuits to stuck-at-1 faults. Stuck-at-1 faults were selected since their impact on a multiplier is on average more severe then stuck-at-zero faults. This is due to the fact that the output part of the 2-bit multiplier truth table contains 14 zeros and 50 ones. Thus, increasing numbers of stuck-at-0 faults force the bitwise correctness towards 50. On the other hand, increasing the numbers of stuck-at-1 faults force the bitwise correctness towards 14.

4.4 Fourth experiment

The last experiment sought to investigate how evolution would be capable of exploiting individual characteristics of given gates. This was done by generating random values for R_e for each gate only once for each run of the evolutionary algorithm. The algorithm would then try to utilize the properties of each individual mMUX in the circuit to solve the problem.

The set of random values was saved with each chromosome as R_c, and the test bench was configured to use the saved random values as the values to be used under test ($R_t=R_c$).

The fitness function was modified by adding the number of redundant nodes to the bitwise correctness (provided the bitwise correctness equalled 64). This was done to investigate if it would be possible to evolve smaller circuits for higher values of m.

5   Results

This section presents the results obtained through the experiments carried out in section 4. The results are briefly discussed within the limits of the scope of this paper.

5.1   First experiment

Evolving chromosomes with bitwise correctness equal to 64 was more computationally expensive for large values of m. Purely digital circuits (m=0) took an average of 2000 generations to evolve. On the other hand, it required an average of more than 30000 generations to evolve circuits with a messiness value of m=0.25.

Even though all chromosomes are evaluated 15 times to distribute the random values creating noise at the gate outputs, the same chromosome may not obtain perfect fitness if a new set of random values are introduced. In such a way, the number of evaluations during evolution can be considered a choice between computational effort and robustness of the evolved circuit (with regards to random values introduced).

The resulting average fitness is shown in figure 5.

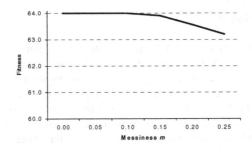

Fig. 5. Average fitness measured through tests

Even though this figure shows a negative trend, it implies that messy circuits are pretty robust to variations in the internal noise within the range m_e used when the particular circuit was evolved, as the drop in fitness was quite small.

Finally, this experiment revealed another property of evolving messy circuits. Figure 6 illustrates how chromosomes evolved with high values of m tend to produce smaller circuits then those produced when messiness is low. This is probably due to the fact that evolution finds a way to reduce the amount of noise the circuit is exposed to. Each new gate means a new noisy value to cope with.

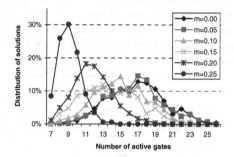

Fig. 6. The number of gates tend to be small for larger values of *m*

5.2 Second experiment

When the evolved circuits were tested in increasingly more noisy environments, a clear trend showed increased robustness for circuits evolved with higher messiness (m_e). This trend is shown in figure 7. The graph shows the deviation from the average fitness over all evolved circuits, when exposed to increasing noise in the environment.

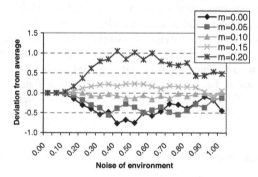

Fig. 7. Average fitness when exposed to more noisy environments

5.3 Third experiment

It was difficult to see any clear trend on whether messy circuits were more tolerant then pure digital ones when they used a small number of gates. It was desirable to investigate whether larger circuits with high messiness would be more fault tolerant. To reliably obtain large circuits (using the maximum number of gates =30) a term had to be added to the fitness that favoured larger circuits. This meant that circuits without a bitwise correctness of 64 had to be accepted. However it was observed that these circuits proved to be largely more fault-tolerant to the stuck-at-1 faults and showed a more graceful degradation when compared to the zero messiness case. This trend is shown in the graph in figure 8. Note that the test bench in this case introduced the faults in the same environment that each chromosome was originally evolved (same value of *m*). So the circuits had to cope with intrinsic randomness associated with the messiness m_c, in addition to the introduced faults.

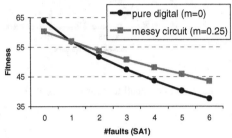

Fig. 8. Difference in degradation of large circuits

An interesting aspect of this graceful degradation is the fact that this tolerance is implicit, since these circuits were never evolved explicitly with fault tolerance to stuck-at faults as a part of the EA. A comparison of the fault tolerance of the evolved digital case (m=0) and conventional circuits was not carried out. However, certain evolutionary systems may inhibit natural mutation tolerance due to the workings of the evolutionary process [14]. Such tolerance is unlikely to have any effect on the EA used in this paper, as a result of the steep hill-climbing nature of the algorithm.

The results where compared to circuits evolved explicitly for fault tolerance (i.e. tolerance to stuck-at fault was included in the fitness function). This comparison showed that explicit fault tolerance could be evolved to display stronger tolerance, but at a higher computational cost.

5.4   Fourth experiment

The fourth experiment revealed that evolving circuits with permanent sets of values R_c for each chromosome, was slightly more computationally demanding for higher values of m.

Figure 9 shows the percentage of evolved circuits having a particular number of gates for various messiness values. The low values of m tended to yield circuits with 7 gates, higher values of m had a wider distribution of evolved circuits.

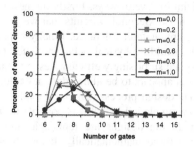

Fig. 9. Distribution of solutions with regards to the number of gates in the evolved circuits

For increasing values of m, a larger amount of the evolved circuits consisted of only six gates, suggesting that only sets of random values with sufficient range allow such small circuit. The small circuits are interesting, since it appears the minimum number of gates (of the two types of multiplexers used) that can be used to solve the problem in a pure digital manner is 7 [6]. An example of a circuit of this type is shown in figure 10. A and B are the two 2-bit numbers being multiplied, while P is the output product. Subscripts on the inputs and outputs refers to the significance of the bits (0 being least significant). A circle at an input illustrates an inverted value. The shown circuit was evolved with m equal to 0.4. It demonstrates the ability of a blind evolutionary process to exploit all the "physical characteristics" of the components. This is reminiscent of Thompson's findings in [13] where he found that robust evolved clocked digital circuits exploited glitches, even though these are shunned in conventional human design.

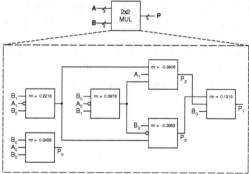

Fig. 10. Example of a small evolved circuit (m=0.4)

6 Conclusions

In this paper a new model of a gate-like component with added random noise was used for initial investigations. The experiments carried out so far indicates that such components are beneficial in several ways. Firstly the new gate-like models naturally offered a robustness to noise. Secondly circuits that were evolved using these messy components exhibited implicit fault tolerance. Finally experiments indicated that in creating a simulateable world, "physical" characteristics (intrinsic random values) could be exploited to create surprisingly efficient designs (the six mMUX 2-bit multiplier). Artificial evolution appears to be quite adept at exploiting such things. It is not easy to imagine a human design process that could exploit such random differences. One advantage of such a simulation is that it is possible to inspect every detail of the "physics". In addition all the designs are replicable and the functionality of the evolved circuit can be verified mathematically. It may be that one cannot design a model with sufficient richness to allow artificial evolution to find really innovative ways of building circuits. In these cases intrinsic evolution may be the only way forward. However this appears to require the use of conventional technology (i.e. FPGAs, programmable-logic). One way forward might be to deliberately disrupt the silicon (perhaps by radiation) in order to enrich the natural physics. Another approach would be to look for novel materials for intrinsic evolution (i.e. evolvable matter arrays [8]). It should also be noted that at present the algorithms that are used in artificial evolution are very far from the sophistication and complexity of real biological evolution processes.

7 Future work

This paper is a preliminary study into an area that to the knowledge of the authors is relatively unexplored. The model of messiness discussed here is really just a starting point. It is likely that more complex models of components would be much more suitable in for use in an evolutionary design process, since the produced circuits would be closer to a real hardware implementation.

Further on, a study into the general principles on how evolution exploits properties of the underlying hardware would be interesting. This paper attempts to initiate thoughts directed at issues that could be important in a search for more suitable media for evolutionary systems.

References

1. A. Aviziensis "Towards systematic design of fault-tolerant systems", IEEE Computer, vol. 30, pp. 51-58, 1997.
2. L.A. Belfore and B.W. Johnson. "The fault-tolerance of neural networks", The International Journal of Neural Networks Research and Applications 1, pp. 24-41 (Jan 1989).
3. T. Higuchi and M. Iwata (eds.). Proceedings of the 1st International Conference on Evolvable Systems: From Biology to Hardware, Lecture Notes in Computer Science (ICES96), Springer-Verlag, Berlin, vol. 1259, 1996.
4. D. Keymeulen, R. Zebulum "Fault-Tolerant Evolvable Hardware Using Field-Programmable Transistor Arrays", in IEEE Transactions on Reliability, vol. 49, no. 3, 2000.

5. J. F. Miller and P. Thomson. "Cartesian Genetic Programming", in R. Poli, W. Banzhaf, W.B. Langdon, J. F. Miller, P. Nordin, T. C. Fogarty (eds.), Third European Conference on Genetic Programming Edinburgh 2000 (EuroGP2000), Lecture Notes in Computer Science, vol. 1802, pp. 121-132, Springer-Verlag, Heidelberg, 2000.

6. J. F. Miller, D. Job, V.K. Vassilev. "Principles in the Evolutionary Design of Digital Circuits - Part I", in W. Banzhaf (ed.), Genetic Programming and Evolvable Machines, Vol. 1, No. 1/2, Kluwer Academic Publishers, Netherlands, pp. 7 - 35, 2000.

7. J. F. Miller, A. Thompson, P. Thomson, T. C. Fogarty (eds.). Proceedings of the 3rd International Conference on Evolvable Systems: From Biology to Hardware (ICES00), Lecture Notes in Computer Science, vol. 1801, Springer-Verlag, Berlin, 2000.

8. J. F. Miller, "Evolvable Hardware: Some directions for the Future", an invited talk at the 2nd NASA/DOD Workshop on Evolvable Hardware, July 2000, available at http://ic-www.arc.nasa.gov/ic/eh2000/slides.html (last accessed 6-4-01)

9. J. Nijhuis, B. Hofflinger, A. Schaik and L. Spaanenburg. "Limits to Fault-Tolerance of a Feedforward Neural Network with Learning", Digest FTCS, pp. 228-235, June 1990.

10. C. Ortega-Sanchez, D. Mange, S. Smith and A. Tyrrell "Embryonics: A Bio-Inspired Cellular Architecture with Fault Tolerant Properties", in W. Banzhaf (ed.), Genetic Programming and Evolvable Machines, Vol. 1, No. 3 Kluwer Academic Publishers, Netherlands, pp. 187 - 215, 2000.

11. A. Thompson. "An evolved circuit, intrinsic in silicon, entwined with physics", in T. Higuch, M. Iwata, W. Liu (eds.), Proceedings of The 1st International Conference on Evolvable Systems: From Biology to Hardware (ICES96), Lecture Notes in Computer Science, vol. 1259, Springer-Verlag, Heidelberg, pp. 390 - 405, 1997

12. A. Thompson. "On the Automatic design of Robust Electronics through Artificial Evolution", in M. Sipper, D. Mange, A. Pérez-Uribe (eds.), Proceedings of The 2nd International Conference on Evolvable Systems: From Biology to Hardware (ICES96), Lecture Notes in Computer Science, vol. 1478, Springer-Verlag, Heidelberg, pp. 13- 24, 1998

13. A. Thompson and P. Layzell. "Evolution on Robustness in an Electronics Design", in J. Miller, A. Thompson, P. Thomson, T.C. Fogarty (eds.), Proceedings of The 3rd International Conference on Evolvable Systems: From Biology to Hardware (ICES00), Lecture Notes in Computer Science, vol. 1801, Springer-Verlag, Heidelberg, pp. 218 - 228, 2000.

14. A. Thompson. "Hardware Evolution - Automatic Design of Electronic Circuits in Reconfigurable Hardware by Artificial Evolution", Springer-Verlag, London, 1998.

15. B.E. Segee and M. J. Carter. "Fault Tolerance of Pruned Multilayer Networks", Digest IJCNN, pp. II-447-452, 1991.

16. M. Sipper, E. Sanchez, D. Mange, M. Tomassini, A. Pérez-Uribe and A. Stauffer, "A phylogenetic, ontogenetic, and epigenetic view of bio-inspired hardware systems", IEEE Transactions on Evolutionary Computation vol. 1(1) pp.83-97, 1997.

17. M. Sipper, D. Mange and A. Pérez-Uribe (eds.). Proceedings of the 2nd International Conference on Evolvable Systems: From Biology to Hardware (ICES98), Lecture Notes in Computer Science, Springer-Verlag, Berlin, vol. 1478, 1998.

18. A. Stocia, D. Keymeulen, R. Zebulum, A. Thakoor, T. Daud, G. Klimeck, Y. Jin, R. Tawel and V. Duong. "Evolution of analog circuits on Field Programmable Transistor Arrays", in J. Lohn, A. Stocia, D. Keymeulen and S. Colombano, Proceedings of The 2nd NASA/DoD Workshop on Evolvable Hardware (EH'00), 2000, CA IEEE Computer Society Press.

19. R. White and F. Miles "Principles of fault-tolerance", in Proceedings of 11th Annual Conference on Applied Power Electronics, IEEE Press, vol. 1, pp. 18-25, 1996.

Evolutionary Design Calibration

Thorsten Schnier and Xin Yao

School of Computer Science
The University of Birmingham
Edgbaston, Birmingham B15 2TT, UK
{T.Schnier, X.Yao}@cs.bham.ac.uk

Abstract. Evolutionary methods are now beginning to be used routinely in design applications. However, even with computing speeds growing continuously, for many complex design problems evolutionary computing times are so long that their use is not practical. Divide and conquer based methods sometimes improve the situation, but in most cases the biggest speed improvement can be gained by adding domain knowledge. Combining evolutionary methods with conventional design methods is one way of doing this. This paper shows how evolutionary computation can be used to improve designs created by conventional design methods. A digital filter design problem is used to illustrate how a conventionally derived design can be further improved by evolutionary calibration. Our experimental results show that the evolutionary calibration algorithm is able to consistently improve the original designs by a considerable margin.

1 Introduction

While evolutionary computation is able to create designs that are superior to conventionally derived designs (Yao 1999a, Schnier, Yao & Liu 2001), it sometimes does so at the expense of very long computation times. Introduction of domain knowledge is one of the major ways runtimes of evolutionary processes can be sped up. This can happen for example in the representation (Schnier & Yao 2000), selection, or genetic operators.

In digital filter design, there are existing conventional methods available for some classes of filters. In such cases, it make sense to use a hybrid approach, where the evolutionary algorithm (EA) is used to 'calibrate' a design created using a conventional design process.

1.1 Limits of the Conventional Design Process

Designing digital filters, especially recursive filters, is not straightforward. For certain design problems with particular characteristics, it is possible to mathematically derive the optimal filter configuration; but in general, no such method exists. Instead, a number of approximation methods have to be developed, usually applicable only for a particular class of design problems.

Y. Liu et al. (Eds.): ICES 2001, LNCS 2210, pp. 26–37, 2001.

There are two limitations with this design method. First, for new problem classes, an approximation approach has to be developed first. For example, in Lu (1999), a process for designing stable IIR filters with equiripple passbands and peak-constrained least-squares stopbands is developed. As the title indicates, the class of filters that the method is applicable to is fairly limited. The development of the approximation requires special technical knowledge. This is different from evolutionary approaches, which can generally be used by non-specialists for larger classes of filter design problems.

The second limitation with the conventional design approach is that, depending on the exact approach taken, the resulting design can be a suboptimal because it follows a deterministic heuristic (or approximate) procedure. For example, the approach taken in Lu (1999) is based on iterative quadratic programming method with linearised constraints using a least-square objective. The resulting design may be suboptimal for three different reasons:

- Linearization of constraints: All constraints have to be formulated as linear inequalities. Constraints that are not initially linear have to be linearised. To ensure that the linearised constraint still excludes all designs that initially violated the constraint, it has to exclude some feasible designs (otherwise the linearised version would have to be identical to the original).
- Objective as a least-square problem: The objective has to be implemented as a weighted least-square function. Often, this is not exactly the same as the actual design goal. For example, in Lu (1999) the square error from the desired behaviour is used as the objective for the quadratic programming, but the maximum deviation from the desired behaviour is actually used to compare the final designs with other filter designs. While the maximum deviation is the 'real' design goal, the accumulated weighted square error is used for the design algorithm.
- The result of the iterated quadratic programming algorithm generally depends on the starting conditions. There is no guarantee of convergence to the global optimum (Lu 1999).

An evolutionary approach to digital filter design has been demonstrated in Schnier et al. (2001). The approach shows some success, but at the cost of a large computational effort. While the authors believe that the approach described in Schnier et al. (2001) can be improved and made more efficient, more reductions in computation time can be achieved by incorporating domain knowledge, whenever available, into the evolutionary design process.

For the design case studied in this paper, the starting point is the conventional design that was obtained through the existing conventional approach (Lu 1999) although the conventional approach can only find a suboptimal design. An EA will then be employed to calibrate the design and produced an improved design. The assumption is that by creating a hybrid approach where conventional design knowledge is combined with an evolutionary process, it is possible to generate results that are superior to the conventionally derived designs with an acceptable computational effort.

1.2   Related Work

Evolutionary digital filter design has been explored in other work. Generally, the representations used are low-level, for example gate level or function block level (Miller 1999, Wilson & Mcleod 1993, Uesaka & Kawamata 2000, Uesaka & Kawamata 1999). This has the advantage that a complete filter is evolved, and additional fitness criteria (e.g. sensitivity to parameter quantisation) can be used. On the other hand, the complexity of the evolved filters is limited, typically to second or third order filters. The work presented here is aimed at constructing filters of 'real world' complexity; the examples shown are order 12 and order 15 filters.

Seeding EAs with designs derived by different methods is not a new idea. For example, it is mentioned in Ibaraki (1997), where a simple 'greedy heuristic' is used to create initial individuals in the knapsack problem. Other examples include Langdon & Nordin (2000) and Louis & Johnson (1997). However, seeding is not actually used very often in hardware design.

The work presented in Baicher & Turton (2000) can be seen as the opposite of the approach used in this paper: use an EA to locate regions of high fitness, and then use conventional iterative optimisation methods like iterated quadratic programming and the Quasi-Newton method to perform a local search for the final result. While this is also a promising direction to pursue in hybrid methods, it is often very difficult to evaluate when "regions of high fitness" has been located since we do not know how high is "high" for a given problem.

1.3   Organisation of the Paper

The rest of this paper is organised as follows: Section 2 describes the filter design problem that we use as an example in this paper. Section 3 explains how a filter is encoded by a chromosome in our EA, how the EA is initialised and how genetic operators are implemented. Section 4 presents our experimental results and discusses our findings. Lastly, Section 5 concludes the paper with some brief remarks on our work.

2   The Design Problem

The design problem is that of defining the transfer function of a linear, infinite impulse response (IIR) digital filter.

In general, digital filter design is a two-step process. In the first step, a mathematical description of the filter fulfilling the design criteria is derived. The first step, generally the more difficult one, produces a polynomial description of the filter, i.e., the transfer function. From this polynomial, it is then possible to derive a hardware implementation using one of a number of well-established methods. Following other filter design papers (Lu 1999), the work reported here only deals with the more difficult first step.

2.1   Mathematical Description of Linear Digital Filters

Any linear digital filter can mathematically be specified by a complex-numbered polynomial function, i.e., the transfer function (Equation 1). This polynomial function, i.e., Equation 2, can be rewritten as the quotient of two product terms with the numerator specifying the zeroes of the polynomial and the denominator specifying the poles. The function usually has a scaling constant (Equation 2). The two descriptions are equivalent. It is easy to transform a pole-zero description to a polynomial description, but not vice versa. The frequency response can be derived from the transfer function by calculating the values for $z = e^{j\omega T}$ where T is the sampling frequency.

$$H(z) = \frac{\sum_{i=0}^{n} b_i z^{n-i}}{z^{(n-r)} \sum_{i=0}^{r} b_i z^{r-i}}. \tag{1}$$

$$H(z) = b_0 * \frac{(z - z_{z0})(z - z_{z1})\dots(z - z_{zn})}{z^{r-i}(z - z_{p0})(z - z_{p1})\dots(z - z_{pi})}. \tag{2}$$

Not all transfer functions can be realized in a hardware filter. Two major requirements have to be observed:

Real coefficients: The coefficients in the polynomial description have to be real numbers. In terms of poles and zeroes, this can be achieved if all poles and zeroes are either real, or exist in conjugate-complex pairs (i.e. $a + jb$ and $a - jb$).

Stability: In a stable filter, a bounded input will always produce a bounded output. A filter is only stable if all poles are within the unit circle, i.e. $\|a + jb\| < 1$. While there are uses for unstable filters in specific applications, most filters are designed to be stable.

More details about filter design can be found in some text books (Antoniou 1993, Parks & Burrus 1987).

2.2   Performance Criteria

The required behaviour of the filter is specified in terms of the frequency response (see Section 2.2). This includes the phase and the magnitude of the output signal relative to the input signal for all the frequencies between 0 and half the sampling frequency.

Filter performance is usually multi-objective. This paper considers two filters. Both of them are low-pass filters. An ideal low-pass filter lets signals pass unchanged in the lower frequency region (passband) and blocks signals completely in the upper frequency region (stopband). In reality, a transition band is located between the passband and stopband.

To minimise distortion of the signal in the passband, two criteria have to be met. The first is that the amplitude of the frequency response in the passband should be as constant as possible. The second criterion is that the phase in the passband has to be as linear as possible. In practice, the so-called 'group delay,' i.e., the first derivative of the phase $\delta\phi/\delta\omega$, is often used. The second criterion can therefore be stated as a constant group delay. This means that all frequencies are delayed by the filter by the same amount of time.

In the stopband, the design goal is generally to attenuate the signal as much as possible. Because the signal is attenuated, the phase and group delay of the signal in the stopband usually become unimportant.

Figure 1 shows a typical lowpass filter. The top half shows the amplitude and the lower half the group delay. The ideal behaviour is shown with thick lines. The 'real' behaviour (thin line) is acceptable as long as it is within the shaded regions.

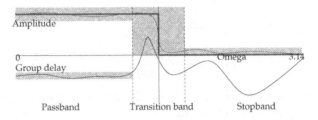

Fig. 1. Design objective and constraints for an example lowpass filter

To facilitate comparison between our and other existing work, this paper follows the criteria used previously (Lu 1999) whenever possible, that is,

1. weighted square error over the amplitude in the passband and stopband,
2. peak amplitude in the stopband,
3. maximum deviation from constant amplitude in the passband,
4. maximum deviation from the goal group delay, and
5. the stability.

The conventional design process adopted by human experts uses criterion (1) as the optimisation criterion and criteria (2) to (5) with predefined values as constraints.

3   The Evolutionary Algorithm

The three major parts of our EA are chromosome representation, population initialisation and genetic search operators.

3.1   Chromosome Representation

The choice of (chromosome) representation for each individual is crucial in the success of any EA (Schnier & Yao 2000). Generally, the representation used in an EA has to be 'appropriate' to the application domain (Fogel & Angeline 1997).

In this section, we will discuss how to represent a transfer function that defines a filter in a chromosome (i.e., a genotype).

As described in Section 2.1, the transfer function is generally given in one of two forms: a polynomial or a pole-zero description of the filter. Because of the direct relationship between the transfer function and frequency response, poles and zeroes in the pole-zero form of the transfer function (Equation 2) can be directly interpreted: a pole near the current frequency amplifies the signal, a zero attenuates it. Since poles and zeroes are complex numbers, their locations in the complex plane can be naturally expressed in polar coordinates. Under such coordinates, the angle directly specifies the frequency at which the pole or zero is active, and the distance from the origin indicates its 'strength'.

Because of this direct relationship to the fitness, a polar coordinate based representation of poles and zeroes promises to be suitable. It has a number of advantages. It can represent all feasible linear IIR filters. It is possible (as shown below) to ensure the feasibility of all phenotypes. Additionally, locality is preserved at least on a local scale. That is, similar genotypes will have similar frequency responses. Because of this, the fitness landscape is expected to be reasonably smooth.

In detail, the transfer function of a filter is represented by a sequence of paired real-value numbers, where each pair indicates the polar coordinates of a pole or zero. An additional pair of real-valued numbers encode the scaling parameter b_0. Each pair of real-valued numbers is called a gene in our EA.

In order to impose the constraints specified in Section 2.1, poles and zeroes need to be either positioned on the real axis (i.e. $IM(z) = 0$), or exist in conjugate-complex pairs when a genotype is mapped into a phenotype (i.e., a transfer function). All poles have to be located within a unit circle in order to ensure filter stability.

For example, a genotype of $N_{p2} + N_{p1} + N_{z2} + N_{z1} + 1$ pairs of real-valued numbers will consist of

N_{p2} **pole pairs:** Each pair of real valued numbers in the genotype represents a complex pole. The conjugate-complex pole is automatically generated by the genotype-phenotype mapping to ensure the filter is feasible. The radius can lie between -1.0 and 1.0, which ensures stability.

N_{p1} **single poles:** For these poles, the angle is ignored. Only the radius is used to determine the position on the real axis. Radius is restricted to between -1.0 and 1.0.

N_{z2} **zero pairs:** Such a pair determine one of conjugate-complex pair of zeroes. The partner is automatically generated. The radius can lie between -1.0 and 1.0, but is scaled in the genotype-phenotype mapping with the factor R_{zMax}.

N_{z1} **single zeroes:** The angle is ignored. The radius is between -1.0 and 1.0 and scaled with R_{zMax}.

Scaling factor b_0: The angle is ignored. The radius is used to scale the polynomial. It is between -1.0 and 1.0 and scaled with S_{Max}.

The scaling employed for zeroes and b_0 means that all pairs of real value numbers have exactly the same range: between -1.0 and 1.0 for radius, and between $-\pi$ and π for the angle. This facilitates evolutionary search without special knowledge about the differences between zeroes, poles, and b_0. The scales R_{zMax} and S_{Max} are fixed in our experiments. Setting them optimally requires domain knowledge (see also Section 5).

3.2   Population Initialisation

The hybridisation of conventional and evolutionary design processes occurs in the creation of the initial population. A large portion of the initial population (e.g. 80%) are created randomly, as in any normal EA. The rest, however, is initialised by mutating the known design generated by the conventional method. The mutation is the same as that used in the evolutionary process. Because of the way the mutation operation is implemented (see Section 3.4), some individuals may not be mutated.

3.3   Selection

Selection is done through tournament selection. Fitness sharing is implemented, where the raw fitness is modified by a factor depending on the number of individuals in the same niche (based on the genotype similarity). The selection for the next generation is elitist. All individuals in the Pareto front survive. The rest of the population is made up of the best of the remaining individuals from the union of current population and offspring.

3.4   Genetic Operators

The crossover operator is a modified uniform crossover. As radius-angle pairs are closely coupled, it does not seem to make sense to allow crossover to separate them. Crossover points are therefore limited to be between these pairs. In other words, two parents can only be crossed over between genes, but not within a gene.

Mutation is based on the Cauchy mutation (Yao, Liu & Lin 1999, Yao & Liu 1997) with a fixed scaling factor η (Yao et al. 1999, Yao & Liu 1997). For each value in a genotype, the mutation probability is 0.1. Because of different ranges allowed, angle mutation is performed differently from radius mutation. When a radius is mutated, the mutation is 'reflected' from the edges of the search space (e.g., if a pole currently has a radius of 0.9 and the mutation is +0.3, it will end up being $(1.0 - 0.2) = 0.8$). When angle is mutated its value is simply 'wrapped around' at $\pm\pi$.

4   Experimental Studies

4.1   Two Test Problems

Both test problems are lowpass filters (Lu 1999) with slightly different numbers of poles and zeroes, cutoff frequencies, and goals for delays and amplitude.

Problem Case 1: $\omega_p = 0.2$, $\omega_a = 0.28$, maximum amplitude deviation $0.1dB$, minimum stopband attenuation $43dB$, group delay $= 11$ samples with maximum deviation 0.35, order 15 with 7 zero pairs, 1 single zero, 2 pole pairs, 1 pole single, 10 poles at the origin.

Problem Case 2: $\omega_p = 0.25$, $\omega_a = 0.3$, maximum amplitude deviation $0.3dB$, minimum stopband attenuation $32dB$, group delay $= 9$ samples with maximum deviation 0.5, order 12 with 6 zero pairs, no single zeros, 5 pole pairs, 1 pole single, 1 poles at the origin.

Genotypes representing individuals require 12 pairs of real numbers for case 1 and 13 pairs of real numbers for case 2. When computing the fitness for the human design in case 1, it was noted that the amplitude curve seemed to be slightly too high. When the value for b_0 was modified from -0.00046047 as given in the paper (Lu 1999) to -0.000456475, the fitness value becomes very similar to that given in the paper (Lu 1999). We think this is caused either by a typo in the published paper (Lu 1999) or the result of rounding errors in the fitness calculations in either the paper or our implementation. We will use the corrected value in all our performance comparisons in this paper.

In all the results given below, 300 samples have been used in the passband and 200 in the stopband. To conduct a fair comparison, fitness values have been computed for the designs given by the human expert (Lu 1999) using exactly the same sampling and fitness computation methods as those used in our evolutionary system. Because the sampling and rounding error issues, the computed fitness values for the filters are similar to but not exactly the same as those reported by the human expert (Lu 1999).

4.2   Results

The filter design problem is formulated as a multi-objective optimisation problem in this paper. As such, it is difficult to identify a single 'best' individual in any population. For statistical comparisons, however, single performance numbers are very useful. For this reason, a weighted sum is computed for each individual, and used in many of the comparisons in this paper. The weights are somewhat arbitrary. They have been chosen to roughly give each criterion a similar weight, corrected for the magnitude of the expected value. As long as the weights are positive an individual that outperforms another in all fitnesses will have a better weighted average independent of the weights. The weighted average therefore proves to be a sufficient indicator of the performance of the algorithm. For a few individual designs, the performance with respect to all three objectives is shown.

Two sets of 30 runs have been performed for each of the two test cases. One set used an algorithm with a seeded initial population as described above. As a comparison set, the other runs were performed using exactly the same algorithm and parameter settings, but with a completely random initial population. Each run was allowed to run for 1000 generations with a population size of 550 individuals.

Table 1 shows the results of the runs. The numbers in the table indicate the weighted fitness for an individuals. The first column shows the performance of the conventional human design. The next two columns show the performance of seeded and unseeded runs. Each column shows the performance of the best design in the best run, the worst run, and average and median of the best individual over the runs.

As another comparison, the last column shows the performance of the best individual in a run using an algorithm similar to that described in Schnier et al. (2001). This algorithm, which uses a combination of clustering and Pareto optimisation, is computationally more intensive but requires no seeding. Apart from using slightly different parameters, the main difference between the algorithm used here and that shown in Schnier et al. (2001) is the fact that tournament selection based on shared fitness was used here. Not enough runs have been performed to include meaningful values for mean, average and worst performance in the last column.

Table 1. Performance of evolutionary calibration and comparison with other methods for the two test cases.

	Conventional	Seeded	Unseeded	Clustering	Case
Best	6.293	4.103	5.092	4.842	1
Mean		4.675	7.907		1
Average		4.63	8.281		1
Worst		5.091	11.85		1
Best	10.099	8.433	7.75	8.741	2
Mean		8.687	11.05		2
Average		8.700	11.13		2
Worst		8.910	16.385		2

What the results show is that it is indeed possible to improve the performance of the conventionally derived design using the evolutionary approach. The improvement is fairly consistent over different runs; between 20% and 35% in the first design case, and between 12% and 16% in the second design case.

The designs created using the evolutionary process without any seeding are more variable. Only a few of the runs produced results with a better combined fitness than the conventional design. For the first design case, even the results of the best run is worse than the worst run using a seeded population. For the second design case, however, there is a surprise: while the designs are similarly variable, the best design has a considerably better fitness than any of the other designs. This indicates one potential limitation of evolutionary calibration, i.e., it is most likely to search around the initial design.

As mentioned previously, the overall combined fitness depends on the weights used to calculate it. It is useful to examine individual fitnesses in order to understand the differences among different filters better. Table 2 gives the actual fitness values according to three different criteria, which were used in generating the combined fitness.

Table 2. Fitness values, according to the three separate criteria, of the best individual from four different approaches. The first test case was used.

Method	Passband Amplitude	Passband Delay	Stopband Amplitude	Case
Conventional	0.103	0.293	0.023	1
Seeded	0.100	0.115	0.019	1
Unseeded	0.119	0.163	0.023	1
Clustering	0.07	0.148	0.026	1
Conventional	0.27	0.437	0.030	2
Seeded	0.266	0.296	0.028	2
Unseeded	0.145	0.263	0.037	2
Clustering	0.31	0.178	0.038	2

Figure 2 illustrates the difference in performance for the first test case. In Figure 2(a) , the amplitude in passband and stopband are shown. The stopband amplitude is noticeably lower in the computer designed case, and the passband amplitude deviates slightly less from a linear amplitude. Similarly, the deviation from the goal delay in the passband is slightly less in the computer generated design (Figure 2(b)).

It is clear from Table 2 that the designs created through evolutionary calibration can improve the conventional design according to all three fitness criteria. In contrast, EAs without any seeding tend to be very good at exploring the whole design space and create designs that have an extremely good fitness value according to a single criterion. Such results confirm our expectation that evolutionary calibration is good at exploitation and fine-tuning an existing design, while other evolutionary approaches (Schnier et al. 2001) are better at exploration and discovering novel designs.

Finally, a comparison of the phenotypes of the designs created in Table 3 shows the amount of calibration done by the EA for the first test case. Although changes are small in values, they are noticeable. For example in the first case,

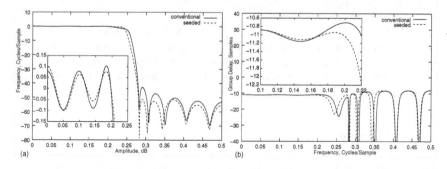

Fig. 2. Performance of conventional design and computer generated design, for the first test case: (a) amplitude, (b) group delay

Table 3. Phenotypes generated by the conventional and evolutionary approaches for the test case 1.

	Evolutionary Calibration	Conventional
Zero Pair	$1.4966 \pm i0.482$	$1.4968 \pm i0.4822$
Zero Pair	$0.9252 \pm i1.250$	$0.8976 \pm i1.2291$
Zero Pair	$-0.9977 \pm i0.206$	$-0.9990 \pm i0.2001$
Zero Pair	$-0.2062 \pm i0.979$	$-0.2022 \pm i0.9800$
Zero Pair	$-0.8521 \pm i0.566$	$-0.8544 \pm i0.5624$
Zero Pair	$-0.3347 \pm i0.947$	$-0.3546 \pm i0.9392$
Zero Pair	$-0.5588 \pm i0.841$	$-0.5986 \pm i0.8131$
Real Zero	-14.3286	18.6313
Pole Pair	$0.0792 \pm i0.555$	$0.0557 \pm i0.5576$
Pole Pair	$0.0295 \pm i0.929$	$-0.0325 \pm i0.9356$
Real Pole	0.2366	0.2062
b0	0.0045598	-0.0004564

the second pole pair have moved from the negative to the positive half plane (the sign changes in b_0 and the real zero cancels each other, because the frequency response is symmetrical in respect to the z-axis). In the second design case, the fifth zero pair have moved the most, from $-1.1449 \pm i0.0360$ to $-1.0793 \pm i0.2636$.

5   Conclusions

This work has shown that an evolutionary system can be used successfully to calibrate designs created using existing conventional design methods. The improvements are consistent over many runs and sufficiently large to be of relevance and importance. Seeding the initial population with the conventional design is a simple and efficient method to combine the conventional and evolutionary design processes. Although filter design was used as the domain of investigation in this paper, the approach and techniques described here are applicable to other design problems.

Evolutionary calibration is very good at exploitation and tuning an existing design. However, it is unsuitable for exploring a large design space and discovering novel designs that are drastically different from the original one. Both evolutionary calibration and unseeded evolutionary design have their own niches in evolutionary hardware design.

Acknowledgement

This research is generously supported by a grant from Marconi Communications, Ltd. We are grateful to Pin Liu and John Evans for their support and comments on this work.

References

Antoniou, A. (1993). *Digital Filters - Analysis, Design and Applications*, 2 edn, Mc-Graw Hill International.

Baicher, G. S. & Turton, B. (2000). Comparative study for optimization of causal IIR perfect reconstruction filter banks, *Proceedings of the 2000 Congress on Evolutionary Computation (CEC 2000)*, IEEE, pp. 974–977.

Fogel, D. B. & Angeline, P. J. (1997). Guidelines for a suitable encoding, *in* T. Bäck, D. B. Vogel & Z. Michalewicz (eds), *Handbook of Evolutionary Computation*, IOP Publishing and Oxford University Press, chapter C1.7:1. Release 97/1.

Ibaraki, T. (1997). Uses of problem-specific heuristics, *in* T. Bäck, D. B. Vogel & Z. Michalewicz (eds), *Handbook of Evolutionary Computation*, IOP Publishing Ltd and Oxford Press, chapter D3.3:1. Release 97/1.

Langdon, W. B. & Nordin, J. P. (2000). Seeding GP populations, *in* R. Poli, W. Banzhaf, W. B. Langdon, J. F. Miller, P. Nordin & T. C. Fogarty (eds), *Genetic Programming, Proceedings of EuroGP'2000*, Vol. 1802 of *LNCS*, Springer-Verlag, Edinburgh, pp. 304–315.

Louis, S. J. & Johnson, J. (1997). Solving similar problems using genetic algorithms and case-based memory, *in* T. Bäck (ed.), *Proceedings of the 7th International Conference on Genetic Algorithms*, Morgan Kaufmann, San Francisco, pp. 283–290.

Lu, W. S. (1999). Design of stable IIR digital filters with equiripple passbands and peak-constrained least-sqaures stopband, *IEEE Transactions on Circuits and Systems II: Analog and Digital Signal Processing* **46**(11): 1421–1426.

Miller, J. (1999). Digital filter design at gate-level using evolutionary algorithms, *in* W. Banzhaf et al. (eds), *Proceedings of the Genetic and Evolutionary Computation Conference (GECCO'99)*, Morgan Kaufmann, San Francisco, CA, pp. 1127–1134.

Parks, T. & Burrus, C. (1987). *Digital Filter Design*, Topics in Digital Signal Processing, John Wiley & Sons.

Schnier, T. & Yao, X. (2000). Using multiple representations in evolutionary algorithms, *Congress on Evolutionary Computation CEC 2000*, Vol. 1, IEEE Neural Network Council, IEEE, pp. 479–487.

Schnier, T., Yao, X. & Liu, P. (2001). Digital filter design using multiple pareto fronts, *The 3rd NASA/DoD Workshop on Evolvable Hardware (EH-2001)* (to be published).

Uesaka, K. & Kawamata, M. (1999). Synthesis of low coefficient sensitivity digital filters using genetic programming, *Proceedings of the IEEE International Symposium on Circuits and Systems*, Vol. 3, pp. 307–310.

Uesaka, K. & Kawamata, M. (2000). Synthesis of low-sensitivity second-order digital filters using genetic programming with automatically defined functions, *IEEE Signal Processing Letters* **7**(4): 83–85.

Wilson, P. B. & Macleod, M. D. (1993). Low implementation cost IIR digital filter design using genetic algorithms, *IEE/IEEE Workshop on Natural Algorithms in Signal Processing*, Chelmsford, U.K., pp. 4/1–4/8.

Yao, X. & Liu, Y. (1997). Fast evolution strategies, *Control and Cybernetics* **26**(3): 467–496.

Yao, X., Liu, Y. & Lin, G. (1999). Evolutionary programming made faster, *IEEE Transactions on Evolutionary Computation* **3**: 82–102.

Yao, X. (1999a). Following the path of evolvable hardware, *Communications of the ACM* **42**(4): 47–49.

Implementation of a Gate-Level
Evolvable Hardware Chip

Masaya Iwata[1], Isamu Kajitani[1], Yong Liu[2],
Nobuki Kajihara[3], and Tetsuya Higuchi[1]

[1] National Institute of Advanced Industrial Science and Technology,
AIST Tsukuba Central 2, 1-1-1 Umezono, Tsukuba, Ibaraki, 305-8568 Japan,
{m.iwata, isamu.kajitani, t-higuchi}@aist.go.jp
[2] The University of Aizu,
Tsuruga, Ikki-machi, Aizu-Wakamatsu, Fukushima, 965-8580 Japan,
yliu@u-aizu.ac.jp
[3] Adaptive Device NEC Laboratory, RWCP,
4-1-1, Miyazaki, Miyamae-ku, Kawasaki, Kanagawa, 216-8555 Japan,
n-kajihara@ab.jp.nec.com

Abstract. Evolvable hardware (EHW) is hardware that can change its
own circuit structure by genetic learning to achieve maximum adaptation to the environment. In conventional EHW, the learning is executed
by software on a computer. However, there are problems associated with
this method, of slow learning speeds and large systems, which are serious
obstacles to utilizing EHW in various kinds of practical applications. To
overcome these problems, we have developed a gate-level evolvable hardware chip, by integrating both GA hardware and reconfigurable hardware
within a single LSI chip. The chip consists of genetic algorithm (GA)
hardware, reconfigurable hardware logic, and the control logic. With this
chip, we have successfully executed GA learning and hardware reconfiguration. In this paper, we describe the architecture, functions, and a
performance evaluation of the chip. We show that its learning speed is
considerably faster than with software.

1   Introduction

Evolvable hardware (EHW) is hardware that can adapt to new environments,
which cannot be anticipated by hardware designers [1], [2]. EHW contrasts with
conventional hardware, which has no provision for adaptive change. Consisting
of reconfigurable devices, EHW is adaptive hardware where the architecture can
be reconfigured through use of genetic algorithms (GAs) [3].

   In conventional EHW, the learning is executed by software on a computer.
However, this method is not always practical due to slow learning speeds and
large system sizes. These are serious problems to utilizing EHW in various kinds
of practical applications. To overcome these problems, we have previously proposed a gate-level EHW chip consisting of GA hardware and reconfigurable
hardware (PLA: Programmable Logic Array) [4]. This was the first EHW chip

Y. Liu et al. (Eds.): ICES 2001, LNCS 2210, pp. 38–49, 2001.

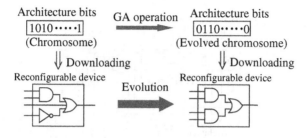

Fig. 1. Conceptual diagram of evolvable hardware (EHW).

where the GA hardware and the reconfigurable hardware were integrated within a single LSI chip. A discussion of the design considerations for the GA hardware and an overview of the chip are provided in [4].

This paper describes the more precise chip architecture of the latest version of the chip. In this version, we have improved most of the circuits and have added some new optional functions to the chip for more adaptive learning. As a result, we have successfully executed the GA operation and hardware reconfiguration within the chip. In this paper, we describe the architecture, functions, and a performance evaluation of the chip. We compare the processing speed to that of a program running on a personal computer, and show that the learning speed of the chip is considerably faster. This chip makes it easier to apply EHW to practical applications that require on-line hardware reconfiguration, because application systems using this chip can be much smaller.

This paper consists of the following sections. Section 2 outlines the EHW concept and its problems. Section 3 explains the developed EHW chip. Section 4 describes the performance evaluation and Section 5 concludes the paper.

2   Evolvable Hardware Chip

2.1   Evolvable Hardware (EHW)

Evolvable Hardware (EHW) modifies its own hardware structure in accordance with environmental changes [1], [2]. EHW is implemented on a reconfigurable device, where the architecture can be altered by downloading a binary bit string, or *architecture bits*. The PLA (Programmable Logic Array) and the FPGA (Field Programmable Gate Array) are typical examples of reconfigurable devices. Appropriate architecture bits for the devices are evolved by the genetic algorithm (GA).

The basic idea of EHW is to regard the architecture bits of a reconfigurable device as a chromosome for a GA, which searches for an optimal hardware structure (see Fig. 1). The GA chromosome, that is the architecture bits, are downloaded onto the reconfigurable device during genetic learning. Therefore, EHW can be considered as on-line adaptive hardware.

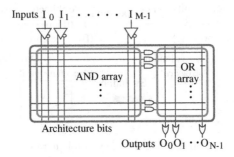

Fig. 2. PLA architecture for EHW.

The reconfigurable device we use is a PLA (Programmable Logic Array) device, which is a popular commercial device (Fig. 2). This architecture consists mainly of an AND array and an OR array.

2.2   Basic Concepts of EHW Chip

In conventional EHW, genetic learning is executed with software on computers. However, this method always requires a personal computer or workstation, which increases the size of the system. Furthermore, the learning speed is slow due to the execution by software. These restrictions reduce the range of applications. For example, myoelectric artificial hands should be the same size as a human hand and weigh less than 700 gram. This means that conventional EHW cannot be applied to myoelectric artificial hands as long as large computers are used to execute the learning. Our solution to this problem is to implement the GA operations with hardware, and to integrate the GA hardware together with the reconfigurable device within a single LSI.

Based on survey and design considerations for the GA hardware, we have designed the GA hardware for compact implementation. Accordingly, we have combined elitist recombination [8] with uniform crossover. We have proposed an EHW chip according to this scheme [4]. In the latest version of the chip we have improved most of the circuits and have added some new optional functions. As a result, we have been successfully in producing a complete chip.

2.3   Applications

In this section we introduce two applications for the EHW chip. First, this chip has been applied to a myoelectric artificial hand (Fig. 3) [4], [5], [6]. The chip is used as an artificial-hand controller to recognize appropriate hand actions using myoelectric signals. Although users of conventional myoelectric hands have to adapt to the hand through a long period of training (almost one month), by using the EHW chip, the learning time for the artificial hand can be reduced to a few minutes because the EHW chip can adapt itself quickly. We have tested the possibility of using this chip for the artificial hand in a simulation, confirming that the myoelectric-signal learning is effective for fast adaptations of hands. We are currently making an experiment system for the adaptive control of an artificial hand using the chip.

Another application of this chip is to the adaptive navigation task for a real-world mobile robot that must track a moving colored ball while avoiding obstacles (Fig. 4) [11]. The adaptation speed of Evolver is two orders of magnitude faster than speeds achieved with classical approaches. In this robot, the EHW is used as control logic, which is continuously updated to adapt to the changing environment. We have tested the possibility of using this chip for the robot in a simulation, where the learning of the robot controller circuit was successfully completed [9].

In this paper, we report a performance-evaluation experiment using training patterns for the myoelectric artificial hand.

Fig. 3. A myoelectric artificial hand.

Fig. 4. The autonomous mobile robot, Evolver.

3 Chip Architecture

3.1 Overview

A block diagram of the EHW chip is shown in Fig. 5. The EHW chip consists of a genetic algorithm (GA) unit, a PLA (programmable logic array) unit as the reconfigurable hardware logic, registers, and the control logic. The merit of this architecture is that it can process two chromosomes in parallel. As shown in Fig. 5, this architecture has two ports for parallel access to the external 2-port RAM. The GA unit and PLA unit also have a parallel-processing architecture for the data stream from the two ports. This architecture was first proposed in [4]. In this chip we have improved most of the circuits and have introduced some new

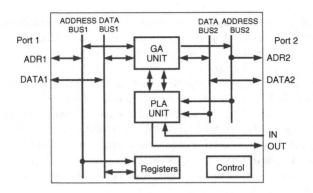

Fig. 5. A block diagram of the EHW chip.

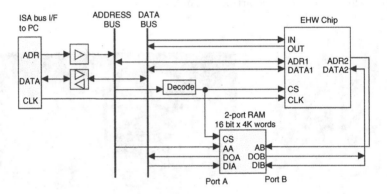

Fig. 6. A block diagram of the EHW chip board.

optional functions. The main functional units and the workflow are explained in the next two sections.

The chip has to be connected to a memory and a CPU. A block diagram of the board for the EHW chip is shown in Fig. 6. A photograph of the board is shown in Fig. 7. The EHW chip is located on the upper side. The board consists of the EHW chip, a 2-port RAM, and an interface controller. The RAM consists of a chromosome memory, a training-pattern memory, and a memory for the fitness value. In our board, we have used an ISA bus controller for the interface with a personal computer. The reported experiment was conducted using this board.

3.2   Functions

GA Unit The GA unit executes the GA learning operations using the steady state GA and elitist recombination [8]. A block diagram of the GA unit is shown in Fig. 8. In this figure, only the data bus is shown. The procedure of the GA unit

Fig. 7. A photograph of the EHW chip board.

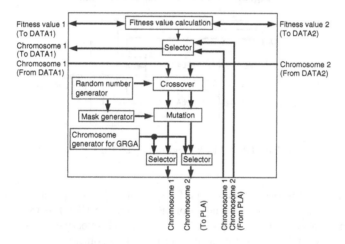

Fig. 8. A block diagram of the GA unit.

is shown in Fig. 9. The GA unit selects two chromosomes from the chromosome memory in parallel in units of 32 bits, and then carries out uniform crossover and mutation on these to make two chromosome segments of 32 bits (Fig. 9 (a)). Uniform crossover is carried out using a random 32-bit string. If a location in the random bit string has a value of '1', then the information corresponding to the same location in the two segments is exchanged. A mutation rate of either 0, 1/256, 2/256, or 3/256 can be selected. The two new chromosome segments are then sent to the PLA. After all chromosome segments have been sent, their fitness values are calculated using training patterns. The fitness values of the two children and the two parents are compared and two chromosomes survive according to the rule described below (Fig. 9 (b)). This GA unit was proposed in [4] as a suitable GA unit for hardware implementation.

When this GA unit compares fitness values, it carries out elitist recombination with a new option for more adaptive learning. In conventional elitist

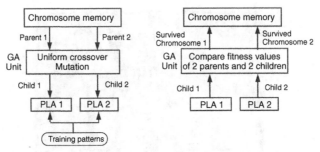

(a) Generation of two children   (b) Keeping two chromosomes.
and evaluation.

Fig. 9. The procedure for the GA unit.

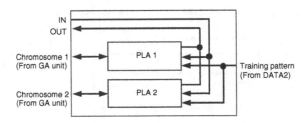

Fig. 10. A block diagram of the PLA unit.

recombination, the best chromosome from among the two parents and the two children always survived. The new option can ensure that the worst parent is always exchanged with the best child. This method is effective when the learning process has a tendency to stick at a local minimum solution.

This GA unit has an option for GRGA (Gene Replacement Genetic Algorithm) [5], which has been proposed to accelerate the genetic search by replacing a part of a chromosome with a bit string, referred to as a chromosome candidate segment. In this chip, the chromosome candidate segment is generated from a training input-output pattern used for the evaluation of circuit candidates. For more details, refer to [5].

This chip has an on-line editing mode for the training-pattern memory. In the previous EHW chip, the training pattern memory could not be edited once the GA unit had begun to operate. Therefore, the training sample patterns had to be made off-line. This chip allows us to change the training-pattern memory during learning to provide on-line learning. This helps to ensure a smooth adaptation process [5].

PLA Unit. A block diagram of the PLA (Programmable Logic Array) unit is shown in Fig. 10. In this figure, only the data bus is shown. There are two PLAs for parallel evaluation of two circuits. These blocks read two chromosomes from the GA unit in parallel in units of 32 bits to implement two circuits in

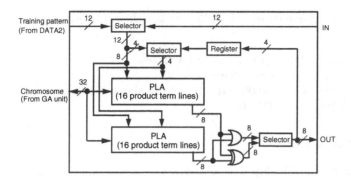

Fig. 11. A block diagram of one PLA.

parallel. The two circuits are then evaluated using the training data. We can select one of two input-output modes: an 8-bit input/8-bit output mode, or a 12-bit input/4-bit output mode.

The PLAs have a new architecture as shown in Fig. 11. We have introduced XOR logic in the final stage of the PLA. In a PLA, 32 product term lines are divided into two groups of 16 lines. Each bit of the output (8-bits) from the two groups can be connected to either an OR, or an XOR gate. If the OR gate is selected, the operation is the same as a conventional PLA. If the XOR gate is selected, then the XOR operation is executed on the each bit of the two outputs from the two groups. If XOR is used with AND and OR gates, the PLA can generate a circuit with less product term lines [10]. This option is useful for circuits that need many product term lines.

Each PLA has an option for selecting a feedback loop from the output to the input. If this option is selected, the upper 4 bits of the PLA output is connected to the upper 4 bits of the PLA input via a register (Fig. 11). This feedback loop is useful when the EHW has to learn a sequential circuit, because the feedback loop can store the state of the circuit.

Random Number Generator. A parallel random number generator using cellular automata [7] was selected for implementation on the EHW chip as this is very popular for GA hardware, and can produce 560 bits of random bit-string at every clock cycle.

RAM on the Board. The EHW chip works with a 2-port RAM on the board. The RAM is divided into three memories: a chromosome memory, a training-pattern memory, and a memory for the fitness value. The chromosome memory stores all the individuals in 16 bits x 2048 words. The chromosome length is 1024 bits, and the population number is 32. This memory has two input/output ports. Using these ports, two chromosomes can be read or written in parallel in units of 16 bits from the GA unit. The training data memory can store a maximum of 256 training data set of 16 bits. The memory for the fitness data stores all the fitness values for all 32 chromosomes with an 8-bit integer value.

Technical Information. The following is technical information for the EHW chip.

- Package: 144pins QFP, 20x20mm. Cell base LSI.
- Circuit size: about 80,000 gates.
  (GA unit: about 20,000 gates, PLA unit: about 46,000 gates.)
- Clock frequency: 33MHz.

3.3 The Workflow

The workflow of this chip is as follows, which is also expressed in Fig. 9.

1. Initialization
   This process initializes the control signal, the register, the random-number generator, and the memory.
2. Selection of two parents
   This process selects two chromosomes as parents. One is selected using the random number, and the other is the best chromosome.
3. Crossover, Mutation
   This process operates uniform crossover and mutation on the two chromosomes. Two new chromosomes are then generated as two children, which are applied to configure the two PLAs.
4. Fitness calculation
   Training patterns are loaded on the PLA inputs. The PLA outputs are then evaluated and fitness values are calculated.
5. Keeping two chromosomes for the next evaluation
   The fitness values of the two parents and the two children are compared and two chromosomes survive for the next evaluation process. The survived chromosomes are written on the chromosome memory.
6. Return to 2.

The crossover and mutation are operated with 32 bits each to manage the chromosome of 1024 bits.

4   Performance Evaluation

4.1   Experiment

We conducted a learning experiment on the EHW chip to evaluate its performance. The EHW board (Fig. 6) was connected to an ISA bus slot on a DOS/V personal computer. The clock frequency of the board is 33MHz. The training patterns were made from the myoelectric artificial hand system described in Sec. 2.3. The input was 8 bits that was encoded from human myoelectric signals [6]. The output was 8 bits. The lower 6 bits correspond to the 6 movements of the hand (open, grasp, supination, pronation, flection, and extension), and the upper 2 bits are not used. The number of training patterns was 252.

Table 1. Comparison of execution times for the EHW chip and the program.

Function	EHW chip (μs)	Program (μs)	(Program)/(EHW chip)
One evaluation	94.8	3670	38.7
Crossover and mutation	12.6	78.9	6.3
Fitness calculation	68.2	3560	52.2
Comparison of fitness values	0.03	0.15	5.0
PLA execution	0.03	13.42	447.3

4.2   Execution Time

To measure the effect of the hardware implementation of the GA operations, we compared the execution time with a GA program for the same algorithm written in C language. The program was executed on a personal computer with an AMD Athlon processor (1.2 GHz). The results are shown in Table 1. The processing time of the EHW chip for one evaluation was 94.8 μs, and that of the program was 3670 μs. These results show that the execution speed of the EHW chip is 38.7 times faster than the program on a personal computer. That means that the learning speed of the EHW chip is considerably faster. For example, the learning time of the EHW chip (252 training data, 10000 evaluations) was 0.948 seconds, whereas that of the GA program was 36.7 seconds. Table 1 also shows the execution time for each function. The fastest function compared with the program is the fitness calculation. The execution speed is 52.2 times faster than that of the program. The main reason for this is that the PLA execution time was 447 times faster than that of the program. The PLA execution is included in the fitness calculation and is called 252 times (the number of the training patterns) in one execution.

There are some possible methods to increase the execution speed still further, for this chip was made primarily for evaluation purposes and there is still some slack in the timing. For example, the processing speed for crossover and mutation could be made at least 3 times faster by improving the timing of the read or write control signals for the RAM and the register. It would also be possible to speed up the system clock.

4.3   Learning

We have measured the correct output rate for the training data with the chip. The maximum correct output rate was 72% at about 10000 evaluations. The mutation rate was 0.0117 (3/256). This rate is about the same as that achieved with a program of the same algorithm at about 10000 evaluations. Although a maximum correct output rate of 85% was achieved on the program at about 100000 evaluations, the number of evaluations with the chip is presently limited to 16384. Increasing the maximum number of evaluations in order to improve the correct output rate is an issue for further research. The correct output rate might increase if the range of each GA operation value becomes wider.

5 Conclusions

We have developed an evolvable hardware chip by integrating the learning hardware for genetic algorithm (GA), and reconfigurable hardware logic (PLA). GA learning and hardware reconfiguration were successfully carried out on the chip. We have described its detailed architecture and functions. We have evaluated its performance in an experiment, and confirmed that the execution speed was about 40 times faster than with a GA program on a personal computer. This chip represents a breakthrough for practical applications that require on-line hardware reconfiguration, because smaller systems are possible using this chip compared to conventional EHW systems.

Acknowledgement

This work is supported by MITI Real World Computing Project. We thank Dr. Otsu and Dr. Ohmaki at AIST, Dr. Shimada of RWCP for their support.

References

1. Higuchi, T., Niwa, T., Tanaka, T., Iba, H., de Garis, H., Furuya, T.: Evolvable Hardware with Genetic Learning. Proc. Simulation of Adaptive Behavior, MIT Press (1993) 417-424
2. Higuchi, T., Iba, H., Manderick, B.: Evolvable Hardware. Massively Parallel Artificial Intelligence, MIT Press (1994) 398-421
3. Goldberg, D.: Genetic Algorithms in Search, Optimization, and Machine Learning. Addison Wesley (1989)
4. Kajitani, I., Hoshino, T., Nishikawa, D., Yokoi, H., Nakaya, S., Yamauchi, T., Inuo, T., Kajihara, N., Iwata, M., Keymeulen, D., Higuchi, T.: A Gate-Level EHW Chip: Implementing GA Operations and Reconfigurable Hardware on a Single LSI. Evolvable Systems: From Biology to Hardware, Lecture Notes in Computer Science, Vol. 1478, Springer Verlag, Berlin (1998) 1-12
5. Kajitani, I., Hoshino, T., Kajihara, N., Iwata, M., Higuchi, T.: An Evolvable Hardware Chip and its Application as a Multi-Function Prosthetic Hand Controller. Proc. 16th National Conference on Artificial Intelligence (AAAI-99) (1999) 182-187
6. Kajitani, I., Sekita, I., Otsu, N., Higuchi, T.: Improvements to the Action Decision Rate for a Multi-Function Prosthetic Hand. Proc. 1st International Symposium on Measurement, Analysis and Modeling of Human Functions (2001) (in press)
7. Hortensius, P. D., et al.: Parallel Random Number Generation for VLSI Systems using Cellular Automata. IEEE Trans. Computers, Vol. 38, No. 10 (1989) 1466-1473
8. Thierens, D., Goldberg, D. E.: Elitist Recombination: an Integrated Selection Recombination GA. Proc. 1st IEEE Conference on Evolutionary Computation (1994) 508-512
9. Liu, Y., Iwata, M., Higuchi, T., Keymeulen, D.: An Integrated On-Line Learning System for Evolving Programmable Logic Array Controllers. Proc. Parallel Problem Solving from Nature (2000) 589-598

10. Debnath, D., Sasao, T.: Minimization of AND-OR-XOR Three-Level Networks with AND Gate Sharing. IEICE Trans. Information and Systems, Vol. E80-D, No 10 (1997) 1001-1008
11. Keymeulen, D., Iwata, M., Kuniyoshi, Y., Higuchi, T.: Online Evolution for a Self-Adaptive Robotic Navigation System using Evolvable Hardware. Artificial Life, Vol. 4 (1999) 359-393

A VLSI Implementation of an Analog Neural Network Suited for Genetic Algorithms

Johannes Schemmel, Karlheinz Meier, and Felix Schürmann

Universität Heidelberg, Kirchhoff Institut für Physik,
Schröderstr. 90, 69120 Heidelberg, Germany,
schemmel@asic.uni-heidelberg.de,
http://www.kip.uni-heidelberg.de/vision.html

Abstract. The usefulness of an artificial analog neural network is closely bound to its trainability. This paper introduces a new analog neural network architecture using weights determined by a genetic algorithm. The first VLSI implementation presented in this paper achieves 200 giga connections per second with 4096 synapses on less than 1 mm^2 silicon area. Since the training can be done at the full speed of the network, several hundred individuals per second can be tested by the genetic algorithm. This makes it feasible to tackle problems that require large multi-layered networks.

1 Introduction

Artificial neural networks are generally accepted as a good solution for problems like pattern matching etc. Despite being well suited for a parallel implementation they are mostly run as numerical simulations on ordinary workstations. One reason for this are the difficulties determining the weights for the synapses in a network based on analog circuits. The most successful training algorithm is the back-propagation algorithm. It is based on an iteration that calculates correction values from the output error of the network. A prerequisite for this algorithm is the knowledge of the first derivative of the neuron transfer function. While this is easy to accomplish for digital implementations, i.e. ordinary microprocessors and special hardware, it makes analog implementations difficult. The reason for that is that due to device variations, the neurons' transfer functions, and with them their first derivatives, vary from neuron to neuron and from chip to chip. What makes things worse is that they also change with temperature. While it is possible to build analog circuitry that compensates all these effects, this likely results in circuits much larger and slower than their uncompensated counterparts. To be successful while under a highly competitive pressure from the digital world, analog neural networks should not try to transfer digital concepts to the analog world. Instead they should rely on device physics as much as possible to allow an exploitation of the massive parallelism possible in modern VLSI technologies. Neural networks are well suited for this kind of analog implementation since the compensation of the unavoidable device fluctuations can be incorporated in the weights.

Y. Liu et al. (Eds.): ICES 2001, LNCS 2210, pp. 50–61, 2001.

A major problem that still has to be solved is the training. A large number of the analog neural network concepts that can be found in the literature use floating gate technologies like EEPROM of flash memories to store the analog weights (see [1] for example). At a first glance this seems to be an optimal solution: it consumes only a small area making therefore very compact synapses possible (down to only one transistor [2]), the analog resolution can be more than 8 bit and the data retention time exceeds 10 years (at 5 bit resolution) [3]. The drawback is the programming time and the limited lifetime of the floating gate structure if it is frequently reprogrammed. Therefore such a device needs predetermined weights, but to calculate the weights an exact knowledge of the network transfer function is necessary. To break this vicious circle the weight storage must have a short write time. This would allow a genetic algorithm to come into play. By evaluation of a high number of test configurations the weights could be determined using the real chip. This could also compensate a major part of the device fluctuations, since the fitness data includes the errors caused by these aberrations.

This paper describes an analog neural network architecture optimized for genetic algorithms. The synapses are small, 10×10 μm^2, and fast. The measured network frequency exceeds 50 MHz, resulting in more than 200 giga connections per second for the complete array of 4096 synapses. For building larger networks it should be possible to combine multiple smaller networks, either on the same die or in different chips. This is achieved by confining the analog operation to the synapses and the neuron inputs. The network inputs and neuron outputs are digitally coded. The synapse operation is thereby reduced from a multiplication to an addition. This makes the small synapse size possible and allows the full device mismatch compensation, because each synapse adds either zero or its individual weight that can include any necessary corrections. Analog signals between the different analog network layers are represented by arbitrary multi-bit connections.

The network presented in this paper is optimized for real-time data streams in the range of 1 to 100 MHz and widths of up to 64 bits. We plan to use it for data transmission applications like high speed DSL[1], image processing based on digital edge data produced from camera images by an analog preprocessing chip [4] and for the fitness evaluation of a field programmable transistor array [5] also developed in our group.

2   Realization of the Neural Network

2.1   Principle of Operation

Figure 1 shows a symbolic representation of a recurrent neural network. Each input neuron (small circle) is linked to each output neuron (large circle) by a synapse (arrow). The output neurons are fed back into the network by a second set of input neurons. The input neurons serve only as amplifiers, while the

---

[1] digital subscriber line

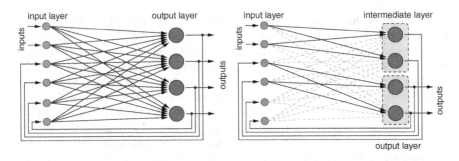

Fig. 1. Left: A recurrent neural network. Right: The same network configured as a two layer network.

processing is done at the output neurons. This architecture allows virtual multi-layer networks by choosing the appropriate weights. On the right of Figure 1 an example is shown for two layers. Synapse weights set to zero are depicted as dashed arrows. A recurrent network trained by a genetic algorithm has usually no fixed number of layers. Of course, the algorithm can be restricted to a certain number of layers, as in Figure 1, but usually it seems to be an advantage to let the genetic algorithm choose the best number of layers. Also, there is no strict boundary between the virtual layers. Each neuron receives input signals from all layers. To avoid wasting synapses if not all the feedback pathways are used, the presented network shares input neurons between external inputs and feedback outputs.

Figure 2 shows the operation principle of a single neuron. The synaptic weights are stored as charge on a capacitor (storage capacitor). The neuron oper-

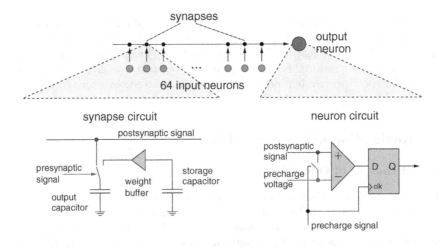

Fig. 2. Operation principle of the neuron.

ation is separated in two phases, *precharge* and *evaluate*. In the precharge phase all the switches in the synapses are set towards the buffer and the precharge signal in the neuron is active. In each synapse the output capacitor is charged via the weight buffer to the same voltage as the storage capacitor. The neuron consists of a comparator and a storage latch. The precharge signal closes a switch between both inputs of the comparator. This precharges the post-synaptic signal to a reference voltage that constitutes the zero level of the network.

In the evaluate phase the sum of all the synapses is compared to this precharge voltage. If the synapse signal exceeds it, the neuron fires. This neuron state is stored in the flip-flop at the moment when the phase changes from evaluate to precharge. In the evaluate phase the synapse switch connects the output capacitor with the post-synaptic signal if the pre-synaptic signal is active. The pre-synaptic signals are generated by the input neurons depending on the network input and feedback information.

This cycle can be repeated a fixed number of times to restrict the network to a maximum layer number and limit the processing time for an input pattern. The network can also run continuously while the input data changes from cycle to cycle. This is useful for signal processing applications.

Figure 3 shows a block diagram of the developed neural network prototype. The central element is an array of 64×64 synapses. The post-synaptic lines of

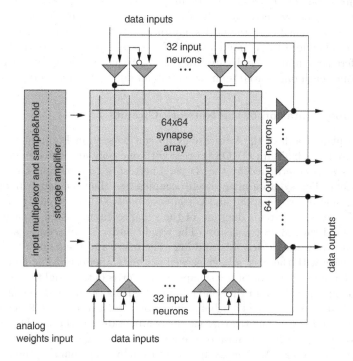

Fig. 3. Block diagram of the neural network prototype.

the 64 output neurons run horizontally through the synapses, while the pre-synaptic signals are fed into the array from the input neurons located below and above. Each input neuron can be configured to accept either external data or data from the network itself for input. This internal data comes alternately from two sources. The odd input neurons receive a feedback signal from an output neuron while the even ones get the inverted output from its odd neighbor. If the even neuron is switched to its odd counterpart, they together form a *differential* input since the binary signal is converted into a pair of alternately active inputs. This is useful for two purposes: if binary coded data is used the number of active input neurons stays always the same, independently of the input data. The second reason is linked to the way the post-synaptic voltage is calculated:

$$V_{postsyn} = \frac{\sum_{i=1}^{64} I_i Q_i}{\sum_{i=1}^{64} I_i C_i} \qquad (1)$$

Q_i is the charge stored on the synapse output capacitor C_i. I_i is the pre-synaptic signal. As a binary value it is either zero or one. The neuron fires if $V_{postsyn} > V_{precharge}$. Not only the numerator, but also the denominator depends on all the input signals. This has the drawback that if one input signal changes, the other weights' influence on $V_{postsyn}$ changes also. Even though it can be shown that in the simplified model of Eq. 1 the network response stays the same, the performance of the real network may suffer. The differential input mode avoids this effect by activating always one input neuron per data input. The capacitance switched onto the post-synaptic signal line becomes independent of the data. Therefore the denominator of Eq. 1 stays the same for any changes of a differential input. The disadvantage is the reduced number of independent inputs since each differential input combines an odd with an even input neuron.

2.2   Implementation of the Network Circuits

A micro photograph of the fabricated chip can be seen in Figure 4. The technology used is a 0.35 μm CMOS process with one poly and three metal layers. The die size is determined by the IO pads necessary to communicate with the test system. The synapse array itself occupies less than 0.5 mm^2. It operates from a single 3.3 volt supply and consumes about 50 mW of electrical power. Figure 5 shows the circuit diagram of the synapse circuit. Both capacitors are implemented with MOS-transistors. The weight buffer is realized as a source follower built from the devices M1 and M2. The offset and the gain of this source follower vary with the bias voltage as well as the temperature. Therefore an operational amplifier outside of the synapse array corrects the weight input voltage until the output voltage of the source follower equals the desired weight voltage which is fed back via M7. The charge injection error caused by M6 depends on the factory induced mismatch that can be compensated by the weight value. M3 is closed in the precharge phase of the network to charge the output capacitor to the weight voltage. M5 speeds up this process by fully charging the capacitor first. Since the output current of the source follower is much larger for a current

Fig. 4. Micro photograph of the neural network chip.

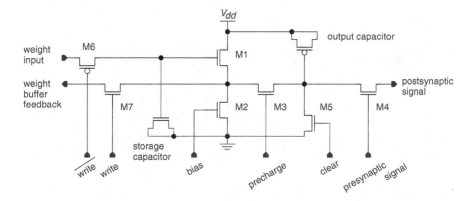

Fig. 5. Circuit diagram of the synapse.

flowing out of M1 instead of into M2 it discharges the capacitor faster than it is able to charge it. The total time to charge the output capacitor to the desired voltage decreases therefore by this combination of discharging and charging. In the evaluate phase M4 is enabled by the pre-synaptic signal of the input neuron connected to the synapse. Charge sharing between all the enabled synapses of every output neuron takes place on the post-synaptic lines. In Figure 6 a part of the layout drawing of the synapse array is shown. Most of the area is used up by the two capacitances. The values for the storage and output capacitances are about 60 and 100 fF respectively. The charge on the storage capacitors must be periodically refreshed due to the transistor leakage currents. In the training phase this happens automatically when the weights are updated, otherwise the refresh takes up about 2 % of the network capacity.

Figure 7 shows the circuit diagram of the neuron circuit. It is based on a sense amplifier built from the devices M1 to M4. In the precharge phase it is disabled (*evaluate* and $\overline{evaluate}$ are set to the precharge voltage). The activated *precharge* and *transfer* signals restore the post-synaptic input signal and the

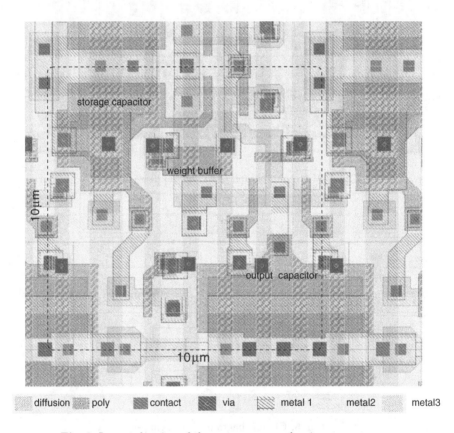

Fig. 6. Layout drawing of the synapse array showing one synapse.

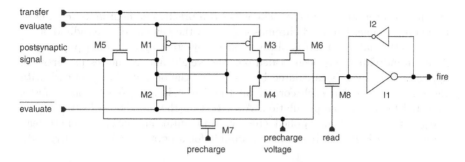

Fig. 7. Circuit diagram of the neuron.

internal nodes of the sense amplifier to the precharge voltage. At the beginning of the evaluate phase the *precharge* signal is deactivated while *transfer* stays on. The potential on the post-synaptic input changes now by the activated synapses. Transistor M5 transfers it onto the gates of M3 and M4. The small differential voltage between the gates of M1/M2 and M3/M4 is amplified by disabling *transfer* and activating *evaluate/evaluate*. At the same moment the synapses switch back to the precharge phase. The sense amplifier restores the signal in about 1 ns to the full supply voltage. With the *read* signal the result is stored in the output latch formed by the inverters I1 and I2. The output of I1 is fed back to the input neurons. The output neuron forms a master/slave flip-flop with the sense amplifier as the master and the output latch as the slave. This results in a discrete-time operation of the network. Together with the feedback the network acts as a kind of mixed-signal state machine. The neurons are the state flip-flops while the synapse array represents the logic that determines the next state. The simulated maximum clock frequency of the network is 100 MHz.

3   Implementation of the Genetic Training Algorithm

The time needed to load the 4096 weight values into the network is about 250 μs. A single test pattern comprised of 64 input bits can be applied in about 100 ns. This makes it feasible to use iterative algorithms needing high numbers of passes. The dependency between a weight value and a neuron output could be highly nonlinear, especially if more than one network cycle is used to implement a multi-layered recurrent network. Therefore a genetic algorithm seems to be well suited to train the network. The network has also built-in hardware support for perturbation based learning [6], an iterative algorithm that needs no knowledge about the transfer function of the network.

The implemented genetic algorithm represents one weight value by one gene. To avoid close hardware dependencies the weight value is stored in a normalized way using floating point numbers between -1 for the maximum inhibitory and +1 for the maximum excitatory synapse. These numbers are converted into the voltage values needed by the analog neural network while translating the genome into the weight matrix. The genes comprising one neuron are combined to a chromosome. Up to 64 chromosomes form the genome of one individual.

The artificial evolution is always started by creating a random population. After an individual has been loaded into the weight matrix the testpatterns are applied. The fitness is calculated by comparing the output of the network with the target values. For each correct bit the fitness is increased by one. This is repeated for the whole population. After sorting the population by the fitness two genetic operators are applied: crossover and mutation. The crossover strategy is depicted in Figure 8. It shows an example for a population of 16 individuals.

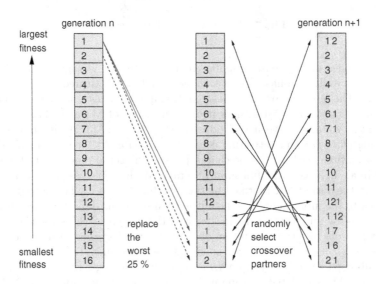

Fig. 8. Crossover pattern used in the genetic algorithm.

The worst 25% of the population are replaced in equal halves by the fittest individual (solid arrows) and the 12.5 % best ones (dashed arrows). 75% of the individuals are kept unchanged. As shown in Figure 8 the crossover partners are randomly chosen. The crossover itself is done in a way that for each pair of identical chromosomes (i.e. chromosomes coding the same output neuron) a crossover point is randomly selected. All the genes up to this point are exchanged between both chromosomes. After the crossover is done the mutation operator is applied on the new population. It alters every gene with equal probability. If a gene is subject to mutation, its old value is replaced by a randomly selected new one (again out of the range $[-1, 1]$).

4   Experimental Results

The network and the genetic training algorithm have been tested with the setup shown in Figure 9. The population is maintained on the host computer. The

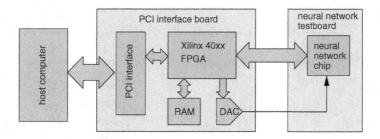

Fig. 9. Testbench used for the evaluation of the neural network.

data for each individual is sent via the FPGA to the neural network using a 16 bit digital to analog converter to generate the analog weight values from the gene data. The testpatterns and the target data are stored in the RAM on the testboard throughout the evolutionary process. They are applied to the individual after the neural network has stored its weights. The FPGA reads the results and calculates the fitness. After the last testpattern the final fitness value is read back by the host computer and the test of the next individual starts. To speed up this process the weight data for the next individual can be uploaded into the RAM while the testpatterns are applied to the current individual. Since the test board is not yet capable of the full speed of the network the number of individuals tested per second is limited to about 150 to 300, depending on the number of testpatterns used.

To test the capability of the genetic algorithm a training pattern was chosen that is especially hard to learn with traditional algorithms like back-propagation [7]: the calculation of parity. While easy to implement with exclusive-or gates, it can not be learned by a single layered neural network. Therefore it also shows the ability of the presented neural network to act as a two-layered network. Figure 10 shows the testpattern definition for an eight bit parity calculation. Since the input data is binary coded, the input neurons are configured for differential input (see Section 2.1). The number of network cycles is set to two and four

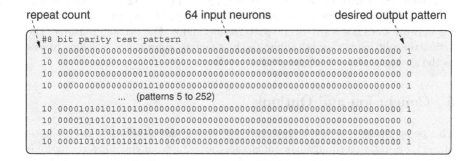

Fig. 10. Testpattern definition for the 8 bit parity.

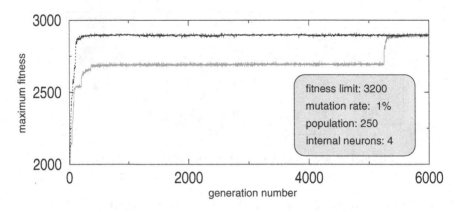

Fig. 11. Example fitness curves from 6 bit parity experiments.

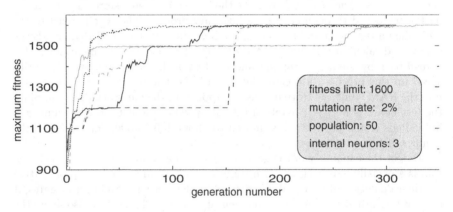

Fig. 12. Example fitness curves from 4 bit parity experiments.

output neurons are fed back into the network. This allows the genetic algorithm to use an internal layer with four neurons. For each testpattern one target bit is defined: the parity of the testpattern. Figures 11 and 12 show plots of the fitness versus the generation number for different parity experiments. At 6 bit the network does not learn all the patterns any more. The random nature of the artificial evolution is clearly visible: the black curve approaches the same fitness as the gray one about 5000 generations earlier.

5   Conclusion and Outlook

This paper presents a new architecture for analog neural networks that is optimized for iterative training algorithms, especially genetic algorithms. By combining digital information exchange with analog neuron operation it is well suited for large neural network chips. Especially, the very small synapse area makes network chips with more than a million synapses possible. The mapping of input data to the network and the effective number and size of the network layers is programmable. Therefore not only the weights, but also a significant part of the architecture can be evolved. The accuracy of the network is not limited by the precision of a single analog synapse since arbitrary synapses can be combined. By implementing this task in the genetic algorithm, the network could automatically adapt its prediction performance to a specific data set.

The presented prototype successfully learned the parity calculation of multibit patterns. This shows that genetic algorithms are capable of training two-layered analog neural networks. At the time of this writing the test setup was limited in its analog precision. This makes it difficult to train binary patterns of 6 or more bits without errors. Also, the genetic algorithm used is a first approach to show the functionality of the system. These limitations will be hopefully overcome in the near future.

References

1. Shibata, T., Kosaka, H., Ishii, H. , Ohmi, T.: A Neuron-MOS Neural Network Using Self-Learning-Compatible Synapses Circuits. IEEE Journal of Solid-State Circuits, Vol. 30, No. 8, (August 1995) 913–922
2. Diorio, C., Hasler, P., Minch, B. , Mead, C.: A Single-Transistor Silicon Synapse. IEEE Transactions on Electron Devices, Vol. 43, No. 11, (November 1996) 1972–1980
3. Kramer, A.: Array-Based Analog Computation. IEEE Micro, (October 1996) 20–29
4. Schemmel, J., Loose, M., Meier, K.: A 66 × 66 pixels analog edge detection array with digital readout, Proceedings of the 25th European Solid-State Circuits Conference, Edition Frontinières, ISBN 2-86332-246-X, (1999) 298–301
5. Langeheine, J., Fölling, S., Meier, K., Schemmel, J.: Towards a Silicon Primordial Soup: A Fast Approach to Hardware Evolution with a VLSI Transistor Array. ICES 2000, Proceedings, Springer, ISBN 3-540-67338-5 (2000) 123-132
6. Montalvo, J., Gyurcsik R., Paulos J.,: An Analog VLSI Neural Network with On-Chip Perturbation Learning. IEEE Journal of Solid-State Circuits, Vol. 32, No. 4, (April 1997) 535–543
7. Hertz, J. Krogh, A., Palmer, R.: Introduction to the Theory of Neural Computation. Santa Fe Institute, ISBN 0-201-50395-6 (1991) 131

Initial Studies of a New VLSI Field Programmable Transistor Array

Jörg Langeheine, Joachim Becker, Simon Fölling,
Karlheinz Meier, and Johannes Schemmel

Address of principle author: Heidelberg University, Kirchhoff-Institute for Physics,
Schröderstr. 90, D-69120 Heidelberg, Germany,
langehei@kip.uni-heidelberg.de
http://www.kip.uni-heidelberg.de/vision.html

Abstract. A system for intrinsic hardware evolution of analog electronic circuits is presented. It consists of a VLSI chip featuring 16 × 16 programmable transistor cells, an FPGA based PCI card and a software package for setup and control of the experiment. The PCI card serves as a link between the chip and the computer that runs the genetic algorithm to produce the configurations for the Field Programmable Transistor Array (FPTA). First measurement results prove chip and system to be working as well as they indicate the tradeoff between performance and configurability. The system is now ready to host a wide variety of evolution experiments.

1 Introduction

While digital hardware is becoming more and more powerful, there are a lot of problems requiring analog electronic circuits. Examples are sensors (e.g. [1]), that will always use some analog front end to measure a physical quantity in an analog manner, analog filters or sometimes (massive parallel) signal processing circuits. For the latter example the use of analog circuitry can result in a better ratio of performance and area and/or power consumption (cf. e.g. [2], [3]). Unlike its digital counterpart the domain of analog design is not blessed with powerful tools simplifying the design process. This is, at least to some extent, due to the tight relationship between the used technology, the chosen layout and the performance of the resulting circuit, which makes the simple reuse of standard building blocks without any adaption virtually impossible. Moreover great care has to be taken in how the specific process parameters can be used to achieve the desired behavior because of the device variations on the actual dice. As evolutionary algorithms are assumed to yield good results on complex problems without explicit knowledge of the detailed interdependencies involved, they seem to be a tempting choice. Accordingly the project described in this paper tries to make a step towards the design automation of analog electronics by means of evolvable hardware.

From the variety of different approaches intrinsic evolution on a fine grained FPAA, namely a Field Programmable Transistor Array (FPTA) designed in

Y. Liu et al. (Eds.): ICES 2001, LNCS 2210, pp. 62–73, 2001.

CMOS technology, is chosen for the following reasons: First, the use of hardware in the loop is expected to be advantageous because it faces the algorithm with the full complexity of the problem including device mismatching as well as any kind of electronic noise inherent to the chip. There is evidence that the presence of different environmental conditions during the evolution process is helpful to evolve circuits that work on different dice under different conditions (cf. [4], [5]). Second, intrinsic evolution is expected to be faster than evolution using software models for the hardware. Third, the use of large scale integration techniques facilitates the design of complex systems. CMOS nowadays is the most widely used and therefore cheapest technology for the design of integrated electronic circuits.

The final goal can be twofold: On one hand, it would be desirable to have a system that can be fed with an abstract problem description, such as a sort of fitness function, and that after some time produces a solution to the problem. Without caring about the details of the implementation the designer merely has to ensure that the circuit is working correctly under all expected conditions. On the other hand it may be useful to analyze the circuits obtained by the hardware evolution process and understand them to such an extent that it is possible to use the extracted circuits or design principles in a different chip, thus using the system as a design tool.

The paper is organized as follows: Section 2 gives an overview over the evolution system. In section 3 the implementation of the Field Programmable Transistor Array is discussed. Finally in section 4 experimental results are given and the expected performance of the chip is discussed, before the paper closes with a summary.

2   The Hardware Evolution System

Figure 1 shows the setup of the evolution system. A commercial PC is used to control the system and as a user interface. The software allows to create and edit circuit configurations for the FPTA chip. A PCI-card serves as the link between the FPTA board that can be plugged into the PCI-card and the computer. A state machine run on the FPGA generates all the necessary digital signals: It creates the signals used to write the configuration to the SRAM of the FPTA and performs the read out of the SRAM. Furthermore the state machine provides the DAC with the necessary data and timing signals to produce the analog input patterns for the FPTA and controls the data conversion of the analog outputs of the FPTA carried out by the ADC. The RAM module on the PCI-card can be used for example to cache the data for the analog input patterns, the output of the ADC and the next individuals to be loaded into the FPTA.

In figure 2 a screenshot of the user interface of the software is displayed. The right window contains 6 × 4 cells of the lower right corner of the transistor array, consisting of a total of 128 P- and 128 NMOS transistors arranged in a checkerboard pattern as denoted by the letters P and N in each of the cells. From this window any circuit can be downloaded to the chip in order to test it. The

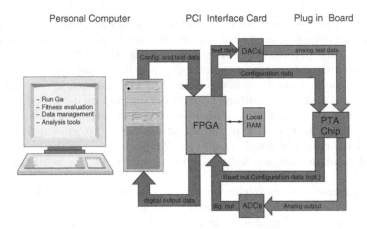

Fig. 1. Schematic diagram of the evolution system.

left window reflects the configuration of the cell 15/15 (cells are identified by their x/y coordinates): Each of the three terminals gate, source and drain of the MOS transistor can be connected to either the supply voltage, ground, or any of the four edges of the cell. Furthermore to enable signals to be routed through the chip any of the four cell edges can be connected to any of the remaining three edges.

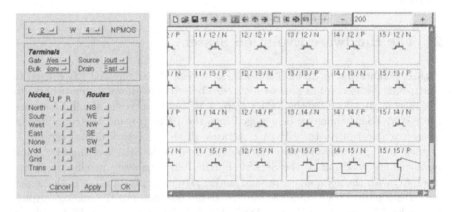

Fig. 2. Screenshot of the circuit editor window of the software: **Left:** Editor to set the connections and W/L values for one cell (here 15/15). **Right:** Editor showing the setup for the measurement of one PMOS transistor in the lower right corner of the chip.

3   Implementation of the FPTA

In order to provide some *primordial soup*, i.e. a configurable hardware device, for the intrinsic evolution a Field Programmable Transistor Array (FPTA) has been designed and manufactured in a $0.6\,\mu$m CMOS process (More information can be found in [6]). Figure 3 shows a micro photograph of the chip whose die size is about $33\,\mathrm{mm}^2$.

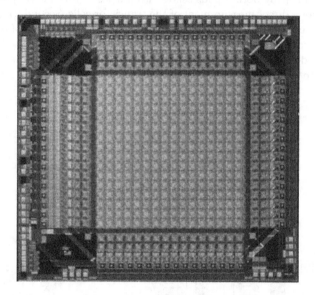

Fig. 3. Micro photograph of the FPTA chip.

The core of the chip consists of an array of 16×16 programmable transistor cells. These cells contain either a programmable P- or NMOS transistor, whose channel geometry can be tuned. The terminals of these transistors can be connected to the four neighboring cells. The signals from the adjacent cells can be routed through the cells.

The choice for this implementation is motivated as follows: First, it was desired to have distinct transistors, that contain the circuit functionality as transistors do in usual designs, in order to simplify the analysis of evolved circuits. Second, the array was designed as homogeneous and symmetric as possible to keep the implementation details of the evolutionary algorithms simple and to enable it to reuse parts of the genome by copying and translating it. However, a single cell was reserved for P- and NMOS transistors respectively to save die area. Third, the transistor geometry can vary in 5 logarithmically graded lengths and 15 linearly graded widths resulting in 75 different aspect ratios in order to obtain a smooth fitness landscape at least for choosing the transistor dimensions.

3.1   Architecture of the Complete Chip

The transistor cell array is surrounded by 64 IO-cells that are connected to the 64 terminals of the 60 transistor cells forming the edges of the array (see fig. 4). The functionality of the IO-cells can be selected by setting their registers. Possible settings are to connect the terminal of the according border cell directly to the analog input or output, directly to the according *array border pad*, leave it unconnected, or to access it via a sample and hold stage. The direct access granted by the array border pads serves two purposes: First, it simplifies debugging and allows direct measurement of the transistor cells. Second, the transistor array can be expanded by bonding together the array border pads of two or more chips. The array border pads are smaller than the standard pads used for analog signals to reduce their capacity and lack the ESD protection circuitry in order to contribute as little distortion as possible to the signals crossing a die border.

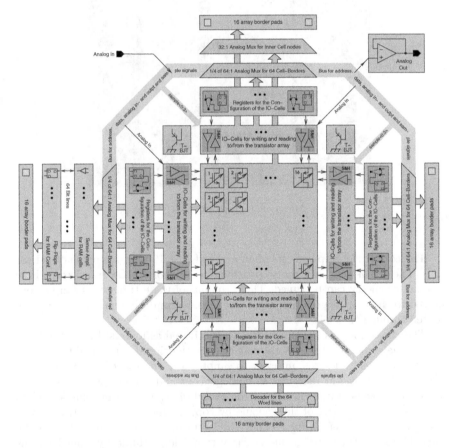

Fig. 4. Block diagram of the PTA chip. Note that the address and data buses are used for all multiplexers and demultiplexers as well as for the programming of the IO-cells.

Each sample and hold stage can be configured to either buffer an input voltage applied to the border terminal or to sample and hold the voltage present at the border cell. The cells configured in the former manner can be used to create complex input patterns from the single analog input. Therefore the sample signals can be taken from four external sample lines. Used as output buffers the sample and hold stages can be utilized to multiplex more than one border cell voltage to the analog output. Moreover they allow the successive read out of different outputs sampled at the same time.

The configuration of the transistor cell array is stored in static RAM cells that are integrated in the transistor cells. Both, read and write access to the SRAM and the configuration of the IO-cells use a 10 bit wide address and a 6 bit wide data bus, that are looped around the chip as shown in figure 4. Each transistor cell contains an operational amplifier that can buffer one out of four possible nodes in the according cell (cf. figure 5). These signals are used to determine voltages and currents inside the transistor cells and can also be multiplexed to the analog output line, which is buffered again before the output signal leaves the chip.

3.2   Architecture of the Transistor Cell

Figure 5 shows the setup of an NMOS cell. At each corner some of the configuration information is stored in a block of static RAM containing 6 bits each. Of the 22 bits used, 6 bits directly control the routing switches that route signals through the cell. Each terminal of the programmable transistor, whose channel geometry is set by 7 bits, can be connected to either power (vdd), ground (gnd) or any of the four edges of the cell, named after the four cardinal points. The remaining two codes of the multiplexers for drain and source are used to leave the terminals floating. For the gate the same code ties the gate terminal to power or ground for P- and NMOS transistors respectively, thus disabling the transistor.

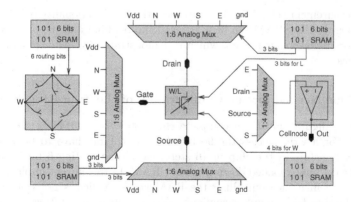

Fig. 5. Block diagram of one NMOS transistor cell.

In order to be able to analyze the behavior of successfully evolved circuits, the voltage at nodes *east, south, drain* and *source* can be read out by means of a unity gain buffer. Thereby all the nodes between adjacent cells can be read out and all currents flowing through the active transistors can be estimated via the voltage drop across the transmission gates connecting them to the cell borders. The layout of a complete transistor cell is shown in figure 6. It occupies an area of about $200\,\mu\text{m} \times 200\,\mu\text{m}$.

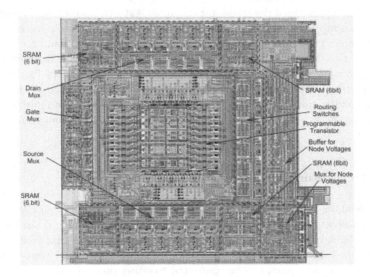

Fig. 6. Layout of one complete NMOS-Cell.

4   Experimental Results

First measurements of the FPTA chip show the full functionality of the transistor cell array: The SRAM can be written to and read out and the programmable transistors behave as expected, which is demonstrated by some transistor characteristics.

4.1   Time Needed for the Configuration of the Transistor Array

For the configuration of the complete chip $256 \times 24 = 6144$ bits have to be written. For a write access the 96 bits for one column have to be written to a row of registers in the chip in 16 steps, each time writing 6 bits. Then one complete column is loaded down into the SRAM. In the current implementation of the state machine controlling the RAM access, which is not optimized for speed, the time for a complete configuration amounts to about 2 ms. From timing measurements however a configuration time of about $70\,\mu\text{s}$ with a more optimized

FPGA configuration can be inferred. As far as the chip is concerned simulation results suggest that even this time can at least be halved. Compared to the expected evaluation times per individual of 1 to 10 ms, this is almost negligible.

4.2 Transistor Characteristics

In order to measure the output characteristic of some of the PMOS transistor cells the configuration shown in figure 2 has been loaded into the chip. The connected border cell terminals are directly routed to the according array border pads such that the transistor cell can be controlled and measured by an HP 4155A semiconductor parameter analyzer.

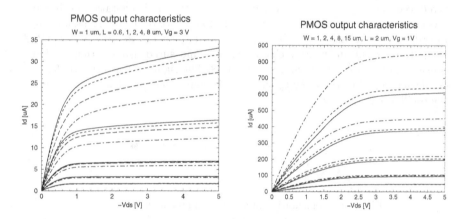

Fig. 7. Output characteristics of programmable PMOS transistors: **Left:** Comparison of PMOS transistors placed at different locations on the chip: Solid: 15/15, dashed: 14/14, long dashed: 9/9 dot-dashed 1/1. **Right:** Comparison of the measured cell 15/15 (solid line), a simulation including all transmission gates (dashed) and one of plain PMOS transistors (dot-dashed).

To compare the output characteristics of different transistors, PMOS transistor cells at different locations on the chip have been measured. For that purpose the terminals of the programmable transistor are always connected to the same pads using the routing capability of the transistor cell array. The results for five different lengths are shown on the left side of figure 7. Apart from looking like transistor output characteristics the curves belonging to the same L value do look similar, but vary in their drain current values. In fact, the output current is the smaller the longer the routing path to the connected border cells. While the relative difference of the saturated drain currents for $L = 0.6\,\mu$m amounts to approximately 32%, it decreases to about 4% for $L = 8\,\mu$m. This is due to the finite resistance of the transmission gates providing the routing, which explains why the effect is more severe for larger currents (i.e. smaller transistor lengths).

In the right half of figure 7 the output characteristic of the PMOS transistor cell in the lower right corner of the chip is compared to the simulation of a plain PMOS transistor as well as to one including the transmission gates used in the measurement. Results are shown for five different transistor widths. While the more precise model of the transistor cell matches the measured curve quite well, the output currents of the transistor cell are always smaller than the ones from the simulation of the plain transistor. Again this is due to the finite resistance of the transmission gates and the discrepancy worsens for higher currents.

4.3   Ring Oscillators

As was already discussed in [6] the bandwidth of any possible circuit in the FPTA is reduced in comparison to the corresponding *direct* implementation in the same process due to the parasitic resistance and capacitance of the transmission gates. In order to get a measure for the maximum frequencies possible in the FPTA the gate delay of an inverter chain has been measured using a ring oscillator consisting of 9 inverters as shown in figure 8. The rightmost inverter buffers the oscillating signal of the circuit, such that it can be measured without changing the oscillator frequency.

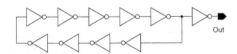

Fig. 8. Implementation of a ring oscillator with 9 inverters.

The circuit was implemented in the FPTA (cf. figure 9) in five different locations, namely all four array corners and the middle of the array. For comparison it was also simulated for different process parameter sets denoting the slowest and fastest as well as the typical behavior of the devices fabricated in the used process.

The aspect ratios used were $14\,\mu\mathrm{m}/0.6\,\mu\mathrm{m}$ and $8\,\mu\mathrm{m}/0.6\,\mu\mathrm{m}$ for the P- and NMOS transistors respectively. Furthermore the oscillator in the lower right corner was measured for an aspect ratio of $2\,\mu\mathrm{m}/8\,\mu\mathrm{m}$ (PMOS) and $1\,\mu\mathrm{m}/8\,\mu\mathrm{m}$

Table 1. Measured period and gate delay of the 9 inverter ring oscillator placed in 5 different locations on the chip and of the 119 inverter ring oscillator. The gate delays are calculated by dividing the period by 18 (238 in case of the 119 inverter ring).

Used location	upper left	upper right	lower right	lower left	middle	average	119 inverters	lower right slowest W/L
Period	148.5 ns	147.5 ns	150 ns	150.5 ns	148.5 ns	149 ns	1.8 μs	6.78 μs
Gate delay	8.25 ns	8.14 ns	8.35 ns	8.36 ns	8.25 ns	8.28 ns	7.56 ns	376.7 ns

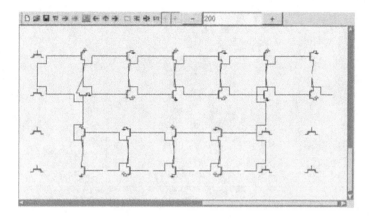

Fig. 9. Implementation of the ring oscillator in the lower right corner of the FPTA.

(NMOS) resulting in a lower oscillation frequency. In addition an oscillator containing 119 inverters occupying the complete transistor array was implemented. The results are listed in table 1. A screenshot of the output signal recorded by an oscilloscope is shown in figure 10.

The ring oscillator was simulated using the exact architecture of the FPTA implementation and an implementation using standard cell inverters. Both simulations were carried out with and without the back-annotated parasitic capacitances of the layout and for three sets of process parameters. While *typical mean* (*tm*) denotes the average set of process parameters, *worst case power* (*wp*) and *worst case speed* (*ws*) refer to the parameter sets marking an upper and a lower bound to the speed of the manufactured devices guaranteed by the manufacturer. The results are listed in tables 2 and 3.

Taking the measured and simulated gate arrays as a measure for the speed of the technology it can be inferred that the loss of speed caused by the overhead for the configurability is about a factor of 100, limiting possible application for the FPTA to frequencies of the order of MHz. Furthermore the fact, that the variation of the observed frequencies is quite small indicates a high level

Table 2. Simulation results for the ring oscillator with 9 inverters. The left part of the table displays the periods and calculated gate delays for simulations with all parasitic capacitances back-annotated from the layout. On the right hand side the simulation results for the pure schematic (without any parasitic capacitances) are given. The abbreviations *tm*, *wp*, *ws*, refer to different parameter sets for the simulations (further explanations see text).

Transistor cell	back-annotated simulation			simulation without parasitics		
simulation	tm	wp	ws	tm	wp	ws
Period	219.6 ns	148.2 ns	365.23 ns	84.5 ns	47.4 ns	161.6 ns
Gate delay	12.2 ns	8.23 ns	20.29 ns	4.69 ns	2.64 ns	8.98 ns

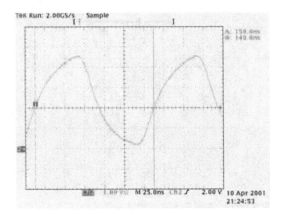

Fig. 10. Screenshot of the output signal of the ring oscillator implemented in the FPTA. One square corresponds to 25 ns and 1 V for the x- and y-axis respectively.

of homogeneity of the array cells. The smaller gate delay extracted from the measurement of the 119 inverters is probably due to the better ratio of the number of cells used as inverter parts to the number of routing cells. Finally the comparison to the implementation with the small aspect ratios shows the range of possible frequency adjustments that can be obtained by simply changing the transistor geometries.

The comparison of measurement and simulation results for the transistor cells yields the following: First, the measured gate delay is significantly smaller than the gate delay extracted from the back-annotated typical mean simulation, although the process parameters accessible from the vendor are closely matching the typical mean parameters. This may be due to the fact that the extraction of the parasitic capacitances yields worst case values. Second, the difference between the gate delay of the simulation without parasitic capacitances and the measured gate delay indicates, that the capacitances introduced by the metal lines are of the same order as the parasitic capacitances introduced by the transmission gates used for connecting the programmed transistors (cf. [6]).

Table 3. Simulation results for a ring oscillator with 9 inverters designed out of digital standard cells. As in Table 2 results are shown for the simulation with and without parasitic capacitances.

Standard cell simulation	back-annotated simulation			simulation without parasitics		
	tm	wp	ws	tm	wp	ws
Period	1.46 ns	942.2 ps	2.47 ns	1.352 ns	853.42 ps	2.34 ns
Gate delay	81.11 ps	52.34 ps	137.3 ps	75.11 ps	47.41 ps	130 ps

5   Summary and Future Plans

A Field Programmable Transistor Array has been fabricated in a 0.6 μm CMOS process. The chip is embedded in a hardware evolution system designed for the intrinsic evolution of analog electronic circuits. First measurements have proven the chip to work. The time for a configuration of the whole chip is extrapolated to be less than 70 μs allowing for testing rates of up to 1000 individuals per second. Time domain measurements suggest that the chip can be used for frequencies in the order of MHz. The evolution system is almost ready to be programmed for first evolution experiments. The next steps are to optimize the system for high throughput rates and extend it to monitor the die temperature and the current used by the transistor cell array itself.

Acknowledgment

This work is supported by the Ministerium für Wissenschaft, Forschung und Kunst, Baden-Württemberg, Stuttgart, Germany.

References

1. M. Loose, K. Meier, J. Schemmel: Self-calibrating logarithmic CMOS image sensor with single chip camera functionality, *IEEE Workshop on CCDs and Advanced Image Sensors, Karuizawa, 1999, R27*
2. J. Schemmel, M. Loose, K.Meier: A 66 × 66 pixels analog edge detection array with digital readout, In: *Proceedings of the 25th European Solid-state Circuits Conference (ESSCIRC'99)*, B.J. Hosticka, G. Zimmer, H. Grünbacher, Eds., pp 298-301, Edition Frontières, 1999.
3. M. Murakawa, S. Yoshizawa, T. Adachi, S. Suzuki, K. Takasuka, M. Iwata, T. Higuchi: Analogue EHW Chip for Intermediate Frequency Filters, In: Proc. 2nd Int. Conf. on Evolvable Systems: *From biology to hardware (ICES98)*, M. Sipper et al., Eds., pp 134-143, Springer-Verlag,1998.
4. Thompson, A, Layzell, P.: Evolution of Robustness in an Electronics Design, In: Proc. 3rd Int. Conf. on Evolvable Systems: *From biology to hardware (ICES2000)*, T. Fogarty, J. Miller, A. Thompson and P. Thompson, Eds., pp 218-228, April 17-19, 2000, Edinburgh, UK. New York, USA, Springer Verlag.
5. A. Stoica, R. Zebulum and D. Keymeulen: Mixtrinsic Evolution, In Proceedings of the Third International Conference on Evolvable systems: *From Biology to Hardware ICES2000)*, T. Fogarty, J. Miller, A. Thompson and P. Thompson, Eds., pp 208-217, April 17-19, 2000, Edinburgh, UK. New York, USA, Springer Verlag.
6. J. Langeheine, S. Fölling, K. Meier, J. Schemmel: Towards a silicon primordial soup: A fast approach to hardware evolution with a VLSI transistor array, In: Proc. 3rd Int. Conf. on Evolvable Systems: *From Biology to Hardware (ICES2000)*, J. Miller et al., Eds., pp 123-132, Springer-Verlag,2000.

An Embryonics Implementation
of a Self-Replicating Universal Turing Machine

Hector Fabio Restrepo and Daniel Mange

Logic Systems Laboratory, Swiss Federal Institute of Technology,
IN-Ecublens, CH-1015 Lausanne, Switzerland,
{HectorFabio.Restrepo, Daniel.Mange}@epfl.ch – http://lslwww.epfl.ch

Abstract. This paper describes a multicellular universal Turing machine implementation endowed with self-replication and self-repair capabilities. In this multicellular artificial organism every artificial cell contains a complete copy of the genome. The mapping of the universal Turing machine onto a multicellular array was made possible thanks to the introduction of a modified version of the W-machine[1].

Keywords: self-replication, self-repair, universal Turing machine, cellular automata, Embryonics.

1 Introduction

Living organisms are complex systems exhibiting a range of desirable characteristics, such as evolution, adaptation, and fault tolerance, that have proved difficult to realize using traditional engineering methodologies. The last three decades of investigations in the field of molecular biology (embryology, genetics, and immunology) has brought a clearer understanding of how living systems grow and develop. The principles used by Nature to build and maintain complex living systems are now available for the engineer to draw inspiration from [7].

The Embryonics (embryonic electronics) project is inspired by the basic processes of molecular biology and by the embryonic development of living beings. By adopting three fundamental features of biology – multicellular organization, cellular division, and cellular differentiation – and by transposing them onto the two-dimensional world of integrated circuits in silicon, we show that properties of the living world, such as self-replication and self-repair, can also be attained in artificial objects (integrated circuits).

Our goal in this paper is to present self-replicating machines exhibiting universal computation, i.e., universal Turing machines. We demonstrate that the dream of von Neumann, the self-replication of such a machine, can be realized in actual hardware thanks to the Embryonics architecture.

---

[1] A W-machine [13] is like a Turing machine, save that its operation at each time step is guided not by a state-table but by an instruction from the following list: **PRINT 0, PRINT 1, MOVE DOWN, MOVE UP, IF 1 THEN (n) ELSE (next)**, **STOP**.

Y. Liu et al. (Eds.): ICES 2001, LNCS 2210, pp. 74–87, 2001.

In Section 2 we present a brief reminder of specialized and universal Turing machines. Section 3 introduces the PICOPASCAL language and a PICOPASCAL interpreter. Sections 4, and 5, present a Embryonics architecture based on a multicellular array of cells and describes the implementation of a self-replicating specialized Turing machine. In Sections 6 and 7 we present the architecture of an ideal and of an actual universal Turing machine able to self-replicate. A discussion of our results follows in the final section (Section 8).

2 Turing Machines

In the 1930's, before the advent of digital computers, several logicians (Kurt Gödel, Alonzo Church, Stephen Kleene, Emil Post, and Alan Mathison Turing) began to think about the theoretical limits of computation. Alonzo Church and Alan Turing independently arrived, through different approaches, at equivalent conclusions. Both solutions described computability, but while Church (1932-34) described it with λ-calculus, Turing's idea (1936) was based on a mathematical model of a machine that could compute any computable functions: the *Turing machine* [11, 3].

Turing machines were first described by Alan Turing in his historic paper, "*On Computable Numbers, with an Application to the Entscheidungsproblem*" [12], which was his answer to the Entscheidungsproblem posed by the German mathematician David Hilbert. Hilbert asked if there existed, in principle, any definite method which could be applied to determine the truth of any mathematical assertion [1].

2.1 Specialized Turing Machines

In his 1936 paper [12], A. M. Turing defined the class of abstract machines that now bear his name: *Turing machines*. A specialized Turing machine (Figure 1), or simply a Turing machine, is a finite-state machine (the *program*) controlling a mobile head, which operates on a tape. The tape, composed of a sequence of squares, contains a string of symbols (the *data*). The head is situated, at any given moment, on some square of the tape and has to carry out three operations to complete a step of the computation (one operation cycle of the finite-state machine). These operations are:

1. reading the square of the tape being scanned;
2. writing on the scanned square;
3. moving the head to an adjacent square (which becomes the scanned square in the next operation cycle).

A Turing machine can be described by three functions f_1, f_2, f_3:

$$Q+ = f_1(Q, S) \tag{1}$$
$$S+ = f_2(Q, S) \tag{2}$$
$$D+ = f_3(Q, S) \tag{3}$$

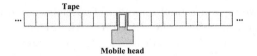

Fig. 1. A specialized Turing machine.

where Q and S are, respectively, the current internal state (of the finite state machine) and the current input symbol (the symbol on the square of the tape being scanned), and where $Q+$, $S+$, and $D+$ are, respectively, the next internal state, the next input symbol, and the direction of the head's next move [8].

The tape can be considered as infinite in both directions. However, we will make the restriction that, when the machine starts operating, the tape must be blank, except for some finite number of squares. With this restriction, we can think of the tape as finite at any particular time but capable of being infinitely extended whenever the machine comes to an end of the finite portion.

When a symbol is printed on the tape, the symbol previously there is erased. Of course, it can be preserved if we print the same symbol that was read. Because the head can move either way along the tape, it is possible for it to return to a previously printed location to recover the information inscribed there. This ability provides the machine a sort of rudimentary memory in a sense that the machine can look up the previous symbols and change them if necessary. Since the tape is as long as desired, this memory is potentially infinite.

At any given time, the read/write mobile head of the Turing machine is positioned on some square on the tape. Furthermore, at any given time, the Turing machine is in one of a finite number of internal states.

A set of *quintuples* can be used to specify what the machine will do for each possible combination of symbol and state. These quintuples have the following form

(*current state, current symbol, next state, next symbol, direction of motion*)

or, equivalently:

(Q, S, $Q+$, $S+$, $D+$)

where the third, fourth, and fifth symbols are determined by the first and second according to the three functions f_1 (1), f_2 (2), f_3 (3) mentioned above.

These quintuples indicate that if a Turing machine is now in the current internal state Q, and the current input symbol is S, the machine will change its current internal state to the next internal state $Q+$, replace the current input symbol on the tape by the next input symbol $S+$, and move the read/write head one square in the given direction $D+$. If a Turing machine is in a condition for which it has no instruction, it halts.

The information contained in the set of quintuples is often represented in the form of a state table, defining the behavior of the machine for each possible combination of symbol and state.

2.2   The Universal Turing Machine

Turing had the further idea of the universal Turing machine (UTM), capable of simulating the operation of any specialized Turing machine, and gave an exact description of such a UTM in his paper [12]. The importance of the universal Turing machine is clear. We do not need to have an infinity of different machines doing different jobs. A single one will suffice [2].

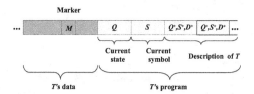

Fig. 2. Universal Turing machine's tape, describing the specialized machine T.

A UTM, U, is a Turing machine with the property of being able to read the description (on its tape) of any other Turing machine, T, and to behave as T would have. The machine U consists of a finite-state machine (the program of U) controlling a mobile head, which operates on a tape. The data on the tape completely describe the machine T to be simulated (the data of T and the program of T, i.e., the three functions $Q+$, $S+$, and $D+$ describing T).

Figure 2 shows the organization of U's tape. To the left is a semi-infinite region containing the data of T's tape. Somewhere in this region is a marker M indicating where T's head is currently located. The middle region contains the current internal state Q and the current input symbol S of T. The right-hand region is used to record the description of T, i.e., the three functions $Q+$, $S+$, and $D+$ for each combination of Q and S.

3   PICOPASCAL

In this section we present the PICOPASCAL language and a possible hardware architecture for a PICOPASCAL interpreter. The PICOPASCAL language and its interpreter have been customized for our embryonics architecture and the UTM implementation described in the next section.

3.1   The PICOPASCAL Language

The PICOPASCAL language consists of a minimal subset of the MODULA-2 language [14]. PICOPASCAL is thus a *high-level* language: it does not make use of explicit addressing and provides a great simplicity of use. PICOPASCAL is, moreover, a *structured* language and thus guarantees, because of its structure, a

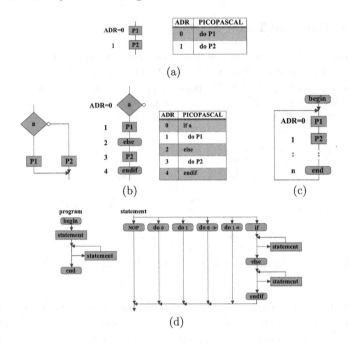

Fig. 3. PICOPASCAL language. (a) Sequence of two assignment instructions: **do** *P1P2*. (b) Choice of either *P1* or *P2*: **if** *a* **then** *P1* **else** *P2*. (c) Non-conditional iteration loop. (d) Syntactic diagram.

rigorous and efficient notation. In conformity with this last feature, PICOPAS-CAL has three fundamental constructs, described below: (1) the sequence, (2) the choice or alternative, and (3) the iteration.

The assignment **do...**, realizing the synchronous transfer of a constant into a register, is a structured program. The *sequence* (or composition) of two such instructions *P1* and *P2*, written **do** *P1P2*, is a structured program, described by the flowchart and by the mnemonic program of Figure 3a. This last notation consists of a linear succession of instructions, displayed in the growing order of addresses *ADR*.

The *choice* (or alternative) of *P1* or *P2*, where *P1* and *P2* are two assignments, is a structured program, written **if** *a* **then** *P1* **else** *P2*. It is represented symbolically by the flowchart of Figure 3b, and realized by the linear succession of the instructions of the corresponding functional diagram and mnemonic program. To facilitate comprehension, and unlike programs written in a low-level language using explicit addresses, there is no jump (notably, to avoid the instruction *P1* when $a = 0$ or the instruction *P2* when $a = 1$): all instructions are read sequentially, from *ADR*=0 to *ADR*=4, and the execution of the assignments *P1* or *P2* depends on the value of a signal *EXEC* (for *EXECUTE*) which, in turn, depends on the value of the test variable *a*. This process will be revisited in detail in the description of the interpreter of the PICOPASCAL language (Subsection 3.2).

The last construct of structured programming, the *conditional iteration* **while** *a* **do** *P1*, is thus not necessary in the PICOPASCAL language. However, since our program must be continually executed, notably to allow self-repair, we allow the loop illustrated by the flowchart of Figure 3c, which in fact introduces a non-conditional iteration on the entire program.

In conclusion, the PICOPASCAL language is described by the *syntactic diagram* of Figure 3d, where we can count ten different terminal symbols (ovals), which make up the instructions of the language: **begin**, **end**, **NOP**, **do** 0, **do** 1, **do** 0→, **do** 1←, **if**, **else**, **endif**. The **NOP** (*No operation*) instruction represents the execution of a neutral operation. Figure 4 shows the operating code (*OPC*) for the instructions of the PICOPASCAL language.

OPC3:0	OPC	Instruction	OPC3:0	OPC	Instruction
0000	0	NOP	1000	8	do 0 →
0001	1	begin	1001	9	do 1 ←
0010	2	end	1010	A	do 0
0011	3		1011	B	do 1
0100	4	else	1100	C	
0101	5	if	1101	D	
0110	6	endif	1110	E	
0111	7		1111	F	

Fig. 4. PICOPASCAL language opcodes.

3.2   PICOPASCALINE:
An Interpreter for the PICOPASCAL Language

Figure 5 suggests a possible hardware architecture to execute the ten instructions of the PICOPASCAL language. From now on, we will refer to this machine as PICOPASCALINE. The instruction **end**, as well as the pseudo-instruction **begin** (not executed), have the same effect: jumping to the instruction at address 0. There exist therefore nine distinct types of instruction to interpret.

To decode the instructions (*OPC3:0*) on the program tape, the PICOPASCALINE consists of the following elements (Figure 5):

- A state register REGISTER storing the current values of the internal and input states Q, and S respectively, with an initial state $Q,S = 01$.
- A register REGISTER storing the values $QL3:0, QC, QR0:3$ of the data tape, with an initial state $QL3:0, QC, QR0:3 = 000010000$.
- A stack STACK characterized by a 1-out-of-3 code (one-hot encoding), with an initial state STACK $= ST3:1 = 001$.
- A decoder DMUX1 controlled by the 4 bits of the operating code $OPC3:0$, which generates the signals controlling the STACK (signals *IF*, *ELSE*, and *ENDIF*).

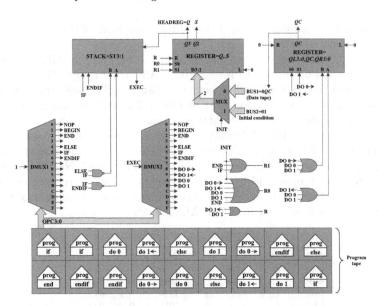

Fig. 5. PICOPASCALINE: PICOPASCAL interpreter for the ten instructions of the language.

- A decoder DMUX2 controlled by the 4 bits of the operating code $OPC3{:}0$ and by the $EXEC$ signal. This decoder generates the signals controlling the (Q,S) and $(QL3{:}0,QC,QR0{:}3)$ REGISTERs (signals $DO\ 0$, $DO\ 1$, $DO\ 0{\rightarrow}$, $DO\ 1{\leftarrow}$, IF, and $ELSE$).
- A multiplexer MUX controlled by the signal $INIT$, which selects one of the two input busses, $BUS1$ coming from the data tape, or $BUS2$ which is a constant used for initialization purposes. At the start of the execution the signal $INIT$ has the value 1 and the (Q,S) REGISTER is initialized, whereas the rest of the execution this variable takes the value 0 and the value QC coming from the $(QL3{:}0,QC,QR0{:}3)$ REGISTER is assigned to the (Q,S) REGISTER.

The signal EXEC controls the execution of the assignment instructions **do** and thus depends on the succession of values of the internal and input states Q and S.

4 Self-Replication of a UTM on a Multicellular Array

Conventional UTMs [8] consist of a finite but arbitrarily long tape, and a single read/write mobile head controlled by a finite-state machine, which is itself described on the tape (Figure 2). In order to implement a UTM in an array of MICTREE[2] artificial cells, we made three fundamental architectural choices (Figure 6):

1. The read/write head is fixed; the tapes are therefore mobile.
2. The data of the given application (the specialized Turing machine to be simulated) are placed on a mobile tape, the *data tape*; this tape can shift right, shift left, or stay in place.
3. The finite-state machine for the given application is translated into a very simple program written in a language called PICOPASCAL; each instruction of this program is placed in a square of a second mobile tape, the *program tape*; this tape just needs to shift left. The transformation of a state table into such a program is directly inspired by the W-machine [13] with the major contribution of avoiding the jumps required by the **if** 1 **then** (*n*) **else** (*next*) instructions.

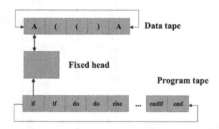

Fig. 6. Universal Turing machine architecture for the parenthesis checker example.

The fixed head, which is in fact an interpreter of the PICOPASCAL language, has to continuously execute cycles consisting of four operations:

1. reading and decoding an instruction on the program tape;
2. reading a symbol on the data tape;
3. interpreting the current instruction, and writing a new symbol on the current square of the data tape;
4. shifting the data tape (left, or right, or not at all) and the program tape (left).

5   An Application: A Binary Counter

In order to test our UTM implementation, we used, as a simple but non-trivial example, a binary counter [8], a machine that writes out the binary numbers 1,

---

[2] the MICTREE (for *tree of micro-instructions*) cell is a new kind of *coarse-grained field-programmable gate array* (*FPGA*) developed in the framework of the Embryonics project. This cell is used for the implementation of multicellular artificial organisms with biological-like properties, i.e., capable of self-repair and self-replication [10, 5].

10, 11, 100, etc. The counter's state table (Figure 7a) has two internal states ($Q \in \{0 \rightarrow, 1 \leftarrow\}$) and two input states ($S \in \{0,1\}$), S being the value of the current square read on the data tape. Depending on the present internal state Q and the present input state S, the specialized Turing machine will [9]:

1. write a new binary value $S+$ (0, 1) on the current square of the data tape;
2. move its data tape to the right ($Q+ = 0 \rightarrow$) or to the left ($Q+ = 1 \leftarrow$), which is equivalent to moving a mobile head to the left or to the right, respectively;
3. go to the next state $Q+(0 \rightarrow, 1 \leftarrow)$.

Figure 7b shows the PICOPASCAL program equivalent to the state table of Figure 7a.

ADR	DATA	PROGRAM
00	5	if (Q)
01	5	if (S)
02	A	do 0 (S)
03	9	do 1← (Q)
04	4	else
05	B	do 1 (S)
06	8	do 0→ (Q)
07	6	endif
08	4	else
09	5	if (S)
0A	B	do 1 (S)
0B	9	do 1← (Q)
0C	4	else
0D	A	do 0 (S)
0E	8	do 0→ (Q)
0F	6	endif
10	6	endif
11	2	end

$Q+,S+$	$S=0$	$S=1$
$Q= 0 \rightarrow$	0→,0	1←,1
$Q= 1 \leftarrow$	0→,1	1←,0

(a)   (b)

Fig. 7. (a) Binary counter state table. (b) PICOPASCAL program equivalent to the state table.

6   An Ideal Architecture for the UTM

A UTM architecture is ideal in the sense that it is able to deal with applications of any complexity, characterized by:

1. a finite, but arbitrarily long data tape;
2. a read/write head able to interpret a PICOPASCAL program of any complexity;
3. a finite, but arbitrarily long program tape.

It must be pointed out that, for any application, the program tape and the read/write head (the PICOPASCAL interpreter) are always characterized by finite and defined dimensions; only the data tape can be as long as desired, as is the case for the binary counter, whose growth is potentially infinite [9].

An ideal architecture, embedding the current example, but compatible with any other application, could be follows (Figure 8):

1. The data tape, able to shift right, to shift left, or hold, is folded on itself. The initial state is defined in Figure 8 by $QL3:0$, QC, $QR0:3 = 000010000$, where QL are the squares to the left of the central square QC, and QR are the squares to the right of QC; the data tape is able to grow to the left of QC, i.e., to the right of $QL0$ ($QL4$, $QL5$, ...) and to the right of QC ($QR4$, $QR5$, ...), as can be appreciated in Figure 8.

2. The fixed read/write head, which is not detailed here, is basically composed of a state register Q,S (storing the current values of internal and input states Q,S, respectively, with an initial state $Q,S = 01$) and a stack $ST1:3$ characterized by a 1-out-of-3 code (one-hot encoding). At the start of the execution of the PICOPASCAL program (i.e., in Figure 7, at address $ADR = 00$), the stack is in an initial state $ST1:3 = 100$. Roughly speaking, each **if** instruction will involve a PUSH operation, each **endif** a POP operation, and each **else** a LOAD operation. When $ST1 = 1$, the **do** instructions are executed. The main characteristic of the stack is its scalability: for any program exhibiting n nested **if** instructions, the stack is organized as a $n + 1$ squares shift register. Both the $ST1:3$ stack and the Q,S register are able to grow to accommodate more complex applications.

3. The program tape is folded on itself; it is able to grow to accommodate more complex applications.

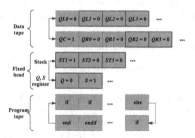

Fig. 8. UTM's ideal architecture.

7   An Actual Implementation of the UTM for the Binary Counter

In order to implement the binary counter application with a limited number of MICTREE artificial cells, we have somewhat relaxed the requirements of the ideal architecture described earlier. Our final architecture is made up of three rows ($Y = 1...3$) and nine columns ($X = 1...9$) organized as follows (Figure 9):

– The 18 instructions of the PICOPASCAL program (Figure 7) are placed in the program tape, using the two lower rows ($Y = 1, 2$) of the array.

- The read/write head is composed of a *ST1:3* stack and of the *Q,S* register ($X = 1...5$, $Y = 3$), while the data tape is implemented by three cells ($X = 6...8$, $Y = 3$) storing 9 bits *QL3:0, QC, QR0:3*.

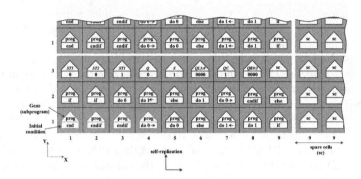

Fig. 9. UTM's actual implementation for the binary counter example on a multicellular array of 27 MICTREE cells.

In order to demonstrate self-repair, we added spare cells in each row, at the right-hand side of the UTM, all identified by the same horizontal coordinate ($X = 9$ in Figure 9). As previously mentioned, more cells may be used not only for self-repair, but also for a UTM necessitating a growth of the tape of arbitrary, but finite, length [9].

Self-replication rests on two hypotheses:

- there exist a sufficient number of spare cells (unused cells at the upper side of the array, at least $3 \times 9 = 27$ for our example);
- the calculation of the coordinates produces a cycle at the cellular level (in our example: $Y = 1 \rightarrow 2 \rightarrow 0 \rightarrow 1 \rightarrow 2 \rightarrow 0$).

Given a sufficiently large space, the self-replication process can be repeated for any number of specimens in the Y axis. With a sufficient number of cells, it is obviously possible to combine self-repair (or growth) towards the X direction and self-replication towards the Y direction (Figure 9).

8   Experiment and Discussion

The UTM was completely implemented and the binary counter fully tested. The values obtained (contents of the *REG* register of cells *Q, QL3:0, QC*, and *QR0:3* at the end of each program execution cycle) correspond exactly to the results presented by Minsky in [8].

The complete genome microprogram describing our artificial organism is composed of 377 16-bit-wide instructions, implying a configuration bit string of 6032 bits.

We tested the self-repair capabilities of our implementation (Figure 9), made possible by the spare column at the right edge of our artificial organism. Using this spare column, our organism is able to tolerate at least one fault in any cell of the array, and up to three faulty cells in the same column [9].

The self-replication of our UTM was tested by doing one copy of original organism. For this test the cellular array contained 6x10 = 60 MICTREE cells (Figure 10).

The largest possible artificial organism implementation with cellular differentiation can reach a size of 16x16 MICTREE cells, a limit imposed by the architecture of the cell. This limitation disappears if we realize a molecular decomposition of the MICTREE cell, allowing us to tailor the cell to meet the requirements of any application [4, 6].

9   Conclusion

In this paper we showed that it is possible to embed a universal Turing machine into a multicellular array based on MICTREE artificial cells, thus obtaining a self-repairing and self-replicating universal Turing machine.

The mapping of the UTM onto our multicellular array was made possible thanks to the introduction of a modified version of the W-machine [13], i.e., an interpreter of the PICOPASCAL language. We showed that an ideal architecture (i.e., an architecture with a semi-infinite data tape) was able to deal with applications of any complexity, (i.e., with a semi-infinite data tape). We also presented an actual implementation in which we relaxed somewhat the characteristics of the ideal architecture in order to use a limited number of MICTREE artificial cells. We slightly simplified our implementation by presenting the example of the binary counter in which the data are binary and where the direction of the head's motion coincides with the internal state (in general functions $Q+$ and $D+$ are independent). A picture of the final implementation is shown in Figure 10.

The property of universal construction raises issues of a different nature, since it requires (according to von Neumann) that a MICTREE cell be able to implement organisms of any dimension. This challenge can be met by decomposing a cell into molecules and tailoring the structure of cells to the requirements of a given application [4, 6].

Acknowledgments

This work was supported in part by grant 21-54113.98 from the Swiss National Science Foundation, by the Consorzio Ferrara Richerche, Università di Ferrara, Ferrara, Italy, and by the Leenaards Foundation, Lausanne, Switzerland. We thank the anonymous reviewers for their helpful remarks.

Fig. 10. Universal Turing machine implementation. This implementation presents six rows and ten columns, which allowed us to test the self-replication of the binary counter.

References

1. A. Hodges. Alan Turing and the Turing Machine. In R. Herken, editor, *The Universal Turing Machine a Half Century Survey*, pages 3–14. Springer-Verlag, Wien, second edition, 1995.
2. D. C. Ince, editor. *Mechanical Intelligence: Collected Works of A. M. Turing*, chapter Intelligent Machinery, pages 107–128. North-Holland, 1992.
3. S. C. Kleene. Turing's Analysis of Computability, and Major Applications of It. In R. Herken, editor, *The Universal Turing Machine a Half Century Survey*, pages 15–49. Springer-Verlag, Wien, second edition, 1995.
4. D. Mange, M. Sipper, A. Stauffer, and G. Tempesti. Towards Robust Integrated Circuits: The Embryonics Approach. *Proceedings of the IEEE*, 88(04):516–541, April 2000.
5. D. Mange, A. Stauffer, and G. Tempesti. Binary Decision Machine-Based Cells. In D. Mange and M. Tomassini, editors, *Bio-Inspired Computing Machines: Toward Novel Computational Architectures*, pages 183–216. Presses Polytechniques et Universitaires Romandes, Lausanne, Switzerland, 1998.
6. D. Mange, A. Stauffer, and G. Tempesti. Embryonics: A Microscopic View of the Cellular Architecture. In M. Sipper, D. Mange, and A. Pérez-Uribe, editors, *Evolvable Systems: From Biology to Hardware*, volume 1478 of *Lecture Notes in Computer Science*, pages 185–195, Berlin, 1998. Springer-Verlag.
7. P. Marchal, A. Tisserand, P. Nussbaum, B. Girau, and H. F. Restrepo. Array processing: A massively parallel one-chip architecture. In *Proceedings of the Seventh International Conference on Microelectronics for Neural, Fuzzy, and Bio-Inspired Systems, MicroNeuro'99*, pages 187–193, Granada, Spain, April 1999.
8. M. L. Minsky. *Computation: Finite and Infinite Machines*. Prentice-Hall, Englewood Cliffs, New Jersey, 1967.
9. H. F. Restrepo. *A Programming Methodology for Configurable Processor Networks*. PhD thesis, Swiss Federal Institute of Technology, Lausanne, Switzerland, 2001. to appear.

10. H. F. Restrepo and D. Mange. *Reconfigurable Computing. Experiences and Perspectives*, chapter MICTREE: A Bio-Inspired FPGA for Embryonic Applications, pages 152–167. Fundaçao de Ensino Eurípides Soares da Rocha, Marília, SP, Brazil, August 2000.

11. B. A. Trakhtenbrot. Comparing the Church and Turing Approaches: Two Prophetical Messages. In R. Herken, editor, *The Universal Turing Machine a Half Century Survey*, pages 557–582. Springer-Verlag, Wien, second edition, 1995.

12. A. M. Turing. On Computable Numbers, with an Application to the Entscheidungsproblem. *Proceedings of the London Math. Soc.*, 42:230–265, 1936.

13. H. Wang. A Variant to Turing's Theory of Computing Machines. *Journal of the ACM*, IV:63–92, 1957.

14. N. Wirth. *Programming in MODULA-2*. Springer-Verlag, Berlin, 1983.

Asynchronous Embryonics with Reconfiguration

Alexander H. Jackson and Andrew M. Tyrrell

Bio-Inspired and Bio-Medical Engineering, The Department of Electronics,
The University of York, Heslington, York, YO10 5DD, UK
{Alex.Jackson, Andy.Tyrrell}@bioinspired.com
http://www.bioinspired.com

Abstract. As embryonic arrays take inspiration from nature they display biological properties, namely complex structure and fault-tolerance. However, hardware implementations have yet to take advantage of a further biological feature at a fundamental level; asynchronous operation. Scalability and reliability are seen as two areas in which embryonic arrays could benefit from asynchronous design. This paper builds upon a previous asynchronous embryonic architecture simulation. The addition of a two-fold reconfiguration strategy that provides fault-tolerance is detailed. The simulation's design is similar to that of a macromodule library that has been implemented using Xilinx Virtex FPGAs, bringing the possibility of truly asynchronous embryonic circuits a step closer.

1   Introduction

Life demonstrates a number of fundamental properties including complex structure, fault-tolerance and asynchronous operation. Some, or all, of these properties are desirable in computing systems. At an abstract level, biologically-inspired systems imitate one or more of these properties within their design. Ontogenetic systems are those motivated by the process of development from a genetic description. In biology, development produces an individual organism, which generally has a complex structure. Although cells within the individual fail, the organism's complex structure provides self-diagnosis and self-repair processes that make it reliable as a whole. Self-diagnosis and self-repair are the equivalent of the detection and recovery stages in fault tolerant systems. In addition, these biological processes are distributed throughout the organism's structure, rather than being under the control of a central unit as typically found in hardware systems. Reliability is paramount where systems are intended for remote or hostile operation and increasingly important for those used generally. These factors imply the need to bestow systems with fault tolerance as their basic components are not entirely reliable.

Conventional digital electronic circuits operate under the synchronous paradigm where stored values are simultaneously updated by a global clock signal. A major difference between the functional units of conventional digital systems and biological cells is the omission of a global signal that coordinates inter-unit communication. In electronic terms they are asynchronous; there is no global clock. Although

Y. Liu et al. (Eds.): ICES 2001, LNCS 2210, pp. 88-99, 2001.
© Springer-Verlag Berlin Heidelberg 2001

asynchronous circuits have been studied since the outset of digital electronics, the simplicity of synchronous techniques has allowed them to dominate.

Embryonic (embryological electronic) arrays [1, 2, 3, 4] draw on the similarities between VLSI array circuits and embryo development to give a fault-tolerant computing architecture. Each cell has an identical processing element with previous architectures having used simple two-input multiplexers or more complex microprogram processors. An embryonic array is essentially a field-programmable gate array (FPGA) bestowed with distributed hardware fault tolerance.

This paper further develops a macromodule-based asynchronous embryonic computing structure [5]. In particular it shows how an asynchronous embryonic array of this type can provide fault-tolerance through hardware reconfiguration. Section 2 briefly covers relevant asynchronous electronic topics. Section 3 considers embryonic systems and describes the structure of an asynchronous embryonic array. Section 4 illustrates how reconfiguration can be included in this array. Following this, section 5 illustrates the results of a simulation through an example circuit. Conclusions are drawn in section 6 together with directions for further work.

2   Asynchronous Electronics

Both synchronous and asynchronous digital systems assume that signals are in one of two possible states. Synchronous systems split time into discrete, regular intervals [6]. The state of the circuit is updated by the clock, with combinational logic processing values and generating the next-state vector between active clock edges. As asynchronous systems do not have a global clock they cannot operate in this way.

Asynchronous systems are typically classified by their timing model, signalling protocol and by the method used for their design and implementation [6, 7, 8]. For brevity, this paper assumes familiarity with the bounded-delay and delay-insensitive timing models, Muller-C gates and macromodules [5, 6, 7, 8, 9, 10].

2.1   Signalling Protocols

Asynchronous circuits generally use an inter-element signalling scheme to act as a protocol layer for all transitions; these are self-timed circuits. Designs based on macromodules, including micropipelines, are among these [5, 7, 9, 10]. Signalling protocol control paths typically use event logic, an event being a logic level transition.

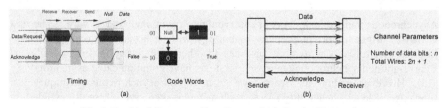

Fig 1. Dual-rail Return-to-Zero Protocol (a) Dual-rail Channel (b)

A dual-rail channel using a request/acknowledge protocol is shown in figure 1. Transfer commences when the sender outputs a data symbol. Two wires are used per data bit, but only one line changes value for every symbol sent; a Gray code [9]. An implicit request is therefore encoded with the data during transfer. Furthermore, even if the two wires have differing delays the correct symbol is received, the dual-rail encoding providing the advantage of delay-insensitive data transmission (in comparison to single-rail systems [6, 7, 8]). The sender can only change output symbol once the receiver acknowledges receipt. Dual-rail channels may use a return-to-zero protocol, also shown in figure 1. Data items are separated by null code words. One line in the pair is set to transmit a logic one, the other to send a logic zero.

2.2   Metastability

Almost all synchronous circuits have inputs, such as those from manually-controlled switches, whose changes may not respect the clock; they are asynchronous. A synchroniser is used to align an asynchronous input with the system signals. A simple way to achieve this is by using a single D-type flip-flop, the asynchronous input being captured by the system clock to give a synchronised output value. For each clock cycle, an unambiguous value for the asynchronous signal would be provided, were it not for metastability [8, 11, 12].

As the input is asynchronous it may change at any time and possibly violate the setup or hold time of the flip-flop. If this occurs the flip-flop output can become metastable, where the output voltage is at neither logic level. The probability of an output remaining metastable decays exponentially with time, the probability and decay rate being dependent upon factors such as the device technology. Synchroniser failure will occur if the synchroniser output is used whilst still metastable, with downstream circuitry possibly interpreting the metastable value differently or itself becoming metastable. It is often implicitly assumed that a single flip-flop can resolve a metastable state within one clock cycle. More sophisticated synchronisers can be built but require a greater amount of logic and introduce latency [12]. Metastability considerations are also needed for certain types of asynchronous logic design, not covered here, where asynchronous arbitration takes place [7, 8].

3   Asynchronous Embryonics

3.1   Benefits

Asynchronous methods may benefit embryonic arrays in a number of ways. The first is that a further important property of biology will be brought to ontogenetically-inspired electronic systems as it has been to other biologically-inspired paradigms. Coarse-grained synchronous embryonic systems have communicated asynchronously [13]. However, embryonic arrays have yet to exploit asynchronous techniques at a fundamental level. Scalability issues, such as clock skew, are inherently avoided by removing the global clock. By making embryonic designs asynchronous, their distributed reconfiguration capability is retained whilst the clock is replaced with localised control. There are other advantages normally associated with asynchronous methods. These include implicit power control and modularity, where systems can be simply interconnected without the need to consider clock matching. Asynchronous designs have switching transitions that are more evenly spread over time, which may lead to reduced electro-magnetic interference [14]. A recent asynchronous processor core is competitive in terms of power, area and performance to its synchronous counterpart [14].

3.2   Design

Figure 2 shows the asynchronous embryonic cell design. The structure closely follows that of previous synchronous architectures where the functional units are multiplexer based [2, 3]. Asynchronous circuits can therefore be realised from binary decision diagrams (BDDs) [1, 2, 3]. The inputs and outputs are referred to by compass point, with each cell requiring eighteen configuration bits. Local routing conveys data from the south input to the functional unit and to both horizontally neighbouring cells. The functional unit output is always routed north, but also to the switch block where it can feed one or more output bus. Alternatively, each output bus can relay the value of any input bus except its partner.

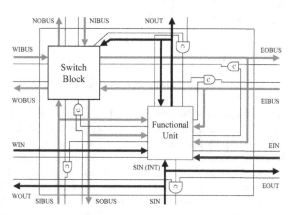

Fig 2. Cell Structure

The fundamental differences from similar synchronous architectures at this level are the need for dual-rail two-bit buses and their acknowledgement lines. Muller-C gates are necessary to combine the acknowledgement of both receivers where a dual-rail line forks. Internally, the switch block uses multiplexers to provide self-acknowledgement when an incoming bus is not in use. This is identical to the technique described below for non-operative lines entering the functional unit. However, the switch block still requires only eight configuration bits.

Macromodules are asynchronous cell library components [5, 7, 9, 10] from which self-timed circuits can be built as the component interfaces are compatible. The functional unit structure (Figure 3) is macromodule based. It closely resembles that of an asynchronous macromodule library that has already been successfully implemented for the Xilinx Virtex FPGA [5]. By using dual-rail encoding delay-insensitivity is achieved between macromodules, an advantage where commercial FPGAs are used for asynchronous implementation.

A return-to-zero protocol is used as this allows the flip-flop to be isolated from the unit output. Since the output encoding logic is simple, and the transistor count for a Muller-C gate is around half that of a D-type flip-flop with preset and clear, the assumption here is that only the flip-flop will incur a fault. By returning to zero between data symbols, symbol transmission can be prevented when an erroneous value has been stored.

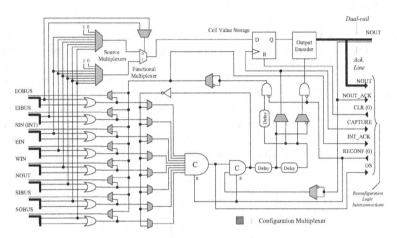

Fig 3. Functional Unit Structure

The configuration determines the inputs used by the functional unit, there being a maximum of three. The functional multiplexer selection is controlled by one of the east buses, whilst data inputs are supplied from fixed values, local connections or south buses. The configuration multiplexers allow non-operating inputs to be ignored by the Muller-C gate, whilst also making them self-acknowledging. This is required, for example, where the functional multiplexer uses *EIBUS* but *EOBUS* carries other data. The cell transmitting data along *EOBUS* must be acknowledged by all potential receivers because of the acknowledge line's Muller-C gates (Figure 2). The configuration multiplexers at the input to the Muller-C gate do not themselves require configuration bits, their setting being derived from that of the source multiplexers. The functional unit has interconnections to the cell's reconfiguration logic, the incoming lines being at logic zero normally. The Muller-C gates initially have a low logic output, but are also reset by the *RECONF* signal. Two Muller-C gates are used so that the *DS* signal is efficiently provided.

During normal operation each OR gate output is asserted when its associated dual-rail line has data. The Muller-C gate combines the OR gate outputs so that its rising transition stores the functional multiplexer output only when all inputs have data. This raised level then acknowledges the senders and activates the encoder to transmit the data symbol. The capture signal goes low once all senders have removed their data and all receivers have acknowledged. This generates a null symbol on the data output and brings the acknowledge output low to complete the communication cycle.

One of the cells within each next-state loop must act to initiate asynchronous operation, otherwise computation will not propagate to the output. The configuration multiplexers can therefore make the cell initially output a data symbol. In addition, the *NOUT* acknowledge line must not be inverted and the connection order of the flip-flop and output encoder must be reversed where a cell initiates computation from within a next-state loop. Configuration multiplexers are used for this purpose and also to alter the level of the outgoing acknowledge where a cell drives it's own inputs, so that the correct protocol is observed.

4   Reconfiguration

4.1   Overview

Hardware fault tolerance can be applied to the array described in the last section through row reconfiguration. Currently only the storage flip-flop is protected from faults. The configuration registers are not protected and all cells contain a complete set of configuration data from the outset [2, 4]. Synchronous designs test for the occurrence of a fault using the clock. Where cell operation is controlled through a state machine [3], an asserted fault status signal moves cells into a non-active state at the clock edge where they remain during reconfiguration. Alternatively, the inactive clock edge is used to store the fault status, this register driving combinational logic to reconfigure the array before the next active clock edge [2]. Therefore, even though in practice the fault status signal is like an asynchronous input (Section 2.2), it has been assumed that its level does not change whilst being stored or equivalently, that a single storage element is sufficiently quick at resolving any possible metastable state.

The asynchronous reconfiguration logic also makes this assumption in addition to assuming that the flip-flop output will remain correct for the remainder of that asynchronous transaction once tested. However, the asynchronous reconfiguration logic is complicated as it cannot use global signals or purely combinational logic for reconfiguration. Synchronous designs [2, 3] can use these approaches as all cells advance simultaneously, whereas asynchronous cells may implicitly operate at different times. As asynchronous logic provides the advantage of global clock removal, the removal of other global signals is pertinent. Although a global reset signal is needed this will not suffer from the same problems as a high-speed clock network and can be routed in a cellular fashion.

4.2   Reconfiguration – Logical Network

The approach to reconfiguration can be split into two phases; logical and physical. Asynchronous finite state machines (AFSMs) are used as part of the logic. These are similar to synchronous FSMs but do not have clock-driven storage elements. Instead state is retained by the delay of feedback loops. Figure 4 shows the reconfiguration logic used during the logical phase. Each data input and output has a pair of reconfiguration lines associated with it; reconfigure in and reconfigure out. The built-in self-test (BIST) logic, currently simulated by an external fail signal, should compare the outputs of duplicated storage flip-flops. The *CAPTURE* signal on the left of figure 4(a) is delayed sufficiently so that the BIST logic can settle following flip-flop storage. This ensures that ANDing the fail output and *CAPTURE* signals will only result in a high logic level output if the stored value is in error. Therefore, following fault detection *SET* causes the logical reconfiguration AFSM output to become high, which in turn puts the cell into reconfigure mode by raising the level of *RECONF*. Figure 3 shows that setting *RECONF* prevents the senders from being acknowledged, the receivers from being sent data and causes the Muller-C gates to be reset. This again requires the signals to be correctly delayed with respect to each other. The *RECONF* signal also drives the reconfigure outputs of the functional unit, which in turn drive those of the cell, shown in figure 4(b). Whilst some of the functional unit outputs can always be driven, those having OR gates at the cell level should only be driven if the functional unit uses that input or output for data; they are configuration dependent. No extra configuration bits are needed here, their setting again being derived from that of the configuration source multiplexers.

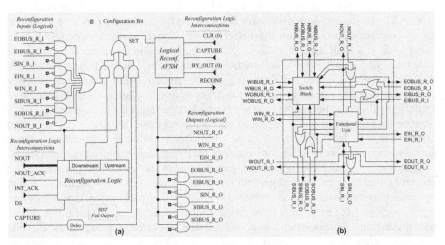

Fig 4. Reconfiguration Logic (Logical) – (a) inside the functional unit (b) within the cell

The above logic results in the reconfiguration signal being passed only to those cells whose functional units receive data from, or pass data to, the faulty cell. That is, the logical reconfiguration network follows the data communication topology of the multiplexer circuit realised by the embryonic array. This is important as only those units immediately upstream and downstream of the faulty cell are in a partially known state. Those downstream are awaiting data whilst those upstream are awaiting acknowledgement. The faulty cell sends neither data nor acknowledgement but indicates the need for reconfiguration.

For a cell downstream of one having a fault, the reconfiguration signal arrives via a reconfiguration input (Figure 4). The cell configuration once more determines those that are active from the data inputs used by the functional unit. The input reconfiguration signal cannot be used to select reconfiguration mode directly, but must be gated as the cell may have receivers that have yet to acknowledge. Where this is the case, the cell will become idle once the send transaction completes. This is known as at least one upstream cell is in reconfiguration mode and so will send no further data. Alternatively a downstream cell may have been sent data, but will not acknowledge as it is in reconfiguration mode (indicated via $NOUT\_R\_I$). This second possibility can arise where an output forms a feedback loop. The downstream reconfiguration logic, which is purely combinational, tests for both these situations and gates the downstream input reconfiguration signal accordingly.

The reconfiguration signal for a cell upstream of one having a fault always arrives through $NOUT\_R\_I$ as the functional unit can only output through $NOUT$. Here also, the cell cannot be placed into reconfiguration mode directly. From figure 3 it can be seen that the upstream cell will have acknowledged its senders just after it sent the data that caused the error. The senders of the upstream cell can therefore revert to a null data output because of the asserted acknowledgement line. The upstream cell must wait for all of its senders to have a null output as switching to reconfiguration mode would otherwise prematurely remove the acknowledgement, violating the communication protocol. As the faulty cell will not have acknowledged the upstream cell it cannot perform any further calculation. Figure 3 shows that DS will become low once all incoming data channels have a null code. This signal is used by the upstream reconfiguration logic, which is again combinational, to gate the reconfiguration input appropriately.

In both upstream and downstream cases the cell will eventually enter reconfiguration mode. In turn, this cell will then propagate the reconfiguration signal to its upstream and downstream cells. Eventually all of the array's active cells will be in reconfiguration mode assuming the array forms a single circuit.

4.3   Reconfiguration – Physical Network

Unlike the logical reconfiguration network, whose topology is controlled by the data flow, the physical network has the fixed compass-point configuration of figure 5. The physical network's purpose is to ensure that the entire array has stopped processing before reconfiguration takes place. The distant interconnection possibilities of the bus network allow the logical communication network topology to be radically different from the physical. However, physically neighbouring cells must be considered from a reconfiguration point of view. The faulty cell is evidently in the row that needs to be removed during reconfiguration. Therefore, when the faulty cell switches to reconfiguration mode its physical reconfiguration logic is activated by the $RECONF$ signal. The physical reconfiguration lines (x_ROW_R and x_COL_R) are configured as a form of distributed AND gate. These lines pass a signal from the west of the array to the east as each column of cells switches to reconfiguration mode. The cells on the periphery of the array have external boundary connections to facilitate this signal propagation. Once all of the cells are in reconfiguration mode the lines pass a signal back from the east of the array to the west. This signal allows the array's cells to safely begin the process of row removal as all cells have stopped computation.

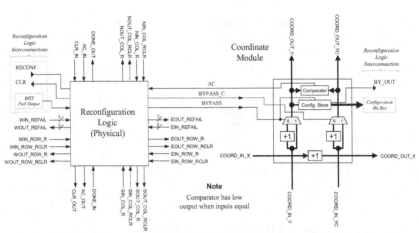

Fig 5. Cell Reconfiguration Logic (Physical) & Cell Coordinate Module

Dual-rail reconfiguration fail signalling lines (x_REFAIL), without acknowledgements, run along the length of each array row, one in each direction. One of these acts as a form of distributed delay-insensitive OR gate that passes from west to east. The western edge of the array passes a false dual-rail code into the first column of cells along the eastbound bus. However, this code will not propagate any further along until the cells have been signalled to indicate that the whole array is in reconfiguration mode. Once signalled, this distributed OR gate becomes active, each cell reading the code from its western neighbour and providing a dual-rail encoded output to the cell to its east. For cells that have not failed the code word is simply passed on. However, for a failed cell a true code is always passed out to the east.

In this way a signal is propagated along the row from west to east, with the array boundary returning the code along the westbound dual-rail line. If any cell has failed in that row a true dual-rail code word will be passed back. This true code activates the physical reconfiguration AFSM of each cell within a row requiring removal. These cells each then act to control reconfiguration for their corresponding column.

The physical reconfiguration AFSM first drives BYPASS high (Figure 5). This causes the normal Y coordinate incrementer to be switched out and also sets the BY_OUT signal, which overrides the configuration selection to make the cell logically transparent from north to south (the bypass multiplexers are not shown in any of the figures). The coordinate change propagates northward to alter the coordinates of all of the cells in the column above the failing row, leading to cells in the next spare row being reconfigured for operation. However, as the cells do not record their previous coordinate, they have no simple way of observing that their location has changed. For this purpose a second Y coordinate datapath exists whose value initially remains unaltered by the cells of the failing row. The comparators can therefore detect this difference and indicate via AC that an address change has occurred.

The AC_x signals run through the array from north to south and act as a distributed AND gate for each column's AC signals. A cell in the row being removed will only receive a high level on its AC_IN line once all of the cells above it in the column have received the address change. The physical reconfiguration AFSM can then set the BYPASS_C signal to switch out the comparison coordinate incrementer.

Once a lowered *AC_IN* signal is detected the *DONE_x* signal path is used to determine if further rows have simultaneously failed. If so, they reconfigure in the same way. Eventually the northernmost cell asserts its *DONE_OUT* signal. This is passed back along the *CLR_x* line to bring the cells out of reconfiguration mode.

Although the cells asynchronous nature would allow them to begin computation at different times, all cells must leave reconfiguration mode before computation starts as otherwise a further fault may occur before the reconfiguration scheme is ready to respond. The AFSMs therefore hold the *DONE* signals, and hence the reconfiguration clear lines, high until all cells have come out of reconfiguration. A further distributed logic network, connected by *x_ROW_RCLR* and *x_CLR_RCLR*, is used to indicate when this has occurred. The array then restarts computation.

5   Simulation Results

The asynchronous embryonic array described has been simulated using VHDL. The simulation is behavioural only at the lowest level, as is that of the previously developed macromodule library. This gives confidence that the array design can be successfully implemented in the future. A two-bit up/down counter has been used before as an example circuit for embryonic arrays [2, 5]. The two-bit counter BDD [2, 15] can be implemented using a two-column, three-row embryonic array.

Figure 6 shows the simulation output for an asynchronous two-bit up/down counter. A two-column, five-row array has been used to allow for reconfiguration. The single asynchronous dual-rail input, *Up/Down*, determines the count direction and also advances the count. The most-significant bit of the counter output is displayed by signal *B*, the least-significant by *A*. All of the dual-rail signals are shown as separate signals and as bus values, although their acknowledgement lines are not displayed. Formatted versions of the input signal and counter output are also given. These show the decoded dual-rail value and do not revert to a null state for clarity. The simulated fail signal, *EXTFAIL*, and configuration register of each cell is also shown, as are the Y coordinates of cells in the first array column.

The counter can be seen to increment as the *Up/Down* input supplies data '1' symbols. A fault is injected to cell (2,1) at time *X* and reconfiguration can be seen to occur shortly after. Reconfiguration can been seen once again at time *Y* with the third physical row being removed from operation, cell (1,3) having been made faulty. Like previous embryonic arrays the asynchronous design produces data output values that do not follow the count sequence during reconfiguration [2], although the correct asynchronous protocol is observed. The array does resume normal operation following reconfiguration, even though the reconfiguration process itself resets the array's state, and hence the count sequence.

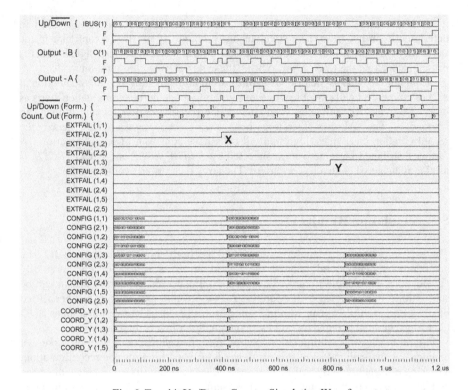

Fig. 6. Two-bit Up/Down Counter Simulation Waveforms

6   Conclusions and Further Work

Asynchronous signalling protocols and metastability have been briefly covered. It has been observed that practical embryonic systems have yet to use asynchronous techniques at a fundamental level. Scalability and reliability have been identified as two areas where an asynchronous approach could be of benefit to embryonic systems, in addition to the normal advantages of asynchronous design.

An asynchronous embryonic architecture based upon dual-rail macromodules has been described. The reconfiguration logic, which provides a fault-tolerance capability, has been detailed. The simulation's design is similar to that of a macromodule library that has been implemented using the Xilinx Virtex FPGA, bringing the implementation of truly asynchronous embryonic circuits closer.

The simulation could be improved by making the address change logic more like that used for physical reconfiguration set and clear. Currently it is assumed that the comparison coordinate logic will have been updated by the time all cells have left reconfiguration mode. Whilst this would be likely in any practical implementation the extra logic and modifications are minor. Future work will include implementation of the design presented using the Xilinx Virtex FPGA. In addition, consideration will be given to the possible problems that metastability may cause fault-tolerant systems.

Acknowledgements

This work is funded through the Engineering & Physical Sciences Research Council, UK, & Xilinx, Inc. "Xilinx" & "Virtex" are registered trademarks of Xilinx, Inc.

References

1. Mange, D. et al. (2000), *"Toward robust integrated circuits: The embryonics approach"*, Proceedings of the IEEE, **88**(4): 516-43.
2. Ortega-Sánchez, C. A. (2000), *"Embryonics: A Bio-inspired Fault-tolerant Multicellular System"*, D.Phil. Thesis, Bio-inspired and Bio-medical Engineering, The Department of Electronics, The University of York: 159.
3. Tempesti, G. (1998), *"A Self-Repairing Multiplexer-Based FPGA Inspired by Biological Processes"*, Ph.D. Thesis, Logic Systems Laboratory, Computer Science Department, Swiss Federal Institute of Technology: 166.
4. Ortega, C. & Tyrrell, A. (2000), *"A Hardware Implementation of an Embryonic Architecture Using Virtex FPGAs"*, Evolvable systems: from biology to hardware, Proceedings of 3rd International Conference (ICES2000), Lecture notes in Computer Science, **1801**, 155-164, Miller, J. et al. (eds.), Springer-Verlag.
5. Jackson, A. H. & Tyrrell, A. M. (2001), *"Asynchronous Embryonics"*, Proceedings of 3rd NASA/DoD Workshop on Evolvable Hardware, IEEE Computer Society.
6. Hauck, S. (1995), *"Asynchronous design methodologies: an overview"*, Proceedings of the IEEE, **83**(1): 69-93.
7. Brunvand, E. (1995), *"Introduction to Asynchronous Circuits and Systems"*, Tutorial, 2nd Working Conference on Asynchronous Design Methodologies, South Bank University, London, UK.
8. Davis, A. & Nowick, S. M. (1997), *"An Introduction to Asynchronous Circuit Design"*, University of Utah Department of Computer Science report, **UUCS-97-013**: 1-58.
9. McAuley, A. J. (1992), *"Four state asynchronous architectures"*, IEEE Transactions on Computers, **41**(2): 129-42.
10. Sutherland, I. E. (1989), *"Micropipelines"*, Communications of the ACM, **32**(6): 720-38.
11. Unger, S. H. (1995), *"Hazards, critical races, and metastability"*, IEEE Transactions on Computers, **44**(6): 754-68.
12. Wakerly, J. F. (1994), *"Digital design principles and practices"*, (2nd ed.), Prentice Hall.
13. Nussbaum, P., Girau, B. & Tisserand, A. (1998), *"Field programmable processor arrays"*, Evolvable Systems: From Biology to Hardware, Proceedings of 2nd International Conference (ICES98), Lecture notes in Computer Science, **1478**, 311-322, Sipper, M. et al. (eds.), Springer-Verlag.
14. Furber, S. B., Edwards, D. A. & Garside, J. D. (2000), *"AMULET3: a 100 MIPS Asynchronous Embedded Processor"*, Proceedings of IEEE International Conference on Computer Design: VLSI in Computers and Processors (ICCD2000), 329-334, IEEE Computer Society.
15. Ortega, C. & Tyrrell, A. (1998), *"Evolvable hardware for fault-tolerant applications"*, IEE Colloquium on Evolvable Hardware Systems (Digest No.1998/233), 30, IEE.

Embryonics: Artificial Cells Driven by Artificial DNA

Lucian Prodan[1], Gianluca Tempesti[2], Daniel Mange[2], and André Stauffer[2]

[1] "Politehnica" University (UPT), Timisoara, Romania
lprodan@cs.utt.ro, http://www.cs.utt.ro
[2] Swiss Federal Institute of Technology (EPFL), Lausanne, Switzerland
name.surname@epfl.ch, http://lslwww.epfl.ch

Abstract. Embryonics is a long-term research project attempting to draw inspiration from the biological process of ontogeny, to implement novel digital computing machines endowed with better fault-tolerant capabilities. For this purpose FPGAs are extremely useful. However, through this project we designed MuxTree, a new coarse-grained FPGA, to implement our embryonic machines. This article focuses on the issues posed by the memory storage and the advances made to achieve more robust memory structures.

Motto. "The nature of an identity lies in its essence." Aristotle

1 Introduction

Present days computing systems are designed from the very beginning to face some challenging problems. One of the main issues is testability, or how to be able to verify that a system functions up to its specs and another is fault tolerance, or how to make the system continue to function properly even while faults are occurring. Though there are techniques developed to satisfy both constraints, engineers have found another source of inspiration, and closer than they thought. The answer could lie in adopting mechanisms tested and refined by nature ever since life began on Earth: bio-inspiration.

A human being is made up of some 60 trillion (60×10^{12}) cells. Key for the survival of the organism is the relentless decoding of the *genome*, a ribbon of 2 billion characters, to produce the necessary proteins [14]. The parallel execution of 60 trillion genomes in as many cells occurs ceaselessly from the conception to the death of the individual. At any given moment, many protecting mechanisms keep an eye on the well operating of the whole organism. Eventual faults, though rare, are in the majority of cases, successfully spotted and repaired. The inspiration of the Embryonics (*embryonic electronics*) project [3, 4, 10, 11] is this astounding degree of parallelism present in nature. Embryonics tries to adapt some of the development processes of multicellular organisms to the purpose of designing novel, robust architectures for massive parallelism in silicon.

It is biology that made possible the miracle of contemplating the successfully operating human organism. It is a miracle indeed to have trillions of cells operating in parallel, forming intricate structures (tissues and organs) only to perform a single goal, that is, the living organism. These are the astounding biological features that make

Y. Liu et al. (Eds.): ICES 2001, LNCS 2210, pp. 100-111, 2001.
© Springer-Verlag Berlin Heidelberg 2001

engineers think about a new way of designing novel electronic systems. A bio-inspired computing system would – theoretically – be capable of online self-testing and self-repairing. The article is structured as follows: Section 2 presents the path of Embryonics from its very beginning, Section 3 and 4 introduce the reader into some of the more peculiar of its features, while the focus is on Section 5 where details of improving the fault tolerance of the memories employed by Embryonics are presented. Finally, Section 6 presents the conclusions and some general guidelines for the future of the project.

2   Overview

2.1   Toward Bio-inspiration

The transition form carbon-based organisms to silicon-based electronic circuits is, of course, far from immediate. Living beings exploit intricate processes, many of which remain undiscovered or unexplained. Therefore, the Embryonics project focuses on two goals [14]:

- *Similarity*: where possible, to develop digital circuits exploit processes similar (but obviously not identical) to those used by living organisms;
- *Effectiveness*: while inspired by biology, the systems we design must remain useful and efficient from an engineer's standpoint.

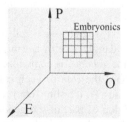

Fig. 1. The POE model and the current position of the Embryonics project.

Therefore the Embryonics project does not try to *imitate* life, but rather to extract some useful ideas from some the most fundamental mechanisms of living creatures.

2.2   The Embryonics Project

Bio-inspired computing systems are categorized by Sipper et al. [13] using their proposed *POE* model, which makes use of three orthogonal axes. Called *phylogenetic* (P), *ontogenetic* (O) and *epigenetic* (E), they define the space inside which all bio-inspired systems, both software and hardware, are situated.

The phylogenetic axis represents the evolution of the genetic program, the reproduction of all living organisms being based on to. The phylogenetic processes exhibit a very low error rate at the individual level and are fundamentally nondeterministic, the source of diversity being mutation and sexual reproduction. Systems represented along the phylogenetic axis are known as *evolvable hardware* or *evolware*.

The epigenetic axis involves online learning through interaction of the systems with the surrounding environment. To the best of our knowledge, only three epigenetic systems exist in living beings: the immune, the nervous, and the endocrine systems. Represented along the epigenetic axis are the bio-inspired systems that are capable of learning, usually under the form of artificial neural networks.

The ontogenetic axis focuses on the development of the single individual from its very own genetic material. Environmentally induced behavior is not considered, thus the main process off the ontogenetic axis being the *growth* of the organism. Characteristics such as replication (self-replication), which can be seen as a special case of growth, and regeneration (self-repair), or recovery after wounds or illnesses, are part of the ontogeny and are extremely attractive for many applications. The Embryonics project presents a consistent view of ontogenetic [2, 12] hardware but it is not necessarily limited to ontogenetic processes. For the moment it can be regarded as situated in the plane defined by the ontogenetic axis and the phylogenetic axis (Fig. 1).

2.3   From Biology to Electronics: Bridging the Gap

With the exception of unicellular organisms (such as bacteria), living beings share three fundamental features [14]:

- *Multicellular organization* divides the organism into a finite number of *cells*, each realizing a unique function.
- *Cellular division* is the process whereby each cell (beginning with the first cell or *zygote*) generates one or two daughter cells. During this division, all of the genetic material of the mother cell, the *genome*, is copied into the daughter cell(s).
- *Cellular differentiation* defines the role of each cell of the organism, that is, its particular function (neuron, muscle, intestine, etc.). This specialization of the cell is obtained through the expression of part of the genome, consisting of one or more *genes*, and depends essentially on the physical position of the cell in the organism.

As each cell contains the genome, that is the whole of the organism's genetic material, the cell's "universality" comes as a consequence. This makes the living organisms capable of self-repair (regeneration, cicatrisation) or self-replication (cloning or budding). These two properties, based on a multicellular tissue, are essentially unique to the living world.

Obviously, the differences between the worlds of biology and of electronics are far too many, preventing us from simply copying the nature in silicon. One difference that is not addressable by present days engineering is that the environment living beings interact with is continuously changing, whereas the environment in which our quasi-biological development occurs is imposed by the structure of the electronic circuits, consisting of a finite (but arbitrarily large) two-dimensional surface of silicon and metal. Taking into account the extant differences, we developed a quasi-biological system architecture based on four levels of organization (Fig. 2), described in detail in previous articles [3, 9, 10]. The particular subject of this article lies at the *molecular level*, the bottom layer of our system, and concerns the implementation of fault tolerant memory structures. We will therefore introduce the other levels and discuss them only insofar as they are useful for a clearer understanding of our subject matter.

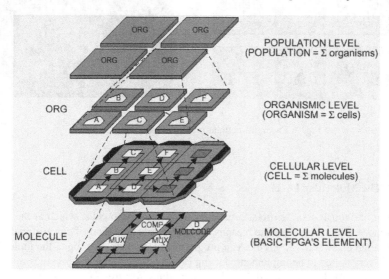

Fig. 2. Embryonics: the 4 levels of organization.

3   The Two Level Organization

3.1   The Cellular Level

As shown in Fig. 3, our artificial organisms are divided into a finite number of cells. Each cell is a simple processor (a binary decision machine), which realizes a unique function within the organism, defined by a set of instructions (program), which we will call the *gene* of the cell. The functionality of the organism is therefore obtained by the parallel operation of all the cells.

Cells are delimited by the existence of a cellular *membrane*, which is in fact an automaton receiving its configuration at initialization time. The information specifying the cellular membrane is known as *polymerase genome*, and it is part of the genome. The dimensions of the cells are programmable, and the mechanism specifying the cellular membrane is called *space divider*. The space divider extends its operations also in the next, more basic level.

Each cell stores a copy of all the genes of the organism (which, together, represent the operative part of our artificial genome, the *operative genome*), and determines which gene to execute depending on its position (X and Y coordinates) within the organism, implementing *cellular differentiation*. In Fig. 3 each cell of a 6-cell organism realizes one of the six possible genes (A to F), but stores a copy of all the genes.

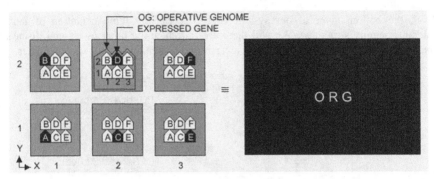

Fig. 3. The multi-cellular organization of an organism.

3.2   The Molecular Level

As seen previously, our artificial organisms decompose into cells, which at their turn decompose into molecules. The reasons for such implementation of our artificial cells are detailed elsewhere [3, 9, 10]. As a programmable substrate of logic, the molecular level is very suited to be implemented using FPGAs. In our case, Embryonics relies on a new type of FPGA, called MuxTree (standing for tree of multiplexers). The molecule is essentially a multiplexer and a D-type flip-flop, linked with the other molecules via a set of programmable connections (Fig. 4). The bit stream configuring the molecule (that is, the connections and the preset value of the flip-flop) is stored into the configuration register CREG.

The unit containing the flip-flop is called the *functional unit* FU, and it provides on-line self-testing and self-repairing. There exist 3 copies of the flip-flop and a simple 2-out-of-3 majority mechanism ensures the fault tolerant operating of the unit.

The switch block SB drives the connections between molecules. Since implementing fault tolerant techniques into the SB would greatly increase the size of the actual implementation of the design, this unit does not provide any such techniques.

4   Memory Structures

Our molecule was initially designed to be able to implement any combinational and/or sequential machines, provided that a sufficient number of molecules were available. But one of the issues raised by our cellular processors was the implementation of the memory required to store the genome program. While capable of fulfilling this role, conventional addressable memory systems present the handicap of requiring relatively complex addressing and decoding logic. Embryonics also had as a disadvantage the fact that one molecule could only store one bit with its functional unit. To store the operative genome we therefore took into consideration other methods.

In living cells, the genetic information is processed *sequentially*. Designing a memory that is inspired by biology suggests a different type of memory, which we called *cyclic memory*. Cyclic memory does not require any addressing mechanism. Instead, it consists of a simple storage structure that circulates synchronously its data in a closed circle, much as the ribosome processes the genome inside a living cell [1].

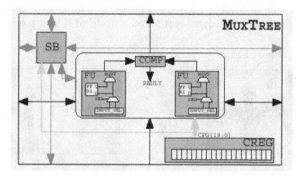

Fig. 4. Internal architecture of the artificial molecule.

The CREG's extended role meant that molecules could operate in two modes, set by special bits in CREG [14]. The first is the *active mode*, in which the molecule acts in its initial way: it makes use of the FU and the SB. The second is the *memory mode*, in which the molecule has two operating sub-modes: short memory (8 bits of CREG are used for data storage) with only SB being active and long memory (16 bits of CREG are used to store data) where neither SB nor FU are active. This was the chosen way to enlarge the storage capabilities of one molecule in order to implement our genome memory.

The molecules operating in either memory modes form rectangular memory structures, the smallest possible such structure being a column composed of only two molecules. There are data output ports at each molecule situated in the top row of the structure. Similar to the way cells are separated from each other by the cellular membrane, we implemented a programmable memory membrane that separates different memory structures. This is to say in the same cell there can exist more than one memory structure, with different shapes and sizes.

To describe the way our cyclic memory operates is not the purpose of this article, it being described in detail in [14]. Until we opted for the cyclic memory architecture the sole purpose of the configuration register CREG was to store the molecule's configuration. For this, an off-line self-testing technique at initialization time was employed. However, adapting the cyclic memory extended its features and there is need to also extend its fault tolerance to suit them, aspects discussed in the next Section.

5   Reliability and Fault Tolerance

The very essence of the Embryonics project is to deliver unprecedented reliability through massive fault-tolerance achieved by bio-inspired design. As difficult as reaching this goal might seem for us, engineers, nature found solutions to fault-tolerance and perfected them throughout hundreds of millions of years. The choice of trying to draw the best of its advantages into the world of silicon seems to be at least worth trying.

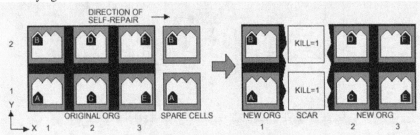

Fig. 5. Self-repair at the cellular level.

Biological entities live continuously under environmental stress. Wounds and illnesses resulting from such stress often cause incapacitating physical modifications. Fortunately, living beings are capable of successfully fighting the great majority of such wounds and illnesses, showing a remarkable robustness through a process that we call *healing*. To reach similar features, a two-level mechanism for self-repair, involving both the cellular and the molecular level is provided in Embryonics. What follows is a description of healing at the cellular and molecular level and of the way the two levels cooperate to produce a higher level of robustness than would be allowed by a single level.

5.1   Self-Repair at the Cellular Level

The redundant storage of the entire genome in every cell is obviously expensive in terms of additional memory. However, it has the important advantage of making the cell *universal*, that is, potentially capable of executing any one of the functions required by the organism. This property is a huge advantage for implementing *self-repair*, the electronic equivalent of biological healing.

Since our cells are universal, the system can survive the "death" of any one cell simply by re-computing the cells' coordinates within the array, provided of course that "spare" cells (i.e., cells which are not necessary for the organism, but are held in reserve during normal operation) are available (Fig. 5).

Self-repair at the cellular level thus consists simply of deactivating the column containing the faulty cell: all the cells in the column "disappear" from the array, that is, become transparent with respect to all horizontal signals. More details about cellular self-repair are provided elsewhere [6, 10, 14].

5.2   Self-Repair at the Molecular Level

Killing a column of processors for every fault in the array represents a penalty we wished to avoid and therefore a certain degree of fault tolerance at the molecular level was introduced. Self-repair in an FPGA implies two separate processes: self-test and reconfiguration. Of these two processes, self-test is undoubtedly the costliest, and we adopted a relatively complex hybrid solution mixing duplication and fixed-pattern testing [9] too complicated to be described here.

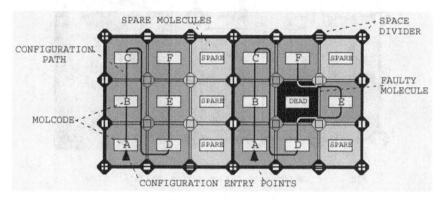

Fig. 6. Self-repair at the molecular level.

On the other hand, the homogeneous architecture of our FPGA simplifies reconfiguration to a considerable extent [5, 7]. Since all molecules are identical, and the connection network is homogeneously distributed throughout the array, reconfiguration becomes a simple question of shifting the configuration of the faulty molecule to its right (similarly to what happens during configuration, as shown in Fig. 6) and redirecting the array's connections. This procedure can be accomplished quite easily, assuming that a set of spare molecules is available. The determination of these spare molecules is in fact one of the most powerful features of our system, since we can exploit the space divider to dynamically allocate some columns as spares. The position and frequency of spares can then be determined at configuration time, and the fault tolerance of our FPGA becomes programmable (and can thus be adapted to the circumstances and the operating conditions).

5.3   Self-Testing Memory Structures

The analysis of Ortega et al. [6] proves that the above strategies of self-repair provide quite a robust architecture. However, when memory structures are employed, the off-line self-testing features implemented in the configuration register (CREG), which is the main memory unit, is insufficient for robust structures of this type. The technique employed to test the CREG was sufficient only for molecules operating in active mode, when the data stored by the CREG remains the same at any given moment. But when molecules are in one of the memory modes, the CREG continuously shifts its data, thus being subject to failure. Therefore the implemented off-line self-testing technique, which takes place at initialization time only does not cover memory data.

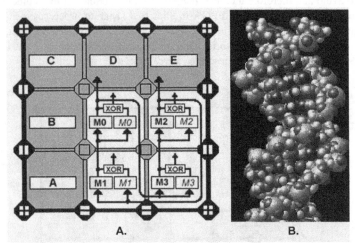

Fig. 7. The new memory implementation (A) and the DNA (B).

The biological DNA, with its two twisted helices (Fig. 7B), provides an intricate means of storing information. As strange as it might seem, information encoded in the DNA takes advantage of techniques used in the designing of fault-tolerant systems: the information is redundant, it is coded in a digital way and it has error control implemented. Each of the two DNA strands is composed of genes, that are strings made of sequences of one of the four bases: A (adenine), T (thymine), C (cytosine) and G (guanine). This is to say that DNA information is stored based on only four characters – the four bases – thus proving the discrete manner of DNA encoding. The error control is achieved by the existence of the second strand, which can be considered as the *complementary* form of the first one. This is because the two strands are in fact linked, and links can only be established between bases A and T, or C and G. Actually, the robustness achieved is so remarkable, DNA can keep its information unaltered in spite of UV radiations, EM fields and other natural stress.

It was discovered that cells have a variety of DNA polymerase enzymes that serve for DNA repair [15]. Since any damage to the DNA would be lethal, biological cells often spend much more energy repairing the DNA than synthesizing it. The correcting process of DNA damage due to environmental effects or proofreading during replication is, unsurprisingly, quite complicated; it assumes the detection of such errors, cutting them out, and then using the remaining good strand as a template for repair synthesis. So even if one strand becomes erroneous, it will be repaired based on the complementary information provided by the second strand.

After all biological features described above, it would seem that Embryonics could also draw some advantages for implementing an error control inside memory structures. But designing a memory that would exhibit on-line self-testing and self-repairing is a difficult and very expensive task. From our standpoint, using even a simple 2-out-of-3 majority function at the configuration register CREG level would almost triple the amount of logic needed for a molecule.

Instead, we could use the two halves of the configuration register in a similar way DNA uses two complementary strands. In memory mode, one molecule offers a maximum of 16 bits of storage space (the *long memory mode*). It is possible to provide fault detection by splitting the CREG in two halves and using them just like the DNA strands: one half stores the complementary data of the other, at any given moment a comparison between the contents of the two halves being made. Because the storage data is continuously shifted [14] there is need only for a few logic gates to be added onto the existing design. A XOR gate implements the comparison process (Fig. 7A). This ensures the detection of any single fault at the expense of storage space. Its correction still remains unsolved. This is somewhat similar to the situation when the biological DNA suffers modifications. In the living cells, correct information is retrieved using the neighborhood around the spotted fault and the complementary strand. In our case, the neighborhood has usually no correlation with the fault, so the data recovery seems not to be possible. In biological terms this is equivalent to non-repairable DNA errors and these typically lead to the death of the cell. The policy considered in the case of errors in the memory molecule is to activate the KILL signal in order to deactivate the entire cell.

For example, instead of storing 16 bits of data, one memory molecule stores 8 bits of data (**M0**) and 8 bits of complementary data (*M0*) as indicated in Fig. 7A. At each clock cycle, the content of both **M0** and *M0* is shifted one position. Specifying the memory mode a molecule may operate in is done through three CREG configuration bits; there are 2 combinations still left unused, there is no problem implementing this new memory mode, which we will call *DNA memory mode*. The presence of the supplementary XOR gate in each molecule, supervising the correct shifting of data, is similar to the existence of the A-T and C-G links between the DNA strands. If an error occurs, the XOR gate will detect it and will forward its output to the KILL signal generator that will trigger hierarchical mechanisms of reconfiguration and re-initialization.

5.4   The KILL and UNKILL Mechanisms

As described in previous articles [9, 10, 14], the robustness of the self-repair mechanism at the molecular level is programmable (through the frequency of columns made of spare molecules). But even so, there are limits to the faults that can be repaired at this level. With the continuous extending of the versatility of molecules, such as introducing the memory modes, the existence of these limits is even more obvious.

The KILL signal is simply generated whenever a non-repairable fault occurs (that is, no more spare molecules are available). It propagates outwards from the non-repairable molecule, rendering al molecules transparent to horizontal signals and thus triggering the cellular-level self-repair briefly described in Subsection 5.1. In the case of a fault being detected inside a memory structure, it being non-repairable, the KILL signal will also be generated and the whole cell will be "killed", meaning that it will cease to operate and will become transparent for the other cells.

In digital electronic systems, the majority of hardware faults that occur during operation are in fact *transient*, that is, they disappear after some time. Based on this observation we might be able to avoid the penalty induced by killing an entire cell due to probably transient errors. This is to say that the parts of the circuit "killed" because of the detection of a fault could potentially come back to "life" after a brief delay. Detecting the disappearance of a fault and handling the "unkilling" at the cellular level proved to be quite a simple task. A killed cell is transparent to the array; being reset, nothing prevents us to resend once again the configuration stream that will restore the functionality of the cell if a sufficient number of detected errors were transient. Since the memory structures are configured in the very same way as the other molecules, the unkilling mechanism does not require to be changed.

6   Conclusions

As a long-term research project, Embryonics is going well on track for many years now. Throughout this time, we have been accumulating considerable experience and were witnesses to the advent of technology that enabled us to actually experiment our ideas. While we adapt continuously to the technological advances, it may be possible that these advances will render some of our specific mechanisms obsolete. But with self-replication [8] being one of the key issues in nanotechnology, we feel that our efforts through the Embryonics project on creating and perfecting bio-inspired computing systems are rewarded.

Though Embryonics is quite at an advanced stage, the technological limits prevent us from experimenting with a great number of cells. A lot of work remains to be done in this direction.

References

1. Barbieri, M.: The Organic Codes: The Basic Mechanism of Macroevolution. Rivista di Biologia / Biology Forum 91 (1998) 481-514.
2. Gilbert, S. F.: Developmental Biology. Sinauer Associates Inc., MA, 3rd ed. (1991).
3. Mange, D., Tomassini, M., eds.: Bio-inspired Computing Machines: Towards Novel Computational Architectures. Presses Polytechniques et Universitaires Romandes, Lausanne, Switzerland (1998).
4. Mange, D., Sanchez, E., Stauffer, A., Tempesti, G., Marchal, P., Piguet, C.: Embryonics: A New Methodology for Designing Field-Programmable Gate Arrays with Self-Repair and Self-Replicating Properties. IEEE Transactions on VLSI Systems, 6(3), (1998) 387-399.
5. Negrini, R., Sami, M. G., Stefanelli, R.: Fault Tolerance Through Reconfiguration in VLSI and WSI Arrays. The MIT Press, Cambridge, MA (1989).

6. Ortega, C., Tyrrell, A.: Reliability Analysis in Self-Repairing Embryonic Systems. Proc. 1st NASA/DoD Workshop on Evolvable Hardware, Pasadena, CA (1999) 120-128.
7. Shibayama, A., Igura, H., Mizuno, M., Yamashina, M.: An Autonomous Reconfigurable Cell Array for Fault-Tolerant LSIs. Proc. 44th IEEE International Solid-State Circuits Conference, San Francisco, CA (1997) 230-231, 462.
8. Sipper, M.: Fifty Years of Research on Self-Replication: an Overview. Artificial Life, 4(3) (1998) 237-257.
9. Tempesti, G.: A Self-Repairing Multiplexer-Based FPGA Inspired by Biological Processes. Ph.D. Thesis No. 1827, EPFL, Lausanne (1998).
10. Tempesti, G., Mange, D., Stauffer, A.: Self-Replicating and Self-Repairing Multicellular Automata. Artificial Life, 4(3) (1998) 259-282.
11. Wolfram, S.: Theory and Applications of Cellular Automata. World Scientific, Singapore (1986).
12. Wolpert, L.: The Triumph of the Embryo. Oxford University Press, New York (1991).
13. Sipper, M., Sanchez, E., Mange, D., Tomassini, M., Perez-Uribe, A., Stauffer, A.: A Phylogenetic, Ontogenetic, and Epigenetic View of Bio-Inspired Hardware Systems. IEEE Transactions on Evolutionary Computation, 1(1) (1997) 83-97.
14. Prodan, L., Tempesti, G., Mange, D., Stauffer, A.: Biololy Meets Electronics: The Path to a Bio-Inpired FPGA. Proc. 3rd International Conference on Evolvable Systems: From Biology to Hardware, Edinburgh, Scotland, UK (2000) 187-196.
15. Terry, T. M.: www.sp.uconn.edu/~bi107vc/fa99/terry/DNA.html

A Self-Repairing and Self-Healing Electronic Watch: The BioWatch

André Stauffer, Daniel Mange, Gianluca Tempesti, and Christof Teuscher

Logic Systems Laboratory, Swiss Federal Institute of Technology,
CH-1015 Lausanne, Switzerland,
andre.stauffer@epfl.ch

Abstract. The Embryonics project is inspired by some of the basic processes of molecular biology, such as the embryonic development of living beings. Transposing these processes in to digital electronic integrated circuits, we design artificial organisms endowed with properties typical of to the living world, such as self-repair and self-healing. In order to illustrate the original features of the Embryonics project, we define the cellular and molecular architecture of a giant artificial organism, the BioWatch. The hardware implementation of a microprogrammed version of our watch exploits a new reconfigurable tissue, the bio-inspired electronic wall or BioWall.

1 Introduction

1.1 Embryonics = embryonic electronics

The *Embryonics* project (for *embryonic electronics*) is inspired by the basic processes of molecular biology and by the embryonic development of living beings [13], [3]. By adopting certain features of cellular organization, and by transposing them to the two-dimensional world of integrated circuits on silicon, we have already shown that properties unique to the living world, such as *self-replication* and *self-repair*, can also be applied to artificial objects (integrated circuits) [4].

Our final objective is the development of very large scale integrated (VLSI) circuits capable of self-repair and self-replication. Self-repair allows partial reconstruction in case of a minor fault, while self-replication allows the complete reconstruction of the original device in case of a major fault [4].

1.2 A bio-inspired watch: the BioWatch

In order to illustrate to the general public the original features of the Embryonics project, we decided to realize a giant electronic watch capable of self-repair and self-healing. In addition to its pedagogical virtues, this project introduces a new concept, the *reconfigurable computing tissue*, that binds tightly together reprogrammable logic circuits (FPGAs), input units (touch-sensitive buttons) and output units (LED-matrix displays).

Y. Liu et al. (Eds.): ICES 2001, LNCS 2210, pp. 112–127, 2001.

Conception, birth, growth, maturity, illness, old age, death: this is the life cycle of living beings. The proposed demonstration will stage the life cycle of the BioWatch from conception to death. Visitors will face a large wall made up of a mosaic of many thousands of transparent electronic modules or *molecules*, each containing a display. At rest, all the modules will be dark. A complex set of signals will then start to propagate through the space (conception) and program the modules to realize the construction of a beating electronic watch (growth). Visitors will then be invited to attempt to disable the watch: on each molecule, a push-button will allow the insertion of a fault within the module (wounding) or its removal from the module (healing). The watch will automatically repair after each aggression (cicatrization). When the number of faults exceeds a critical value, the watch dies, the wall plunges once more into darkness, and the complete life cycle begins anew.

The first part of this paper is devoted to a macroscopic description of the Biowatch project. Section 2 describes the two architectural features of our artificial organism: multicellular organization (the organism consists of an array of identical physical elements, the cells) and cellular differentiation (each cell contains the complete blueprint of the organism, that is, its genome, and specializes depending on its position within the array). While maintaining the correct time, our multicellular organism is capable of self-repair (it can automatically replace one or more faulty cells) and self-healing (it can automatically recover one or more cells, should the fault disappear).

The microscopic structure of the cell relies on three fundamental features: multimolecular organization (the cell is itself decomposed into an array of physically identical elements, the molecules), molecular configuration (the boundaries between cells and the position of spare molecules are defined thanks to a cellular automaton), and fault detection within each molecule leading to the self-repair of the cell (through the replacement of the faulty molecules). This structure is described in the second part of this paper (Section 3). The current implementation of our reconfigurable computing tissue, the *BioWall* (*bio-inspired electronic wall*), forms the core of Section 4, which constitutes in fact a description of the molecular level of our system.

This paper presents a firmware implementation of the bio-inspired electronic watch based on an array of small processors. A fully hardware implementation of this watch is reported elsewhere [10].

2 BioWatch's macroscopic description: an artificial organism

In the framework of electronics, the environment in which our quasi-biological development occurs consists of a finite (but as large as desired) two-dimensional space of silicon. This space is divided into squares or *cells*. Since such cells (small processors and their memory) have an identical physical structure, i.e. an identical set of logic operators and of connections, the cellular array is homogeneous.

As the program in each cell (our artificial genome) is identical, only the state of the cell, i.e. the contents of its registers, can differentiate it from its neighbors.

Our artificial organism is designed to count and display hours, minutes, and seconds, from 00h00'00" to 23h59'59". The input signal used for synchronizing the units of seconds is delivered by wireless broadcast.

2.1   Multicellular organization

Multicellular organization divides the artificial organism (*ORG*) into a finite number of cells (Figure 1), where each cell (*CELL*) realizes a unique function, defined by a sub-program called the *gene* of the cell. The same organism can contain multiple cells of the same kind (in the same way as a living being can contain a large number of cells with the same function: nervous cells, skin cells, liver cells, etc.). Moreover, each cell is associated with some *output state*.

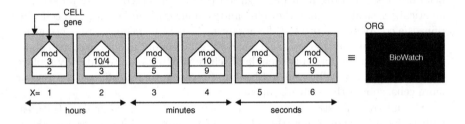

Fig. 1. Multicellular organization of the BioWatch organism. In this example, BioWatch displays 23h59'59".

Our *BioWatch* is thus a one-dimensional artificial organism implemented with six cells and featuring four distinct genes ("mod 10" for counting the units of seconds or minutes, "mod 6" for counting the tens of seconds or minutes, "mod 10/4" for counting the units of hours depending on the value of the tens of hours, and "mod 3" for counting the tens of hours). The output state is the current value of the elapsed time and varies from 0 to 9 (for units of seconds, minutes, and hours), from 0 to 5 (for tens of seconds and minutes), and from 0 to 2 (for tens of hours).

2.2   Cellular differentiation

Let us call *operative genome* (*OG*) a program containing all the genes of an artificial organism, where each gene is a sub-program characterized by a set of instructions and by the position of the cell (its coordinates X, Y). Figure 1 then shows the operative genome of BioWatch, with the corresponding horizontal (X) coordinate; the vertical (Y) coordinate can be ignored in this particular unidimensional case. Let then each cell contain the entire operative genome *OG*

(Figure 2a): depending on its position in the array, i.e. its place in the organism, each cell can interpret the operative genome and extract and execute the gene which defines its function.

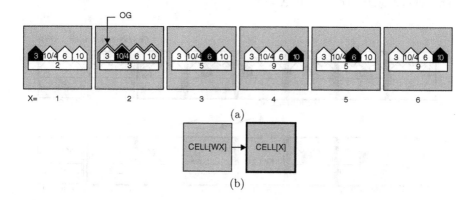

(a)

(b)

Fig. 2. Cellular differentiation of the BioWatch organism. (a) Global organization; OG: operative genome (genes and coordinates). (b) Central cell $CELL[X]$ with its west neighbor $CELL[WX]$.

In summary, storing the whole operative genome in each cell makes the cell universal: given the proper coordinates, it can execute any one of the genes of the operative genome and thus implement *cellular differentiation*. In our artificial BioWatch, any cell $CELL[X]$ computes its coordinate X by incrementing the coordinate WX of its neighbor immediately to the west (Figure 2b). Any cell $CELL[X]$ can thus be formally defined by a set of modulo-counting subprograms (its operative genome OG) and by its coordinate X. In the case of BioWatch, we have the operative genome of Figure 3.

```
OG: operative genome

X = WX+1
case of X:
    X = 1: count_mod_3      (tens of hours)
    X = 2: count_mod_10/4   (units of hours)
    X = 3: count_mod_6      (tens of minutes)
    X = 4: count_mod_10     (units of minutes)
    X = 5: count_mod_6      (tens of seconds)
    X = 6: count_mod_10     (units of seconds)
end case
```

Fig. 3. The operative genome OG of the BioWatch organism (first part).

2.3 Self-repair and self-healing properties

In order to implement *self-repair* and *self-healing*, we need to add *spare cells* (*SC*) to the right of the original unidimensional organism (Figure 4). These cells are defined by the coordinate $X = 7$.

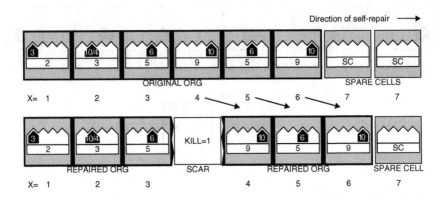

Fig. 4. Self-repair of the 6-cell BioWatch organism with two spare cells and one faulty cell; SC = spare cell.

The existence of a fault is detected by a $KILL$ signal which is produced at the cellular level (see Section 3). The state $KILL = 1$ identifies the faulty cell which is deactivated (column $X = 4$ in Figure 4). All the functions (X coordinate and gene) of the cells at the right of the column $X = 3$ are shifted by one column to the right. Obviously, this process requires as many spare cells, to the right of the array, as there are faulty cells to repair (two spare cells tolerating two successive faulty cells in the unidimensional example of Figure 4). It also implies that the organism must have the capability of bypassing the faulty cell and to divert to the right all the required signals (such as the operative genome and the X coordinate, as well as the data busses).

The disappearance of the $KILL$ signal ($KILL = 0$) means that the faulty cell (column $X = 4$ in Figure 5) has recovered all its functionalities at the cellular level and can be reactivated. All the coordinates and the genes of the cells at its right are then shifted by one column to the left and our organism possesses two spare cells again. While performing the self-repair and self-healing processes, the BioWatch maintains the correct time (23h59'59" in Figures 4 and 5).

3 BioWatch's microscopic description: the artificial cells

In each cell of every living being, the genome is translated sequentially by a chemical processor, the *ribosome*, to create the proteins needed for the organism's survival. The ribosome itself consists of molecules, whose description is an important part of the genome.

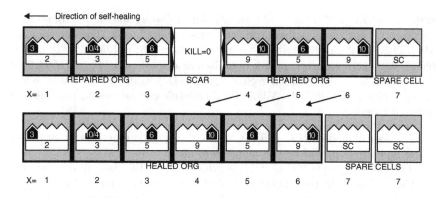

Fig. 5. Self-healing of the 6-cell BioWatch organism after a faulty cell is recovered.

As mentioned previously, in the Embryonics project each cell is a small processor or *binary decision machine*, sequentially executing the instructions of our artificial genome, the operative genome OG (Figure 3). The need to realize organisms of varying degrees of complexity has led us to design an artificial cell characterized by a *flexible architecture*, that is, itself configurable. It will therefore be implemented using a new kind of fine-grained field-programmable gate array (FPGA). We will illustrate the use of this FPGA through the design of the current cell of our BioWatch, a *shift binary decision machine*.

3.1 Multimolecular organization

Each element of our FPGA is then equivalent to a *molecule*, and an appropriate number of these artificial molecules allows us to realize application-specific processors. We will call *multimolecular organization* the use of many molecules to realize one cell.

A consequence of our choices is that we require a methodology to generate, starting from a set of specifications, the configuration of our FPGA, consisting of a homogeneous network of molecules, defined by an identical architecture and a usually distinct state (the *molecular code*, or $MOLCODE$).

To fulfill this requirement, we have selected a particular representation: the *ordered binary decision diagram* (OBDD) [1], [2], [7]. This representation, with its well-known intrinsic properties such as canonicity, was chosen for two main reasons.

1. It is a graphical representation which exploits well the two-dimensional space and immediately suggests a physical realization on silicon.
2. Its structure leads to a natural decomposition into molecules realizing a logic test, easily implemented by a multiplexer.

The reconfigurable molecule, henceforth referred to as MUXTREE (for *multiplexer tree*), consists essentially of a programmable multiplexer (with one control variable), a D-type flip-flop, and a switch block allowing all possible connections between two horizontal and two vertical long-distance busses. The behavior of a MUXTREE molecule, described in detail elsewhere [6](pp. 135-143), [4], is completely defined by a molecular code organized as a 20-bit data $MOLCODE19 : 0$, itself stored in a *configuration register* CREG. Depending on the value of a 1-bit *operation code OPC*, the MUXTREE molecule can be used in two basically different modes.

1. The *logical* mode ($OPC = 0$), in which all the logic functions (multiplexer, D-type flip-flop) and all the connections are activated.
2. The *memory* mode ($OPC = 1$), in which part of the configuration register CREG (the eight bits $MOLCODE7 : 0$ or the sixteen bits $MOLCODE15 : 0$) becomes a shift register that will be used in the implementation of the shift memory of the processor.

Our artificial cell embeds a special kind of binary decision machine, a *shift binary decision machine* [6](pp. 259-265), with a shift memory capable of storing 128 12-bit micro-instructions for the operative genome OG. Assembling eight such cells allows us to realize the current version of BioWatch (Figure 4). The specifications of the organism are described in the program of Figure 3. The operative genome OG consists of four distinct genes whose execution depends solely on the X coordinate. A fifth gene is needed only to identify the cells which are *spare cells* (SC).

The shift binary decision machine is specially designed to fit into an array of MUXTREE molecules: due to the difficulty of embedding a classic random access memory (RAM) in such an array (mainly due to the excessive number of molecules needed for decoding the RAM address), the actual program memory, or *shift memory*, consists of shift registers implemented using the MUXTREE molecules in memory mode.

Figure 6 shows the final BioWatch cell, connecting the shift binary decision machine SBDM and the shift memory SMEM. This structure involves the following hardware resources:

- a 7-bit current time register REG($Q1 : 7$);
- a 7-bit duplicate time register REG($D1 : 7$);
- a 3-bit coordinate register REG($X2 : 0$);
- a 1-bit test register REG(T);
- a 1-bit healing register REG(H);
- a 1-bit ripple count circuit RCC(RC).

The 7-bit coding of the time data is chosen in order to insure full compatibility with a 7-segment display. The operative genome OG also includes a control part that regulates the operations executed by the registers (Figure 7). As shown in the figure, these operations depend on the values of the test and healing bits of the cell itself (T, H), of the first neighboring cell to the west (WT, WH), and of the first neighboring cell to the east (EH). In the normal time counting mode ($WH = 0$, $H = 0$, $WT = T'$), the duplication of the current time is effected

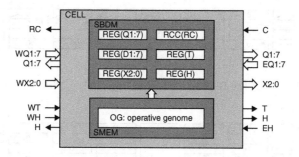

Fig. 6. Block diagram of the BioWatch artificial cell; SBDM: shift binary decision machine; SMEM: shift memory.

at each program execution cycle (Figure 8). When a faulty cell is detected by its right-hand neighbor ($WH = 0$, $H = 0$, $WT = T$), a multicycle self-repair process (Figure 9) sequentially copies the duplicate time and thus allows our BioWatch to maintain the current time. When the faulty cell recovers ($H = 1$), a multicycle self-healing process (Figure 10) shifts the current time sequentially back to the left.

```
OG: operative genome

case of WH,H,EH:
  WH,H,EH = 00-:
  case of WT,T:
    WT,T = 00:    Q = D
                  T = 1
    WT,T = 01:    D = WQ
    WT,T = 10:    D = WQ
    WT,T = 11:    Q = D
                  T = 0
  end case
  WH,H,EH = 011:  D = EQ
                  Q = D
                  H = 0
  WH,H,EH = 1--:  H = 1
end case
```

Fig. 7. The operative genome OG of the BioWatch organism (second part).

Based on the OBDD decomposition of its resources, the MUXTREE implementation of the BioWatch artificial cells leads to the 24 × 25 array of molecules shown in Figure 11. As can be seen in the center of the figure, the 7-segment display of the current time data is part of the array. This is done in order to take full advantage of the reconfigurable computing tissue introduced in Section 4. The seven lower rows of the MUXTREE implementation constitute the 128 × 12-

Q	2	3	5	9	5	9	0	0
D	0	2	3	5	9	5	9	0
T	1	0	1	0	1	0	1	0
H	0	0	0	0	0	0	0	0

Fig. 8. Duplication of the BioWatch's current time (when $T \neq WT$ then D becomes WQ).

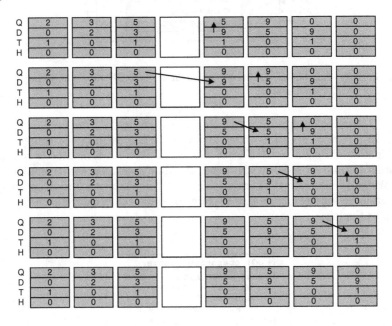

Fig. 9. Time-preserving self-repair process of the BioWatch organism (when $T = WT$ then Q becomes D).

bit shift memory SMEM needed to store the 126 12-bit word instructions of the complete operative genome (Figures 3 and 7).

To sum up, *multimolecular organization* divides finally the BioWatch artificial cell, our shift memory binary decision machine, into a finite number of molecules (600), where each molecule is defined by a unique configuration, its molecular code $MOLCODE19:0$. The *ribosomic genome* (RG) is then the sum of the 600 $MOLCODE$s of the 600 active MUXTREE molecules.

3.2   Molecular configuration

The information contained in the $MOLCODE$ defines the logic function or the memory contents of each molecule. To obtain a functional cell, i.e. an assembly of MUXTREE molecules, we require two additional pieces of information, defining

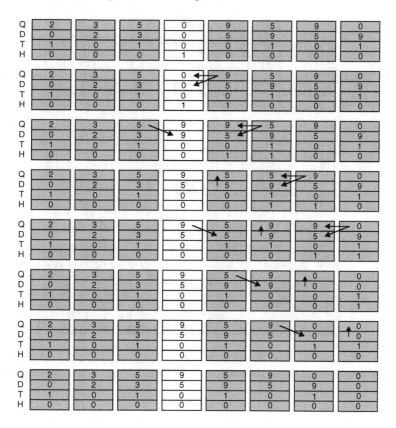

Fig. 10. Time-preserving self-healing process of the BioWatch organism (when $WH, H, EH = 011$ then D and Q become EQ).

the physical position of each molecule within a cell and the presence and position of the spare columns required by the self-repair mechanism (Subsection 3.3).

The mechanism which we have adopted consists of introducing in the FPGA a regular network of automata (state machines) called *space divider* [6], [11], [12]. We call *polymerase genome PG* the sequence of states that needs to be applied to the lower left hand automaton in order to divide the entire space into cells. If the states C, V, and H represent respectively corner, vertical, and horizontal boundaries, the sequence:

$$PG = C, V * [h - 1], H * [w - 1], C, ... \tag{1}$$

defines a cell h molecules high and w molecules wide. In this relation, where the notation $X * [n]$ represents the state X repeated n times, the presence of spare columns is indicated by replacing one or more occurences of H by S. The

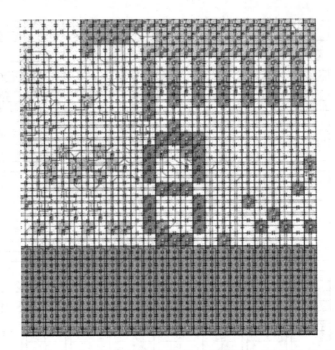

Fig. 11. The 24 × 25 = 600 MUXTREE molecule array implementing the BioWatch's artificial cell; the seven lower rows of molecules constitute the 128×12-bit shift memory.

BioWatch artificial cell shown under construction in Figure 12 with two columns of spare molecules results therefore from a cycle of the following states:

$$PG = C, V * 24, H * 7, S, H * 8, S, H * 8, S, C, ... \qquad (2)$$

The details of the design of the space divider are described elsewhere [4].

3.3   Molecular fault detection

The specification of the molecular self-repair system must include the following features:

- it must operate in real time, in response to the activation of the push-button included in each molecule;
- it must preserve the memorized values, that is, the state of the D-type flip-flop and the $MOLCODE$ of each molecule;
- it must assure the automatic repair at the cellular level;
- in case of multiple faults, it must generate a global signal $KILL = 1$ that activates the suppression of the cell and starts the self-repair process of the

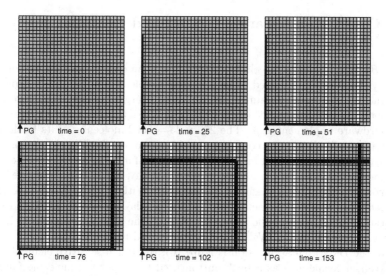

Fig. 12. The BioWatch's space divider (height=25, width=27, 1 spare column out of 9).

complete organism, or $KILL = 0$ which activates the self-healing process (Subsection 2.3).

To meet the specifications, and in particular the requirement that the hardware overhead be minimized, our self-repair system exploits the programmable frequency and distribution of the spare columns (Subsection 3.2) by limiting the reconfiguration of the array to a single molecule per line between two spare columns (Figure 13). This choice allows us to minimize the amount of logic required for the reconfiguration of the array, while keeping a more than acceptable level of robustness.

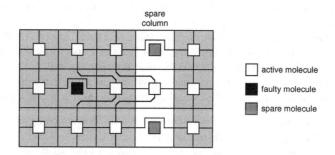

Fig. 13. The self-repair mechanism for an array of MUXTREE molecules.

4   Conclusion: toward a bio-inspired electronic wall (BioWall)

The *BioWall* (bio-inspired electronic wall) [5] is an ongoing project in our laboratory. This wall is intended to be a reconfigurable computing tissue capable of interacting with its environment by means of a large number of touch-sensitive elements coupled with two-color LED displays. Figure 14 shows a prototype of its hardware implementation. The $200 \times 25 = 5000$ molecules of its final implementation will allow us to configure the six cells of our BioWatch organism and two extra spare cells. Each molecule is made up of a transparent touch-sensitive element, a two-color 8×8 dot-matrix LED display, and a reconfigurable Xilinx Spartan XCS10XL FPGA circuit (Figure 15). Within the molecule, the transparent touch-sensitive element and the LED display are physically joined by an adhesive film. As each of the molecules provides the same connections to its four direct neighbors, the BioWall is homogeneous and fully scalable.

In the BioWatch application (Figure 16), the touch-sensitive element of the BioWall molecule acts as a push-button used to render the molecule faulty or healthy again. The LED display shows the boundaries, the spare columns and the current time of the cell, as well as the faulty or healthy state of the molecule. The configuration of the Xilinx FPGA implements all the MUXTREE molecule specifications in the BioWwall molecule.

The programmable robustness of our system depends on a redundancy (spare molecules and cells) that is itself programmable. This feature is one of the main original contributions of the Embryonics project. It makes possible to program (or reprogram) a greater number of spare molecules and spare cells for operation in hostile environments (e.g., space exploration). A detailed mathematical analysis of the reliability of our systems has been carried out at the University of York [8], [9].

With respect to this design process, the programming of the molecular array of MUXTREE elements, our reconfigurable tissue, takes place in the following order.

1. The polymerase genome (PG) is injected in order to set the boundaries between cells and define the spare columns.
2. The ribosomic genome (RG), i.e. the string of $24 \times 25 = 600$ *MOLCODEs*, is injected in order to configure the BioWatch cell and to fix the final architecture of the shift binary decision machine inside the cell.
3. The operative genome (OG) is stored within the shift memory of each cell in order for it to execute the counting, repairing, and healing specifications.

Echoing biology, we have faced complexity by decomposing the organism into cells and then the cells into molecules. This decomposition implies three configuration steps: the polymerase genome organizes the space by defining the cells' boundaries, the ribosomic genome defines the architecture of each cell as an array of molecules, and finally the operative genome makes up the program that will be executed by the cellular processors to accomplish the required task. The Latin motto "divide and conquer" maintains its relevance even today.

Fig. 14. The present BioWall prototype consisting of about $80 \times 25 = 2000$ molecules (Photograph by A. Herzog).

Acknowledgments

This work was supported in part by the Swiss National Foundation under grant 21-54113.98, by the Leenaards Foundation, Lausanne, Switzerland, and by the Villa Reuge, Ste-Croix, Switzerland.

References

1. S. B. Akers. Binary decision diagrams. *IEEE Transactions on Computers*, c-27(6):509–516, June 1978.
2. R. E. Bryant. Symbolic boolean manipulation with ordered binary-decision diagrams. *ACM Computing Surveys*, 24(3):293–318, 1992.
3. D. Mange, M. Sipper, and P. Marchal. Embryonic electronics. *BioSystems*, 51(3):145–152, 1999.

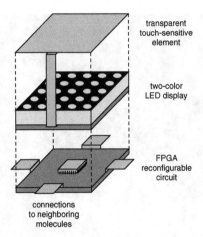

Fig. 15. The basic molecule of the reconfigurable computing tissue.

Fig. 16. The BioWall implementation of the 8-cell BioWatch (Computer graphic by E. Petraglio).

4. D. Mange, M. Sipper, A. Stauffer, and G. Tempesti. Toward robust integrated circuits: The Embryonics approach. *Proceedings of the IEEE*, 88(4):516–541, April 2000.

5. D. Mange, A. Stauffer, G. Tempesti, and C. Teuscher. Tissu électronique reconfigurable, homogène, modulaire, infiniment extensible, à affichage électro-optique et organes d'entrée, commandé par des dispositifs logiques reprogrammables dis-

tribués. Patent pending, 2001.

6. D. Mange and M. Tomassini, editors. *Bio-Inspired Computing Machines*. Presses polytechniques et universitaires romandes, Lausanne, 1998.

7. C. Meinel and T. Theobald. *Algorithms and Data Structures in VLSI*. Springer-Verlag, Berlin, Germany, 1998.

8. C. Ortega and A. Tyrrell. Reliability analysis in self-repairing embryonic systems. In A. Stoica, D. Keymeulen, and J. Lohn, editors, *Proceedings of The First NASA/DOD Workshop on Evolvable Hardware*, pages 120–128, Pasadena, CA, 1999. IEEE Computer Society.

9. C. Ortega and A. Tyrrell. Self-repairing multicellular hardware: A reliability analysis. In D. Floreano, J.-D. Nicoud, and F. Mondada, editors, *Proceedings of the 5th European Conference on Artificial Life (ECAL'99)*, Advances in Artificial Life. Springer-Verlag, Berlin, 1999.

10. A. Stauffer, D. Mange, G. Tempesti, and C. Teuscher. Biowatch: A giant electronic bio-inspired watch. In A. Stoica, D. Keymeulen, J. Lohn, and R. Zebulum, editors, *Proceedings of The Third NASA/DOD Workshop on Evolvable Hardware*, Long Beach, CA, 2001. IEEE Computer Society.

11. G. Tempesti. *A Self-Repairing Multiplexer-Based FPGA Inspired by Biological Processes*. PhD thesis, Computer Science Department, Swiss Federal Institute of Technology Lausanne, 1998.

12. G. Tempesti, D. Mange, and A. Stauffer. Self-replicating and self-repairing multicellular automata. *Artificial Life*, 4(3):259–282, 1998.

13. L. Wolpert. *The Triumph of the Embryo*. Oxford University Press, New York, 1991.

Shrinking the Genotype: L-systems for EHW?

Pauline C. Haddow[1,*], Gunnar Tufte[1,*], and Piet van Remortel[2,*]

[1] The Norwegian University of Science and Technology,
Department of Computer and Information Science,
O.S. Bragstadsplass 2E, 7491 Trondheim, Norway
[2] Vrije Universiteit Brussel,
COMO – Department of Computer Science,
Pleinlaan 2, 1050 Brussels, Belgium
pauline@idi.ntnu.no, gunnart@idi.ntnu.no,pvremort@vub.ac.be

Abstract. Inspired by biological development where from a single cell, a complex organism can evolve, we are interested in finding ways in which artificial development may be introduced to genetic algorithms so as to solve our genotype challenge. This challenge may be expressed in terms of shrinking the genotype. We need to move away from a one-to-one genotype-phenotype mapping so as to enable evolution to evolve large complex electronic circuits. We present a first case study where we have considered the mathematical formalism L-systems and applied their principles to the development of digital circuits. Initial results, based on extrinsic evolution, indicate that our representation based on L-systems provides an interesting methodology for further investigation. We also present our implementation platform for intrinsic evolution with development, enabling on-chip evaluation of grown solutions.

1 Introduction

The field of evolvable hardware promises many possibilities within optimisation and exploration of new circuit designs apart from one missing factor – scalability. Scalability is the property of a method or solution to keep on performing acceptably when the problem size increases. In our case, acceptable performance may be said to be a non-exponential resource increase and performance decrease. Our problem domain is the evolution of electronic circuits. We have seen in recent years that small electronic circuits may be evolved successfully [1,2,3,4]. However, complex circuits are still beyond our reach.

The scalability problem is due to the resource-greedy nature of evolutionary techniques. Generally in EHW, a one-to-one mapping has been chosen for the genotype-phenotype transition. This means that the genotype needs to include all the information required for the phenotype i.e. configuration data to program the device. The larger the circuit to be evolved, the more logic and routing information required. As the complexity of the genotype increases, so increases the computational and storage requirements.

A solution to this problem is to shrink the genotype in some way. This may be stated as the genotype challenge, finding new forms of representation for

---

* The authors names are given in alphabetical order

Y. Liu et al. (Eds.): ICES 2001, LNCS 2210, pp. 128–139, 2001.

evolving complex circuits. In this work we are interested in investigating ways in which development may be combined with genetic algorithms so as to enable smaller genotypes to evolve complex circuit designs.

Turning to biology, the biological development stages of pattern formation, morphogenesis, cell differentiation and growth (see section 3) provide us with pointers as to the principles that should be included in a design methodology for artificial development. Many alternative methodologies are possible. As a first case study, we consider L-systems [5] which may be said to be based on differentiation and growth through rules for changes and growth. Our goal is to adapt L-systems to developing digital circuits within the constraints of our technology, a virtual EHW FPGA [6] and then map our virtual EHW FPGA to Xilinx Virtex. A genetic algorithm will be used in combination with L-systems to evolve a final solution using intrinsic evolution. The work presented herein is a first attempt at using L-systems in combination with a genetic algorithm to achieve growth on a digital platform.

The paper is laid out as follows. In section 2 we introduce other work in the field and our motivation for selecting the given representation. Since much of this work is focused on biological development we give a brief introduction to biological development in section 3. L-systems are described in section 4. Our virtual EHW FPGA is briefly described in section 5. In section 6, we discuss and present our adaptation of L-systems to digital circuits. Section 7 presents our experimental platform and section 8 presents our initial results. Ongoing and further work is described in section 9.

2   Background and Motivation

To solve the genome complexity issue and enable evolution of large complex circuits, the need to move away from a one-to-one genotype-phenotype mapping is becoming generally accepted. We can expect newer features such as incremental evolution [7],'divide and conquer' [8] or growth [9] to become more common in the work within the field of EHW so as to attack the representation problem.

In the past two to three years, researchers in the field of both software and hardware evolution have begun to look again at biology to gain better insights into improving existing techniques.

Inspired by biological development before differentiation, a new family of fault tolerant FPGAs are being developed at York [10]. Each cell can take over the functionality of a faulty cell as it possesses a complete copy of the "genome". To achieve the degree of flexibility required when any cell can have any function, reconfigurable technology may be said to be a requirement. This is especially the case since the cellular structure varies as a function of the application [11].

Cellular encoding [12] is a well-defined formal approach based on biological development. This approach may be used to build complex systems by encoding the cell types, timing of cell division and changes in links involving the cell [13] This methodology is a variation of genetic programming [14].

The work of Mjolsness et al [15] focuses on software modelling of morphogenesis in plants to better understand the morphogenesis stage of development. The long term goal of this work is not just single electronic circuits but more towards much more complex systems such as self-sustaining space industry.

One of the reasons for this interest in biology is due to a move from optimisation to exploration [16]. That is from improving existing solutions to finding novel solutions. If we want to explore for novel solutions one approach is to provide *knowledge-poor representations* and give evolution a greater area of freedom to explore for solutions [16]. In a knowledge-poor representation we allow evolution to find solutions that we haven't thought of ourselves. Using knowledge-rich representations we lead evolution to solutions where we think they can be found and in this way limit the search space of evolution to our own specified search space. Using context insensitive L-systems for the representation may be said to be a form of knowledge-poor representation.

To enable growth, one solution is to follow the working of DNA and combine rules into our representation. That is, the genotype includes rules telling how, where and which gene i.e. component in a component representation, should grow to develop a solution [17]. In DNA, instructions fire and suppress other instructions in an increasingly more complex network of activations. L-systems are a mathematical formalism based on this concept.

In this work our goal is to shrink the genotype. Shrinking the genotype effectively moves the complexity problem over to the genotype-phenotype mapping thus increasing the complexity of the mapping. However, in [6], we proposed a solution to this complex mapping problem which includes splitting the mapping process into two stages using a virtual EHW FPGA as the bridge between the genotype and the phenotype. The mapping from the virtual EHW FPGA to a physical FPGA is a simpler mapping as both architectures are based on FPGA principles. The development process itself is the first stage of the mapping. Instead of developing to our complex organism (phenotype) we are developing to a simpler organism the *intertype* i.e. the configuration data for our virtual EHW FPGA.

3   Biological Development

Biological development enables complex organisms to be built in a robust way. What are the features of this reliable system design?

An initial unit, a cell, holds the complete building plan (DNA). It is important to note that this plan is generative – it describes how to build the system, not what the system will look like. Units have internal state, can communicate locally, can move, spawn other units or die. Groups of units may also exhibit group-wise behaviour i.e. a group state.

The global developmental stages from the zygote (fertilised egg) to the multicellular organism, although interdependent and not strictly sequential, may be categorised as *pattern formation; morphogenesis; cell differentiation and growth.*

During *pattern formation* cells are organised in different regions according to their cell types (set of cells with the same gene activity pattern) in order to become distinct parts (body segments) of the eventual organism. One could say that this is a group-wise change of state due to activation of certain codes after local communication or reaction to globally available signals.

Some cells may change shape (expand/contract) exerting a force on other cells. This process is termed *morphogenesis* and is crucial in the formation of general shape in the organism.

The *differentiation* process is where cells become structurally and functionally different from each other. This includes both intra-cellular factors (cell lineage) and inter-cellular interactions (cell induction).

The final step is *growth*. This is the true enlarging of the almost completely formed organism. This is achieved by multiple cell divisions and expansions. During growth, programmed cell death or *apoptosis* can help generate special structures like fingers and toes from continuous sheets of tissue.

4 L-systems

An L-system is a mathematical formalism used in the study of biological development. One of the main application areas is the study of plant morphology. Phenotypes are branching structures attained through the derivation and graphical interpretation of the development process, described by a set of rules. The rules describe how the development should grow and specialise and interpretation of the development itself enables morphogenism to be studied.

An L-system is made up of an alphabet, a number of ranked rules and a start string or axiom. Applying a rule means finding targets within the search string which match the rule condition. This condition is a pattern on the left hand side (LHS) of the rule. This condition is also a string and the string is made up of elements from the alphabet. Firing the rule means replacing the targets, where possible, with the result of the rule i.e. the right hand side (RHS) of the rule. A more complete description of rule replacement may be found in section 6.2.

Firing of the rules continues until there are no targets found for any rule or until the process is interrupted. In the context of genetic algorithms, the axiom and the associated rules are the genotype and the developed string is the phenotype.

5 Virtual EHW FPGA – Sblock Architecture

The Virtual EHW FPGA contains blocks – *sblocks*, laid out as a symmetric grid where neighbouring blocks touch one another. There is no external routing between these blocks except for global clock lines. Input and output ports enable communication between touching neighbours. Each sblock neighbours onto sblocks on its 4 sides. Each sblock consists of both a simple logic/memory component and routing resources. The sblock may either be configured as a logic or

memory element with direct connections to its 4 neighbours or it may be configured as a routing element to connect one or more neighbours to non-local nodes.
By connecting together several nodes as routing elements, longer connections
may be realised. A detailed description of the sblock architecture may found
in [6] and a discussion around evolution requirements leading to this structure
are given in [18].

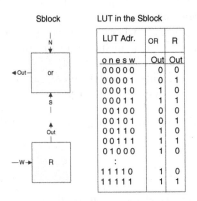

Fig. 1. Sblock Configuration : An OR gate (OR) and Routing from the West (R)

The main element of an sblock is its 5 input LUT. Configuration of the
sblock consists of configuring the response to the inputs of the LUT. With a
5-input LUT, 32 configuration bits are required. The sblocks are programmed
as either logic or routing blocks. Figure 1 illustrates the response for a logic
– two input OR gate (OR), and a routing – west router (R) sblock. For logic,
the output response programmed reflects the active inputs and the functionality
of the LUT. Looking at the LUT contents it can be seen that the OR gate of
the north and south inputs is achieved by ignoring the west and east inputs. For
routing, only one input is active, the others are don't care. In the example given,
a west routing sblock is indicated as the response matches the west input.

The configuration string of the sblock architecture consists of all the sblock
configurations (32 bits), starting at the top left hand corner and crossing the
grid a column at a time. That is, for a 16x16 grid, the configuration string will
consist of 32x16x16 bits.

6   Development

As stated in section 2, a knowledge-poor representation may free evolution to
explore for solutions in a large search space. Our interpretation of knowledge-
poor in an L-system context, is that the rules do not have any chosen meaning
with respect to improving connectivity or logic. Also the rules are not context-
sensitive, relying on neighbouring sblock information. During evolution, the rules
themselves evolve.

The application of the rules to the developing intertype has been designed with determinism in mind. As such, the individual representation provides a reliable representation of the resulting phenotype.

We have added technology constraints where our intertype technology is the virtual EHW FPGA. Following the principles of biological development and treating an sblock as a cell, then growth steps are limited to 32 bit sblocks i.e. a complete cell. As any other developing organism we wish to introduce shape and our goal is the shape of an sblock architecture – a grid. Therefore growth is limited to the grid size chosen for our virtual EHW FPGA. Change rules both effect connectivity of the architecture – in biological terms communication between cells, as well as functionality – specialisation.

6.1 Change and Growth Rules

There are two types of rules: change and growth rules. Figure 2 illustrates part of the rule list used in the experiments of section 8. Change rules have a RHS string of equivalent length to their LHS string. This is to avoid any growth due to the application of a change rule. These rules are ranked from the most to the least specific, as shown. The growth rules are given a random priority and ranked accordingly. In this example change rules have been ranked before growth rules. It may be noted that since these rules are random generated and then evolved, more than one rule may have the same LHS. A don't care feature, typical of digital design, has been introduced to the change rule concept. Don't cares are represented by a 2, as shown. To retain determinism don't cares are only allowed on the LHS of a change rule.

```
 1 - 21200110101212 -> 10110100001110
 2 - 000012110111    -> 100001110011
 3 - 010121121222    -> 100100010100
 4 - 2220021022      -> 0111001110
 5 - 01001 -> 0010111001110110011000000000100111
 6 - 10111 -> 1001011100011101010100011001010110
 7 - 01011 -> 1010110011111110111011100100011100
 8 - 10101 -> 0101111011100110010000101011101010
 9 - 00011 -> 1111101100100001100101100101101010
10 - 01011 -> 1011110100101001001011011110000101
11 - 01110 -> 1101000100100011010101011001000001
12 - 01100 -> 1111110010110011100010101010111001
```

Fig. 2. Change and Growth Rules

Through change rules, the contents of one or more sblocks are changed. A change rule targets the locations in the intertype where the string on the LHS matches. Firing the rule means replacing this string at the different targets with the RHS of the rule. The LHS string may be found within a single sblock or overlapping two or more sblocks as illustrated in figure 3.

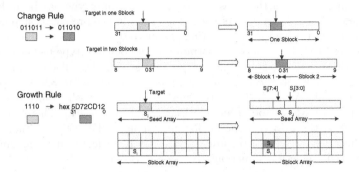

Fig. 3. Applying Change and Growth Rules

As illustrated in figures 3, each sblock has an associated seed. If a seed in the array matches the LHS of a growth rule then this means that the associated sblock is a target for the growth rule. If there is no available space to grow into i.e. the sblock is surrounded by configured sblocks, then there will be no growth at this point. Firing the growth rule means placing an sblock in the first free location according to the priority rule : north, south, west and then east. In introducing a new seed it is important that determinism is maintained. Therefore, the new seed is given the first 4 bits of the sblock which was targeted and its bits 5 to 8 are used to replace its own seed. As shown, the new sblock is configured by the RHS of the growth rule.

6.2   Rule Firing Sequence

To ensure fairness, we propose firing the rules in batches. The rules are still ordered from the most to the least specific but all rules will have a chance of firing before a given rule can fire again.

Figure 4 illustrates this firing sequence. Once any rule in the batch has fired on a target then no other rule can fire on the same target area. As such, these target areas are reserved for the remainder of the current firing batch, as il-

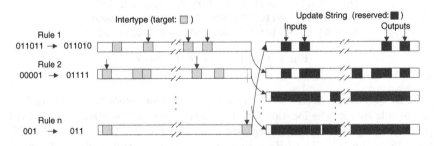

Fig. 4. Rule Firing Sequence

lustrated in what is termed the update string. The default value of the update string is the configuration of the input and output sblocks.

We assume, as shown, that rule 1 (the top ranked rule) has found targets in the intertype i.e. matching strings to its LHS. However, checking the update string, one of its targets can't fire as the target overlaps an input sblock. The rule fires on the other targets in the normal way by writing its RHS string over the target bits at each target point. In addition, it sets the equivalent bits in the update string. Rule 2 finds 5 targets but only 2 are free. Again the rule fires and sets the respective bits in the update string. The last rule of the ranked list finds one target and therefore can fire on that target. The update string can now be reset to its default value and the firing process continues beginning again with the most specific rule. In the example illustrated in figure 2, growth rules are ranked after change rules and therefore don't effect the firing of the change rules within a given batch. However, their effect will be seen at the next batch where the update string will reflect the size of the grown intertype. The firing process may be stopped either when there are no more rules to fire or when the developed string meets some other chosen condition.

7   Experimental Platform

Figure 5 illustrates the workings of our experimental platform. The genetic algorithm (GA) is implemented using GAlib. Our genotype may consist of one or more 32 bit axioms and a ranked list of rules. The GA works on a population of individuals where each individual consists of one or more axioms and the ranked list of rules. Genetic operations applied to individuals affect both its axiom(s) and its list of rules.

The individual is then placed in the fitness module in its genotype form, as shown. To start the evaluation phase, the individual is sent to the development process.

The development process is given the constraints of the required design. These include the technology constraint – sblock grid size, and the design features – the number and position of inputs and outputs. The sblocks specified in

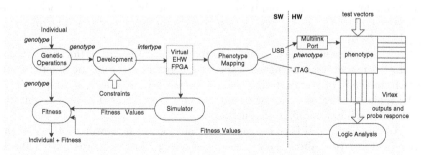

Fig. 5. Platform for Intrinsic Evolution with Development

the design features are protected from further changes during the development process. The axiom(s) is allocated to an sblock location(s). The development process may then begin firing rules, according to the method described in section 6.2 and, as such, develop the genotype to the intertype string.

A simulator is available to generate fitness values based on the intertype representation. This simulator enables initial studies of the interplay between the L-system rules and the GA to be conducted at the intertype level. Fitness values generated are fed back to the fitness module.

The next stage involves mapping the virtual sblock FPGA design (intertype) to a circuit description for Xilinx Virtex FPGA (phenotype). The phenotype mapping is a translation process from the configuration data expressed in the intertype, to Virtex configuration data format. The intertype and the phenotype are configured in columns. In the intertype each sblock's 32 bits is a continuous string, enabling interpretation and growth of the sblock architecture in 32 bit blocks. However, in the Virtex configuration, configuration data is split into frames where configuration of a single CLB is split over several frames. Mapping involves translation between these two formats. In addition, all outputs from the mapped sblock grid are routed to their respective edges so as to be available on the Virtex pads.

To down-load the configuration data to our Virtex chip we have tested out two solutions using Xilinx interfaces. The first is a serial configuration using the USB port of the PC, through Xilinx's multilink port. The second, which is slightly faster uses a JTAG parallel cable.

For fitness evaluation we need feedback of fitness values from the implemented phenotype. The fixed location of the inputs and outputs during the development process means that test vectors may be applied and responses monitored. A number of internal probes on the Virtex chip are also available to provide additional information for fitness evaluation. Responses are fed through the logic analyser, to generate fitness values for fitness evaluation.

8   Experimentation and Evaluation

The experimental goal may be expressed as follows. The "circuit" has no specified function. From a single axiom in the centre of the sblock grid, an sblock solution consisting of north, south, west or east routing modules is grown. The maximum grid size is 16 by 16.

The experiments were conducted using the development platform in extrinsic evolution mode. That is, fitness values were generated from the simulator and not from the Virtex chip itself. The GA parameters were as follows: population 300; roulette wheel selection with elitism; crossover 0.8 or 0 (2 cases); mutation 0.1 and maximum number of generations 150.

Although our goal is to evolve routing sblocks, fitness evaluation is also provided information about how close non-routing blocks are to routing blocks. That is low-input sblocks are credited more than high input sblocks except for the case of zero-input sblocks. These are given a low weighting as they do not offer any routing possibilities. Fitness is expressed in equation 1.

$$F = 15*R + [6*(C-R) - (6*6in + 5*5in + 4*4in + 3*3in + 2*2in + 1*N1in)] \quad (1)$$

where R is the number of routing sblocks; C the number of configured sblocks; Xin the number of sblocks with X inputs and $N1in$ the number of inverter sblocks. A pure router sblock has only one input and no inverter function.

Little growth was achieved in the initial experiments before the rules stopped firing. To aid growth and ensure that firing continued, we both increased the ratio of growth to change rules and made the change rules less specific i.e. shorter LHS. A significant improvement in growth was seen but the results presented no clear trend and seemed almost random.

Studying the results more closely, it was obvious that the present setup had inherent epistasis properties. To further test our assumptions, we removed the crossover operator which we believed was forcing the results to jump around in the fitness landscape. As a result the population gradually found better and better solutions and the average fitness improved gradually. Figure 6 illustrates one of our runs where a typical pattern of fitness improvement is seen. Decreasing number of inputs is illustrated from dark gray to white. White boxes with a cross are router modules. Black boxes indicate either zero-input sblocks or non-configured sblocks i.e. not grown. In general our solutions achieved 30 to 50% routers in less than 100 generations. Limitations in our current simulator have not allowed us to run enough generations to study further improvements.

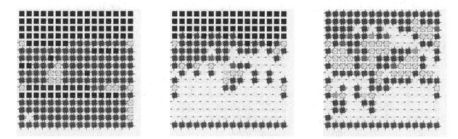

Fig. 6. Configured sblocks after 3, 23 and 57 generations

The achievement of a trend towards increasing routing modules indicates that the combination of the GA and L-systems can at least approach a solution, although the solution achieved may be said to be closer to hill climbing than evolution.

We discussed in section 2 the idea of a knowledge-poor representation. However, our experimental results indicate that our current representation might not be as knowledge-poor as we intended. Figure 6 shows that most router blocks achieved are west router blocks, indicating that west router blocks were favoured over other router blocks. In an earlier set of experiments where zero-inputs were omitted from the fitness calculation, zero-inputs were favoured over the poorly weighted high-input blocks resulting in grids with a large number of zero-inputs. In both cases, our fitness function did not give special credit to the favoured configurations. As such, this tendency towards a specific type of sblock must lie in our representation i.e. in our rules

9   Ongoing and Future Work

As stated, the goal of our work is to shrink the genotype by using some form of growth to an intertype representation and map to a phenotype representation for fitness evaluation. The platform status is that we can test out individuals and obtain detailed feedback for fitness evaluation. However, running generations for a reasonable population is limited due to the slow Virtex interface. For intrinsic experimentation a better interface is essential and is the focus of our attention at present. The new interface will have the advantage that it will allow us to take more control over placement and routing.

We have presented our representation as an adaptation of L-systems. A number of assumptions have been made based on how the development might proceed and how the many different factors might interact. Issues such as, to name but a few: how many axioms? resolving overlapping growths, where to grow from on the chip, how to apply crossover to rules, seed size required and many more are unresolved. As such, we are not in a position to say whether this representation will lead to the possibility to evolve complex circuits. As such, much experimentation is needed to investigate this representation further. However, intuitively we feel that an L-system based representation is worth further investigation.

References

1. A. Thompson. Evolving electronic robot controllers that exploit hardware resources. In *The 3rd European Conference on Artificial life (ECAL95)*, 1995.
2. A. Thompson. An evolved circuit, intrinsic in silicon, entwined with physics. In *1st International Conference on Evolvable Systems, ICES*, Lecture Notes in Computer Science, pages 390–405. Springer, 1996.
3. J.R. Koza et al. Automated synthesis of analog electrical circuits by means of genetic programming. *IEEE Transactions on Evolutionary Computation*, 1(2):109–128, July 1997.
4. T. Higuchi et al. Evolvable hardware and it's application to pattern recognition and fault-tolerant systems. In E. Sanchez and M. Tomassini, editors, *Towards Evolvable Hardware: the Evolutionary Engine ering Approach*, volume 1062 of *Lecture Notes in Computer Science*, pages 118–135. Springer, 1996.
5. Aristid Lindenmayer. Mathematical Models for Cellular Interactions in Development. *Journal of Theoretical Biology*, 1968.
6. P.C. Haddow and G. Tufte. Bridging the genotype-phenotype mapping for digital FPGAs. In *to appear in the 3rd NASA/DoD Workshop on Evolvable Hardware*, 2001.
7. I. Harvey et al. Why evolutionary robotics? *Robotics and Manufacturing: Recent Tends in Research and Applications*, 6:293–298, 1996.

8. J. Torresen. A divide-and-conquer approach to evolvable hardware. In *2nd International Conference on Evolvable Systems, ICES,* Lecture Notes in Computer Science, pages 57–65. Springer, 1998.

9. P.J. Bentley and S. Kumar. Three ways to grow designs: A comparison of embryogenies for an evolutionary design problem. In *Genetic and Evolutionary Computation Conference (GECCO '99),* pages 35–43, 1999.

10. C. Ortega and A. T yrell. A hardware implementation of an embyonic architecture using virtex FPGAs. In *Evolvable Systems: from Biology to Hardware, ICES,* Lecture Notes in Computer Science, pages 155–164. Springer, 2000.

11. L. Prodan et al. Biology meets electronics: The path to a bio-inspired FPGA. In *Evolvable Systems: from Biology to Hardware, ICES,* Lecture Notes in Computer Science, pages 187–196. Springer, 2000.

12. F. Gruau and D. Whitley. Adding learning to the cellular development of neural networks: Evolution and the baldwin effect. *Journal of Evolutionary Computation,* 1993.

13. H. Kitano. Building complex systems using development process: An engineering approach. In *Evolvable Systems: from Biology to Hardware, ICES,* Lecture Notes in Computer Science, pages 218–229. Springer, 1998.

14. J. Koza. *Genetic Programming.* The MIT Press, 1993.

15. E. Mjolsness et al. Morphogenesis in plants: Mo delling the shoot apical meristem and possible applications. In *The first NASA/DoD Workshop on Evolvable Hardware,* pages 139–140, 1999.

16. P.J. Bentley. Exploring component-based representations – the secret of creativity by evolution. In *Fourth Intern. Conf on Adaptive Computing in Design and Manufacture,* 2000.

17. P.J. Bentley and S. Kunar. Three ways to grow designs: A comparison of embryongenies for an evolutionary design problem. In *GECCO,* pages 35–43, 2000.

18. P.C. Haddow and G. Tufte. An evolvable hardware FPGA for adaptive hardware. In *Congress on Evolutionary Computation (CEC00),* pages 553–560, 2000.

Multi-layered Defence Mechanisms: Architecture, Implementation and Demonstration of a Hardware Immune System

D.W. Bradley and A.M. Tyrrell

Department of Electronics, University of York,
Heslington, York, England,
dwb105,amt@ohm.york.ac.uk,
www.bioinspired.com

Abstract. Biology uses numerous methods to keep the body in good working operation. If one line of defence is breached, the next uses a different approach. Very rarely are all lines of defence evaded. This paper analyses the body's approach to fault tolerance using the immune system and shows how such techniques can be applied to hardware fault tolerance. A fault detection layer inspired by the process of self/nonself discrimination used to detect bacterial infections in the body is created. The hardware immune system is then demonstrated to show how such a layer in hardware can provide fault detection.

1 Introduction

Defence mechanisms began when life did [2]. The fabulously complex vertebrate defence system has evolved over hundreds of millions of years to give rise to what we now call the immune system. It is a complex, distributed system with no centralised control. When analysed and segmented a multilayered line of defence is visible, with each layer providing a different method of fault detection and removal. Existing approaches to hardware fault tolerance adhere to this approach, in many situations without even realising the similarities with nature. The immune system performs its job so remarkably well that it is often referred to as a second brain [19]. This paper applies this technique to electronic hardware in an approach called immunotronics, meaning immunological electronics.

This paper analyses biology's way of layering defence mechanisms in section 2 and compares them to those currently used within hardware fault tolerance in section 3. The paper continues with a discussion of artificial immune systems in section 4 to demonstrate some of the successful applications of immune inspired techniques. Section 5 then shows the implementation of the artificial immune system for hardware fault detection. Results are presented in section 6 to demonstrate the systems ability to detect faults in an 'immunised' system. The paper is concluded in section 7.

Y. Liu et al. (Eds.): ICES 2001, LNCS 2210, pp. 140–150, 2001.
© Springer-Verlag Berlin Heidelberg 2001

2   Biological Protection Mechanisms

Biological protection mechanisms are divided into specific and nonspecific components [15]. Nonspecific refers to the basic resistance to infection that an individual is born with and is therefore instantly available to defend the body. Specific, or acquired immunity is developed during the growth of an individual through a series of operations that create a 'bank' of immune system cells to counteract invading bacteria and viruses. Figure 1 shows the organisation of the layers. The outermost and simplest nonspecific layer provides a physical barrier to infection at the atomic layer. The next layer creates physiological characteristics inhospitable for invaders using elements such as temperature and acidity. The phagocytic layer uses roaming scavenger cells that sweep the body clean of easily detectable foreign microorganisms by engulfing them. The specific response is provided in the form of the acquired immune system. This is further divided into two layers of humoral and cell-mediated immunity. Cell-mediated immunity deals with intra-cellular viral infection and humoral with extra-cellular viral infections. This paper concentrates on the processes involved with humoral immunity.

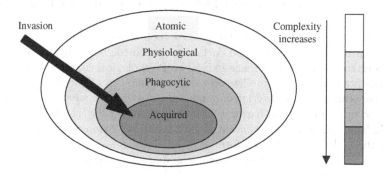

Fig. 1. The layering of defence mechanisms in the body

The humoral layer creates its defence mechanisms through a centralised development stage using the thymus. One form of immature immune cell, known as the helper T cell (CD4) migrates to the thymus where it undergoes a process of clonal deletion [12]. Cells of the body pass through the thymus and any T cells that match a self cell with a certain degree of specificity are destroyed under a process called programmed cell death. Recent theoretical predictions have estimated this matching specificity to be 15 continuous amino acid receptors over the surface of a cell [18]. Upon maturation only T cells that fail to match self cells with this level of specificity and hence match invading microorganisms are left.

Microorganisms with common surface receptors are easily detected. Many bacteria have different surface receptors or have evolved capsules that enable them to conceal these receptors and therefore avoid detection [15]. B cells are a type of antigen presenting cell and are used to detect nonself cells that avoid detection by macrophages. The antigen presenting cells that bind to suspected invaders migrate to lymph nodes - the regions of immune response initiation, where protein fragments of the suspect cells are presented to the helper T cells. If the corresponding T cells exist then a response is initiated and B cells begin the manufacture of antibodies to neutralise the invader. The importance of approximate matching is very apparent here. The human genome simply does not have the storage space to encode for all immune cells to counteract every potential invader. Instead the generation of B and T cells creates a library of cells that initiate a potential response and then proliferate using a form of rapid evolution [12]. In this process the structure of surface receptors on the B cells gradually becomes more specific to the invader currently attempting to attack [14]. Humoral immunity does not directly destroy the invading microorganism, rather it inhibits its proliferation by blocking binding regions with antibodies before the invader can use up valuable resources in the body. Further immunological components guide the destruction and removal of the invading microorganisms [12].

3   Fault Prevention and Detection Layering in Hardware

Hardware systems already use layering to good effect for fault tolerance. Systems can be encased in screened enclosures to provide both physical protection and electromagnetic interference reduction. Physiological characteristics use environmental controls to stabilise operating temperatures. The phagocytic layer in nature causes the destruction and removal of invaders, or what we may treat as faulty components within hardware. Similarities exist with the use of n-modular redundancy [1], although at more of a system level. Stronger similarities exist with the biologically inspired form of cellular electronics known as embryonics [17] [16]. The development of a combined embryonic and immunotronic system has been considered in [3].

4   Artificial Immune Systems

Two computational models of the immune system have been used as methods for problem solving - the negative selection algorithm [7] and the immune network model [13]. The immune network model was developed by the theoretical immunologist Neils Jerne using the hypothesis that the immune system forms an interconnected network of antibodies that decrease and increase in concentration as both antigen and other antibody concentrations change. This technique has been applied to process monitoring [11] and robot control [10].

The negative selection algorithm developed by Forrest and Perelson is a probabilistic detection method that generates a set of data R, that fails to match any of the self strings S, that are to be protected. A match is said to occur if two strings are identical in at least c contiguous positions as illustrated by figure 2.

```
String 1:  0  0  1  0  1  0  1  1  1  0
                                            Match
String 2:  1  0  1  0  1  1  0  1  0  0
                                            Match length c = 4
String 3:  1  0  1  0  1  1  0  0  0  0
                                            No match
String 4:  1  0  1  1  1  0  0  0  1  0
```

Fig. 2. The negative selection process. Strings 1 and 2 match in at least $c = 4$ contiguous positions. Strings 3 and 4 fail to match in at least $c = 4$ contiguous positions.

The negative selection algorithm works upon the basis of self/nonself discrimination that is fundamental to the operation of the immune system. This technique, or forms based upon it, have been successfully used in computer security [6], virus protection [7] and anomaly detection in time series data [4]. The technique for generation of data strings, or *tolerance conditions* that fail to match self is based upon the centralised maturation of T cells in the thymus. Applying the random generation of data to the creation of tolerance conditions is computationally expensive as a significant number of random strings are thrown away due to a match with self strings. However this need only be done once prior to system operation. The detection process is technically much simpler and requires searching to determine if a match occurs in c contiguous locations between data extracted from the system in operation and the stored tolerance conditions. Figure 3 illustrates this. For the hardware immune system self is defined as a valid system state and nonself as an invalid state.

Random generation of tolerance conditions, a process similar to the generation of T cells in the immune system does have weaknesses. Many of the generated tolerance conditions R can match similar subsets of nonself strings. If the number of tolerance conditions needs to be limited then a poor coverage of nonself conditions is likely. The generation time also increases exponentially with respect to the number of tolerance conditions R required. As the number of gen-

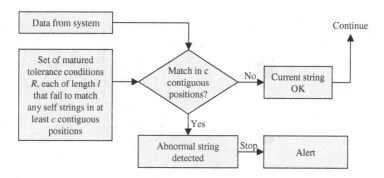

Fig. 3. The matching cycle for self/nonself discrimination - Nonself is detected instead of self, hence negative selection

erated tolerance conditions increases, there is less chance of a tolerance condition being generated that has not previously been generated and that matches one of the nonself strings not yet covered by other tolerance conditions. D'haeseleer [5] developed a method to solve these limitations called the greedy detector generating algorithm. The greedy algorithm works in linear time with respect to the size of R by generating a two dimensional array of all possible c length strings in all possible positions along each nonself string with each string scored against the total number of nonself strings that it matches in the corresponding positions. The highest scoring strings are extracted first, combined to create a complete tolerance condition and the scores adjusted to create a diverse set of tolerance conditions with as little overlap as possible in the nonself strings detected. It is then possible to trade off matching probability against the storage requirements. The greedy detector generator is better suited to the requirements of the hardware immune system as it is desirable to detect as many nonself strings with the smallest number of tolerance conditions. There are likely to be instances where nonself strings are not detectable at all with a specified match length due to certain match length binary sequences matching both self and nonself strings. The algorithm is described in [5] with a detailed analysis of undetectable strings.

5   Implementation of a Humoral Fault Detection Layer

The hardware immune system layer has been developed using a finite state machine representation of the system to be protected (figure 4). This enables self and nonself conditions to be easily extracted from valid and invalid state transitions respectively. Under normal and correct operation only self transitions t_{qx} can occur. A fault may lead to the state machine moving to a nonself state e_x, or a self state q_x, but *not* the correct self state. Monitoring the transitions between states labelled t_{ex} on figure 4 allows both types of faults to be detected.

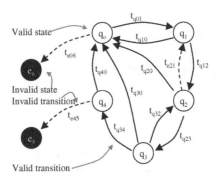

Fig. 4. Generic state machine with valid (self) and invalid (nonself) states and transitions

The hardware immune system is created in four steps:

1. **Data gathering using the immunising test bench**. The hardware and software testbench is used to collect self data from the finite state machine undergoing the immunisation process. This is built using a combination of a graphical based interface on the PC and a programmable hardware testbench constructed on a Xilinx Virtex XCV300 FPGA [9]. The finite state machine is inserted between control inputs and data collection outputs. Inputs are randomly generated until either all self strings are collected, or a predefined coverage is reached.

2. **Tolerance condition generation and analysis**. The self strings are loaded into the greedy detector generator and a complete set of tolerance conditions are generated for all potential match lengths from $1 \leq c \leq l$, where l is the length of both the strings to be protected and the tolerance conditions being generated. A *complete* set enables detection of all detectable nonself strings (see section 4). An analysis of the collected data then enables the lowest failure probability to be determined for the available storage space. Section 6 discusses this for a sample counter state machine.

3. **Configuration of the hardware immune system**. The selected tolerance conditions are then downloaded into the hardware immune system which acts as a wrapper around the finite state machine being immunised.

4. **In service operation of the system with immune wrapper**. During operation of the system the inputs and current state of the finite state machine are extracted and passed through to the hardware immune system. The hardware immune system uses a partial matching content addressable memory[1] to search through all tolerance conditions at the same time to determine the validity of the extracted string. If no match is found in at least c contiguous positions then the system is allowed to continue operation. If a match is found then a potential fault is flagged.

The addition of a hardware immune system layer creates protection organised into the layers shown in figure 5.

6   Results

Fault detection using the hardware immune system is demonstrated using a 4-bit decade counter. 10-bit self strings are created from the concatenation of two user inputs: Count enable and reset, with a 4-bit current state and a 4-bit next state. Taking all permutations of inputs with valid states gives 40 valid self strings to be protected and $984 = (2^{10} - 40)$ nonself strings to be detected.

Self data was collected directly from the hardware through random input injection until all forty valid strings were collected. An average 200,000 input

---

[1] A content addressable memory permits parallel searching by data rather than by address. A match is obtained in one clock cycle irrespective of the depth of the memory. Memory possessed by the natural immune system is also believed to be content addressable [8]

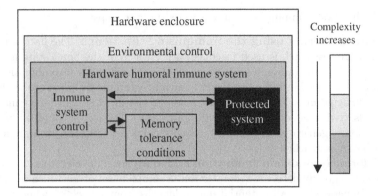

Fig. 5. The hardware immune system wrapper within the additional layers of defence

combinations were needed on each repetition demonstrating the inefficiency, although effectiveness of this initial approach[2]. The integration of pre-defined test sequences would reduce this significantly.

Generation of the tolerance conditions requires two user parameters: the match length c and the number of tolerance conditions required R. String length l is extracted from the self data. The greedy algorithm was used to generate a *complete* set of tolerance conditions where possible, for all match lengths varying from $1 \leq c \leq 10$. Figure 6 shows the percentage of detectable faults for match lengths $4 \leq c \leq 10$ against the minimal number of tolerance conditions required to achieve such coverage. An increase in the number of tolerance conditions has no effect on the percentage of nonself strings detected. Attempting match lengths $c \leq 3$, it was not possible to generate any tolerance conditions that matched nonself and failed to match self. For a match length $c = 4$ only 6 unique tolerance conditions are needed to detect all detectable nonself strings (584 out of 984). With a match length $c = 7$, 103 unique tolerance conditions can detect 913 out of 984 nonself strings. A match length of 10 quite correctly provides 100% coverage as there is a one to one mapping between nonself strings and tolerance conditions (the set of tolerance conditions is equivalent to the set of nonself strings).

Figure 7 confirms the theoretical analysis provided in section 4 concerning the choice of algorithm for generation of the tolerance condition. The plot shows the improvement in fault detection capabilities over a range of 5 to 50 tolerance conditions. The shorter the match length, the greater the improvement in number of nonself strings detected by the greedy detector generator. The greedy detector extracts the tolerance conditions that match the most nonself strings

---

[2] Collection of self strings is inefficient in terms of number of input iterations required. The total time required is dependent upon the speed of the host computer and clock speed on the FPGA testbench. Using a PIII 500Mhz PC and a FPGA test harness running at 1Mhz requires typically 2-4 seconds to generate all self strings.

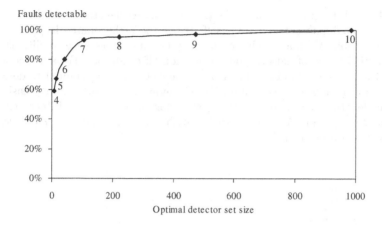

Fig. 6. Optimal number of tolerance conditions required to cover all detectable nonself strings for match length $4 \leq c \leq 10$.

first, compared to the random generator that extracts tolerance conditions on a purely random basis. Furthermore, the random generator creates different tolerance conditions that can match similar subsets of nonself strings.

The next step in the immunisation cycle from section 5 requires the choice of match length to be made. Figure 8 shows the probability of failing to detect a nonself string P_f, against the number of tolerance conditions N_R from 5 to 220. As storage space and therefore the number of tolerance conditions increases the

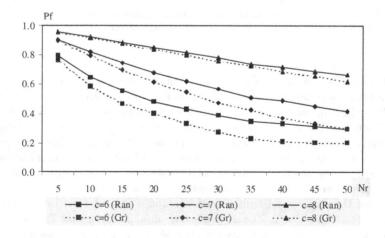

Fig. 7. A comparison of random and greedy detector generator algorithms for variation in probability of match failure (Pf) against number of tolerance conditions (Nr) for match length $6 \leq c \leq 8$.

ideal match length c increases. As the lower match lengths tend towards their limit of maximum detectable nonself strings the higher match lengths become more suitable. With a match length $c = 4$, 60 bits (6 tolerance conditions, each of length 10-bits) of data can detect 584 nonself conditions. For a match length $c = 4$ this is the minimal number of tolerance conditions required to detect all detectable nonself strings. Maintaining 6 tolerance conditions and increasing the match length to $c = 7$, only 120 nonself strings can be detected. Increasing the number of tolerance conditions to 100 with a match length $c = 7$ enables 913 nonself conditions can be detected.

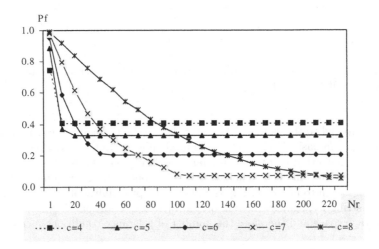

Fig. 8. Probability of match failure (Pf) against number of tolerance conditions (Nr) for match length $4 \leq c \leq 8$ using greedy detector algorithm.

The results have so far considered the total percentage of nonself strings detectable at any time during the operation of the state machine and the combined hardware immune system. The first instance of a fault occurring is likely to lead to a faulty current state, with the existing inputs and previous state conditions holding previously created valid data. In the case of the counter this would result in the last 4-bits of the state string signalling a faulty condition. This invalid state will propagate on successive clock cycles to the previous state, which may consequently lead to a further faulty states. This is considered by figure 9 which compares the single cycle fault detection abilities against total fault detection for a match length $c = 7$. The graphs shows an almost constant 7% increase in match failure for detecting faults within a single clock cycle of the finite state machine. As the match length increases the difference decreases. Given $c = 8$ the difference is typically 2%. As the match length increases individual tolerance conditions detect progressively fewer nonself strings.

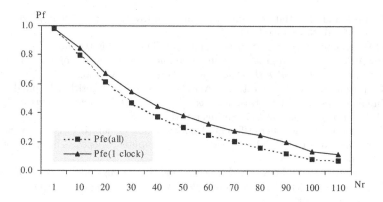

Fig. 9. Probability of match failure (Pf) against number of tolerance conditions (Nr) for all faults and faults detectable in single clock cycle. Match length $c = 7$.

7   Conclusion

The development of a immune system layer for providing further protection against faults has shown that taking inspiration from natural immunity can achieve successful results. The hardware immune system in its current form allows us to detect the presence of faults to a varying degree of probability depending upon the match length used and the number of tolerance conditions created. The process could be completely automated for use in remote applications where the functionality of the system may be changed, reconfigured or even evolved. If the structure of the system changes the immunisation cycle can just be run again to extract self data from the system. At present the hardware immune system only provides fault detection, in essence what the humoral component of the human immune system performs almost to perfection. The human immune system incorporates additional elements to achieve repair through the destruction of the offending cells or molecules. The nearest analogy with biologically inspired fault tolerance for the inclusion of this feature is embryonics. The development of a combined approach will increase the effectiveness of a complete biologically inspired multilayered defence mechanism even further.

Acknowledgements

This work has been supported by the Engineering and Physical Sciences Research Council, UK and Xilinx, Inc.

References

1. A. Avizienis. Fault-Tolerance: The Survival Attribute of Digital Systems. *Proceedings of the IEEE*, 66(10):1109–1125, October 1978.

2. G. Beck and G.S. Habicht. Immunity and the Invertebrates. *Scientific American*, pages 42–46, November 1996.

3. D.W. Bradley, C. Ortega-Sánchez, and A.M. Tyrrell. Embryonics + Immunotronics: A Bio-Inspired Approach to Fault Tolerance. In *Proceedings of 2nd NASA/DoD Workshop on Evolvable Hardware*, pages 215–223, July 2000.

4. D. Dasgupta and S. Forrest. An Anomaly Detection Algorithm Inspired by the Immune System. In D. Dasgupta, editor, *Artificial Immune Systems and Their Applications*, pages 262–277. Springer-Verlag, 1998.

5. P. D'haeseleer. Further Efficient Algorithms for Generating Antibody Strings. Technical Report CS95-3, Department of Computer Science, University of New Mexico, 1995.

6. S. Forrest, S.A. Hofmeyr, A. Somayaji, and T.A. Longstaff. A sense of self for Unix processes. In *Proceedings of 1996 IEEE Symposium on Computer Security and Privacy*, 1996.

7. S. Forrest, A.S. Perelson, L. Allen, and R. Cherukuri. Self-Nonself Discrimination in a Computer. In *Proceedings of the 1994 IEEE Symposium on Research in Security and Privacy*, pages 202–212, Los Alamitos, CA, 1994. IEEE Computer Society Press.

8. C.J. Gibert and T.W. Routen. Associative Memory in an Immune-Based System. In *Proceedings of the 12th International Conference on Artificial Intelligence AAAI-94*, pages 852–857, 1994.

9. Xilinx Inc. Virtex data sheet, 1999. http://www.xilinx.com/partinfo/virtex.pdf.

10. A. Ishiguro, T. Kondo, Y. Watanabe, and Y. Uchikawa. Immunoid: An Immunological Approach to Decentralized Behaviour Arbitration of Autonomous Mobile Robots. In H.M. Voight et al., editors, *Parallel Problem Solving from Nature IV*, volume 1141 of *Lecture Notes in Computer Science*, pages 666–675. Springer-Verlag, September 1996.

11. A. Ishiguro, Y. Watanabe, and Y. Uchikawa. Fault Diagnosis of Plant Systems using Immune Networks. In *Proceedings of the 1994 IEEE International Conference on Multisensor Fusion and Integration for Intelligent Systems*, pages 34–42. IEEE, 1994.

12. C.A. Janeway and P. Travers. *Immunobiology, the Immune System in Health and Disease*. Churchill Livingstone, 3rd edition, 1997.

13. A.K. Jerne. Towards a network theory of the immune system. *Ann. Immunol. (Inst. Pasteur)*, 125C:373–379, 1974.

14. W.S. Klug and M.R. Cummings. *Concepts of Genetics*. Prentice Hall, 4th edition, 1994.

15. J. Kuby. *Immunology*. W.H. Freeman and Company, 3rd edition, 1997.

16. D. Mange, M. Sipper, A. Stauffer, and G. Tempesti. Toward Robust Integrated Circuits: The Embryonics Approach. *Proceedings of the IEEE*, 88(4):516–541, April 2000.

17. C. Ortega-Sánchez, D. Mange, S. Smith, and A. Tyrrell. Embryonics: A Bio-Inspired Cellular Architecture with Fault-Tolerant Properties. *Genetic Programming and Evolvable Machines*, 1(3):187–215, July 2000.

18. J.K. Percus, O.E. Percus, and A.S. Perelson. Predicting the size of the T-cell receptor and antibody combining region from consideration of efficient self-nonself discrimination. In *Proceedings of the National Academy of Science*, volume 90, pages 1691–1695. Oxford University Press, 1993.

19. G.W. Rowe. *Theoretical Models in Biology*. Oxford University Press, 1994.

Ant Circuit World: An Ant Algorithm MATLAB™ Toolbox for the Design, Visualisation and Analysis of Analogue Circuits

Morgan R. Tamplin and Alister Hamilton

Department of Electronics and Electrical Engineering, University of Edinburgh,
King's Buildings, Mayfield Road, Edinburgh EH9 3JL, Scotland,
Morgan.Tamplin@ee.ed.ac.uk, Alister.Hamilton@ee.ed.ac.uk

Abstract. We present a MATLAB™ toolbox which applies an ant algorithm to design simple analogue circuits. Other researchers have applied earlier algorithms inspired by real social insect colonies to travelling salesman and vehicle routing problems with good success. This implementation suggests some features specific to producing analogue circuits. In addition, the toolbox provides the facility to display and assess the progress of the simulation. It will be extended to work with a broad range of circuit component models, multiple ports, and task-specific ants.

Introduction

We present a MATLAB™ toolbox which applies an ant algorithm for designing simple analogue circuits. The algorithm contained in this toolbox is inspired by those which have appeared previously [8,6,5], but contains features unique to designing analogue circuits, in particular those based on Palmo cells [13,9].

Our toolbox is in early development, but so far contains models of a virtual circuit test board and ant environment. This board can be populated with arbitrary numbers of simulated Palmo cells that integrate a pulse-width encoded signal as described in [13,9], input and output ports, and simple mobile agents known as ants. Also included are facilities for visualising this virtual world and the progress of its inhabitants, and for reporting basic statistics on the progress of the algorithm. A description of this system follows with brief explanations of its background, early runs, and expectations for the future.

Ant Colony Optimisation

Ant Colony Optimisation (ACO) was inspired by investigations into the collective search capabilities of real ant colonies [1]. It represents one aspect of a growing interest in optimisation techniques informed research into many kinds of social insects and decentralised systems in general [4,14].

Individually, real ants are extremely limited in their sensing and decision-making abilities, but collectively a colony of ants is proficient at finding the

Y. Liu et al. (Eds.): ICES 2001, LNCS 2210, pp. 151–158, 2001.
© Springer-Verlag Berlin Heidelberg 2001

shortest route to a goal in an irregular environment. Ants have been observed at the beginning of a search (e.g. for food, building materials) to explore their environment almost at random, but to eventually converge on a useful pattern by communicating with each other through changes they make to their environment (i.e. stigmergy). As they explore, the ants leave a chemical pheromone trail behind them. This trail serves two purposes. First it guides ants back to their home from their goal. As ants return to their home, they continue to lay pheromone along their path. The first ants to return to the nest inevitably leave behind a double dose of pheromone which increases the probability of future exploring ants to choose the same path. A second mechanism, pheromone decay, prevents ants from becoming stuck following sub-optimal or obsolete paths. Concentrations of chemical pheromone not updated by exploring ants decrease naturally over time, encouraging new routes to be taken up by the colony.

These natural mechanisms first found computational equivalents in examples such as Ant System [8]. In this programme, an early example of ACO, ants build a solution to a travelling salesman problem (TSP) by searching arcs that connect cities presented as nodes on a graph. Their goal is to find the shortest route around all cities, visiting each city only once. As the ants search the graph, they choose the next node probabilistically according to a local heuristic that indicates the quality of that portion of the route combined with a number that represents a pheromone concentration. The heuristic provides the ants with an initial form of guidance before the colony leave enough pheromone behind to guide its movements. As ants move about the graph, they read the existing pheromone values and write their own values when they act. As pheromone concentrations increase around the graph, ants rely more on them and less on the heuristic to guide their search. Ant System performed comparably to other decentralised biologically-inspired optimisation methods for the TSP, like genetic algorithms, and so adapted versions have been applied to the quadratic assignment problem [12], network routing [5], and other problems. While Eric Bonabeau et al [1] have written about applying ant algorithms to a wide range of benchmark cases, others have applied the method to varied real world problems such as vehicle routing [2] and combinational logic circuits [3]. These have typically inspired novel task-related alterations of the original ant algorithm.

Palmo Cell Circuits

The Palmo cell is our basic unit for producing examples of emergent circuits, previously as evolvable hardware [10], and now using an ant algorithm. Working Palmo cells were originally produced at Edinburgh University [13] for conditioning neural network input signals, and have characteristics that make them suitable to emergent circuit design [9,10]. They are collections of reconfigurable integrators implemented in analogue VLSI that integrate pulse-width encoded signals. One or more of these cells can be employed together to create filters of increasing order [10]. They can encode signals of a broad range of frequencies from analogue, digital, and mixed sources [9].

We use a model of the Palmo cell to demonstrate emergent analogue circuits [10]. The model replicates the Palmo cell on an intermediate level that represents the cell as a computational unit, its inputs and outputs, gain value and integrating components. Signals in this model are defined as real number values that are added to or subtracted from real number integrated values according to the signal's sign and the circuit's interconnect. This choice foregoes the unnecessary and computationally intensive details of an accurate SPICE circuit model, but preserves a sense of circuit structure for analysis and eventual transfer to a real chip.

Ant Circuit World

The Ant Circuit World is our ACO algorithm for the design of analogue circuits. Ant Circuit World models three classes of object:

- a bounded, coarse-grained world environment that contains
- virtual circuit components (including input and output ports), and
- ants, a collection of simple exploring agents with limited sensing, computational, and movement abilities.

The world is a matrix of discrete squares. Each square can hold a physical resident, such as an ant, a component or a port, and a "non-physical resident", or pheromone value written to and read from the world by ants. The dimensions of the world are determined by the number of components it contains. The number of components, the world's dimensions, pheromone decay and update rates are set at the beginning of a run and do not change. The world controls each run, or epoch, divided into days. A day increments each time the entire swarm of ants has attempted or completed a single move.

The component class models circuit components, currently limited to Palmo cells and ports. Ports are typically placed at opposite ends of the world. They are the starting point and goal for the ants' paths. At circuit assessment they serve as the input and output of the circuit test signal. The system currently uses only one input and one output port.

Components are distributed regularly in a single row along the length of the world. They occupy world space and instruct visiting ants' construction of a circuit interconnect according to the component's specialised task. Currently, the only component we have modelled, a Palmo cell, contains two inputs and one output each. It also contains a gain value, K, that affects the integration of signal values passed through the cells at circuit analysis. This K value is initially chosen at random from its maximum range of 0.000 to 1.000. Palmo cells change their K value at random each time an ant passes through them. The legal range from which to choose K decreases equally around the current value of K, proportionately to pheromone value increases of the squares on which the Palmo cell resides. As an individual cell contributes to the construction of increasingly better circuits, the pheromone value of its square rises, thereby restricting the potential change of its gain value.

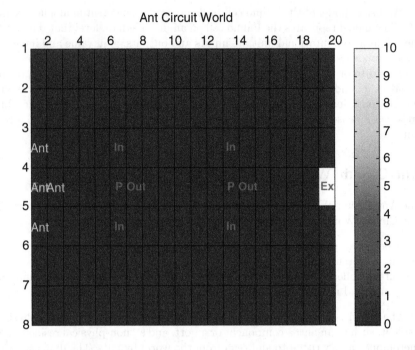

Fig. 1. Graphical representation of Ant Circuit World's starting state with two Palmo cells, and ants at the entrance port. The world is a uniform dark colour indicating it contains no pheromone. The exit is a lighter colour to indicate it has been set to maximum pheromone concentration.

It is hoped that this restriction reduces the damage that might occur to a well developed circuit by changing an already suitable K value too much, but provides some room for variation to prevent the algorithm from stagnating. Any changes to components' legal K-value ranges and the world's pheromone values occur when ants complete their tour of the world and reach the exit port. Provided an ant carries with it a circuit interconnect, it is assessed, scored and an apropriate level of pheromone added to the trail the ant followed. This trail of pheromone will guide future ants' paths depending on its strength.

Ants record the entrance of each palmo cell they pass through, which later becomes a circuit interconnect (if the ant finishes its tour by reaching the exit port), along with the gain values of each palmo cell visited. They also record a complete trail through the world for the purpose of pheromone update. Ant objects determine their movement according to a random proportional transition rule that uses the level of pheromone value one square beyond the ant's current location on the four main compass points (North, East, South, and West). This choice is limited to four directions to simplify the transition rule and reduce the calculations required to move the ant.

The pheromone values surrounding an ant occupy regions of 0 to 180 degrees around their corresponding compass points, proportionate to the amount of pheromone they possess. For instance, a square to the left or East of an ant containing the maximum pheromone concentration would cover the range 90-270 degrees on a corresponding compass representation. A square containing half the maximum pheromone concentration would occupy only 135-225 degrees on the same compass. The former square would consequently have a greater chance of capturing the random number generated from the total compass range of 0-360 degrees for the ant's move than the smaller region representing a lower concentration. When a random direction appears that points to two regions of overlapping pheromone strengths on the compass or falls outside any region, then the transition rule chooses the largest or closest region to the random number.

Physical residents on candidate squares, such as other ants, the entrance port, cell outputs or walls, incur a severe penalty in the calculation of the square's transition desirability to discourage ants from attempting to move toward them, resulting in a lost move. Ants that attempt an illegal move stay put, but their age increments. Ants that reach their maximum age before completing a circuit "die out" and are restarted at the entrance port with an empty interconnect, further constraining the effect of illegal moves on the course of the algorithm. Entrances to Palmo cells and the final exit port, however, attract ants toward them.

Ants have room for exploring the world between components, giving them the possibility of passing through the first, and possibly each consecutive component they reach, due to probabilistic fluctuations in their transition rule, or skipping one or all of the components on their way to the exit goal port. This extention also encourages competitive interaction between ants. The grid representation only allows one agent per square during its search, meaning agents that attempt to move to an occupied square risk losing a move without gaining any immediate benefit.

Increasing space for exploration and providing for interaction between agents is an extension to ant algorithms that is expected to increase the cost of moving ants through the world, ultimately increasing the computational cost and time to achieve a complete circuit solution. However, greater emphasis on exploration and interaction is hoped to increase the production of novel solutions and encourage behaviours, such as backtracking and visiting cells multiple times, necessary for designing analogue circuits that depend on feedback.

Whenever an ant carrying a circuit interconnect of at least one component reaches the exit port, it is analysed and scored before being replaced by a new ant at the entrance port. An error function produced from the difference between ideal circuit behaviour and the behaviour of the test circuit determines the ant's score and the pheromone values applied to the ant's trail. This error value can be achieved by introducing a set of vectors for representative sinusoidal frequencies as in [10], or by fast fourier transform (FFT) analysis of the frequency spectra of the circuit's output for a step input. The ant's score determines the amount

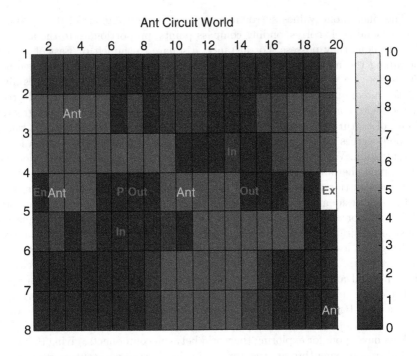

Fig. 2. Graphical representation of Ant Circuit World during an epoch. Lighter shades indicate higher pheromone concentrations.

of pheromone value the world will add to the ant's trail of visited squares, and determines whether the ant and its circuit will be preserved.

We used the same "free netlist" encoding for the interconnect in our ant algorithm as in our previous GA [10], adapted from [15] (and which, coincidentally, resembles the encoding from [3] presented at the same conference). The interconnect for a complete circuit is a row of integers that determine the input connections and internal characteristics of each component. For each Palmo cell, the interconnect represents the source of the cell's inputs, followed by its gain value. For instance, a low pass filter circuit containing a single two-input Palmo cell with one input connected to the signal input and the other connected to its own output would appear as (0 1 0.046) [10]. The circuit input port is arbitrarily set to zero, but the connection between zero and any component is a product of the algorithm. The final real number is the cell's gain value. Unlike genetic algorithm implementations, which employ a fixed length encoding, this encoding used with our ant algorithm will vary in length according to the number of components an ant visits before it exits. Given constraints on the ant's maximum age, allowing the encoding to vary in length presents a useful opportunity for ants to produce the most compact solutions they are capable of.

At the completion of a run, or epoch, the world returns a collection of unique circuit solutions with the best scores, along with summary statistics of the ants' progress through the world.

Continuing Work

We wish to further test this toolbox by comparing its results to those achieved producing low pass filters with a genetic algorithm in [10]. Next steps would then be to add additional components (e.g. transistors, capacitors, resistors, etc.) to the Ant Circuit World's environment to tackle an increased range and complexity of possible circuit problems. Extensions could also include the use of multiple signals, and employ colonies containing ants specialised to follow mulitple, independent pheromone trails linked to multiple signal inputs. The same extension could also extend this toolbox to explore the design of mixed signal circuits.

Our MATLAB™ toolbox represents a first step toward the application of ant algorithms to the design of reconfigurable analogue circuits. Unlike previous implementations of ant algorithms [7,8,1], which placed explicit restrictions on ants backtracking or visiting nodes multiple times, this toolbox provides specialised features to encourage the emergence of feedback loops necessary to the design of analogue circuits and the adjustment of cell parameters during system runtime. To the best of our knowledge it is the first example of an attempt to apply an ant algorithm to analogue circuit design and, apart from a system designed for the creation of combinational logic circuits [3], the only other current example of ACO for circuit design. More generally, representing circuit design problems as ant algorithms overcomes necessary limitations on the number of component nodes and circuit types reported in previous examples [11] of evolvable hardware.

References

1. E. Bonabeau, M. Dorigo, and G. Theraulaz. *Swarm Intelligence: From Natural to Artificial Systems*. Oxford University Press, New York, 1999.
2. B. Bullnheimer, R. F. Hartl, and C. Strauss. An Improved Ant System Algorithm for the Vehicle Routing Problem. Technical Report POM-10/97, Institute of Management Science, University of Vienna, Vienna, Austria, 1997.
3. C. A. C. Coello, R. L. Zavala, B. M. Garcia, and A. H. Aguirre. Ant Colony System for the Design of Combinational Logic Circuits. In *Third International Conference on Evolvable Systems: From Biology to Hardware*, number 1801 in Lecture Notes in Computer Science, pages 21–30. Springer Verlag, April 2000.
4. D. Corne, M. Dorigo, and F. Glover. *New Ideas in Optimization*. MacGraw-Hill Publishing Company, Berkshire, England, 1999.
5. G. DiCaro and M. Dorigo. AntNet: Distributed Stigmergic Control for Communications Networks. *Journal of Artificial Intelligence Research (JAIR)*, 9:317–365, 1998.
6. M. Dorigo and G. D. Caro. The Ant Colony Optimization Meta-Heuristic. In *New Ideas in Optimization*, pages 11–22. MacGraw-Hill Publishing Company, Berkshire, England, 1999.

7. M. Dorigo, V. Maniezzo, and A. Colorni. Positive Feedback as a Search Strategy. Technical Report 91-016, Dipartimento di Elettronica, Politicnico di Milano, Italia, 1991.

8. M. Dorigo, V. Maniezzo, and A. Colorni. The Ant System: Optimization by a Colony of Cooperating Ants. *IEEE Transactions on Systems, Man, and Cybernetics*, (Part B 26(1)):29–41, 1996.

9. A. Hamilton, K. Papathanasiou, M. R. Tamplin, and T. Brandtner. Palmo: Field Programmable Analogue and Mixed-Signal VLSI for Evolvable Hardware. In *Second International Conference on Evolvable Systems: From Biology to Hardware*, number 1478 in Lecture Notes in Computer Science, pages 335–344. Springer Verlag, September 1998.

10. A. Hamilton, P. Thomson, and M. R. Tamplin. Experiments in Evolvable Filter Design Using Pulse Based Programmable Analogue VLSI Models. In *Third International Conference on Evolvable Systems: From Biology to Hardware*, number 1801 in Lecture Notes in Computer Science, pages 61–71. Springer Verlag, April 2000.

11. J. D. Lohn and S. P.Colombano. Automated Analog Circuit Synthesis Using a Linear Representation. In *Second International Conference on Evolvable Systems: From Biology to Hardware*, number 1478 in Lecture Notes in Computer Science, pages 123–133. Springer Verlag, September 1998.

12. V. Maniezzo and A. Colorni. The Ant System Applied to the Quadratic Assignment Problem. *IEEE Transactions on Knowledge and Data Engineering*, 1998.

13. K. Papathanasiou. *Palmo: a Novel Pulsed Based Signal Processing Technique for Programmable Mixed-Signal VLSI*. PhD thesis, Department of EEE, University of Edinburgh, Edinburgh, Scotland, 1998.

14. M. Resnick. *Turtles, Termites, and Traffic Jams: Explorations in Massively Parallel Worlds*. The MIT Press, Cambridge, Massachussetts, 1994.

15. P. Thomson. Circuit Evolution and Visualisation. In *Third International Conference on Evolvable Systems: From Biology to Hardware*, number 1801 in Lecture Notes in Computer Science, pages 229–240. Springer Verlag, April 2000.

Human-Like Dynamic Walking for a Biped Robot Using Genetic Algorithm

Jin G. Kim[1], Kyung-gon Noh[2], and Kiheon Park[3]

[1] School of Electrical Engineering
Inha University, Inchon, Korea
john@inha.ac.kr
[2] Dept. of Automation Engineering
Inha University, Inchon, Korea
g2001370@inhavision.inha.ac.kr
[3] School of Electrical and Computer Engineering
Sungkyunkwan University, Suwon, Korea
khpark@skku.ac.kr

Abstract. This paper presents the smooth walking trajectory for a biped robot using genetic algorithm. Suitable velocities and accelerations at the via-points are required for dynamic smooth walking since the incorrect via-points data can cause the discontinuity on the trajectory and the unstable walking motion as a result. Optimal via-points data can be found by minimizing the sum of deviation of velocities and accelerations as well as jerks. Using genetic algorithm, we obtained the continuity on the entire trajectory interval and the energy distribution during the walking. In conclusion, it is shown that the proposed genetic algorithm guarantees a satisfactory smooth and stable walking through the experiment on the real biped robot.

1 Introduction

Many researches for biped robots have been done over the last 30 years because of its human-like walking behavior and the adaptability in various terrains. These researches can be classified into two groups on the whole. One is a group to study walking motion control such as control for dynamic walking [1], walking pattern control using force or torque sensors [2], and stability analysis in state-space [3]. The other is a group to study walking stability analysis such as stabilization with additional balancing joints, which restrict GCM (ground projection of the center of mass), ZMP (zero moment point) or FRI (front rotation indicator) to remain within the supporting foot region during the walking [4-10]. To achieve dynamic walking for the biped robot, we use the ZMP as stabilization index.

Furthermore, suitable trajectories are considered for smooth walking and good trajectory tracking. For trajectory generation, the approach in 3-dimensional Cartesian space is more preferable to that in joint space in that adaptability in surrounding environment, easiness for checking contact with the ground, and avoidance for obstacles. So far, the fifth-order polynomial interpolation for the time has been used for trajectory planning including at least 4 via-points; starting, final and two mid via-

Y. Liu et al. (Eds.): ICES 2001, LNCS 2210, pp. 159–170, 2001.

points for avoiding obstacles. The velocities and the accelerations at these via-points are considered as control parameters for determining the shape and characteristic of a trajectory. Generally, a trajectory is generated on the interval between two via-points and the whole trajectory could be acquired by repeated time shifting after the iteration on each interval and their combination. Via-points data such as velocities and accelerations at the via-points can affect the relation and connection between the neighboring intervals. Therefore, the incorrect information might cause the indifferentiability on the trajectory and discontinuous walking motion in consequence.

In this paper, dynamic walking stability of biped robot IWR-III, developed by Control & Robotics Lab. in Inha Univ., is investigated, and trajectory optimization method using genetic algorithm is proposed to minimize the sum of deviation of velocities and accelerations and to reduce the jerk. Using this optimized trajectory, we obtained the continuity on the entire trajectory interval, the energy distribution through the optimum velocities and accelerations, and the smooth human-like walking for biped robot. Smooth and stable dynamic walking is shown through the experiment on the real biped robot.

2   Mathematical Model of Biped Robot

2.1   Mathematical Model

IWR-III has three degrees of freedom in each leg and two degrees of freedom in balancing joints that consist of a roll and a prismatic joint. Fig. 1 represents the real biped system IWR-III and its corresponding mass model with nine rigid bodies.

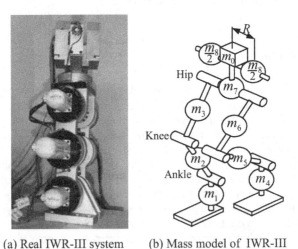

(a) Real IWR-III system   (b) Mass model of IWR-III

Fig. 1. Photo and mass model of IWR-III

For dynamics analysis, let each vector be defined as shown in Fig. 2 on the reference coordinate frame. By applying D'Alembert's principle, the equation of motion at the arbitrary point P is given by

$$\sum_{i=0}^{8} m_i (\vec{r}_i - \vec{P}) \times (\vec{\ddot{r}}_i + \vec{G}) + \vec{M}_T = 0 \tag{1}$$

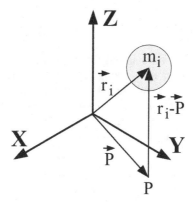

Fig. 2. Vector representation.

where m_i is the mass of the particle i, $\vec{r}_i$ is the position vector of the particle i. $\vec{P}$ is the position vector of the point P on the X-Y plane, $\vec{G}$ is the gravitational acceleration, and $\vec{M}_T$ is the compensational total moment acting on the point P. By modifying the equation (1), the components of $\vec{M}_T$ can be obtained [11].

$$M_{Tx} = -\sum_{i=0}^{8} m_i (\ddot{z}_i + G_z)(y_i - y_{zmp}) + \sum_{i=0}^{8} m_i (\ddot{y}_i + G_y) z_i$$

$$M_{Ty} = \sum_{i=0}^{8} m_i (\ddot{z}_i + G_z)(x_i - x_{zmp}) - \sum_{i=0}^{8} m_i (\ddot{x}_i + G_x) z_i \tag{2}$$

$$M_{Tz} = \sum_{i=0}^{8} m_i (\ddot{x}_i + G_x)(y_i - y_{zmp}) - \sum_{i=0}^{8} m_i (\ddot{y}_i + G_y)(x_i - x_{zmp}) + m_8 R^2 \ddot{\theta}_z$$

where R, represented in Fig.1, in the third equation in (2) is a distance from the center of balancing joint coordinate to the one of the masses of the balancing roll which is assumed to have two masses considering moment of inertia, $\ddot{\theta}_z$ is the angular acceleration with respect to (w.r.t) the erecting z-direction, and x_i, y_i, and z_i are the positions of mass m_i w.r.t the world coordinate. Also x_{zmp} and y_{zmp} are the components of ZMP on the X-Y plane.

The angular acceleration $\ddot{\theta}_z$ of balancing roll joint is obtained when the motion of balancing joints is derived from above first two equations in equation (2). The third in equation (2) means the spin moment caused by motions of legs, trunk and the rotation of balancing roll joint. If M_{Tz} is larger than the friction of walking floor, the robot can't go straight any more. We assume, in this paper, the spin moment between the supporting plane and the sole of the robot is negligible.

2.2   Balancing Motion

Suppose the trajectories of the leg and the desired ZMP are given, the motion of balancing joints that stabilize the walking of a robot is determined by two balancing equations for sagittal and lateral direction.   They can be obtained by arranging equation (2) for the balancing mass m_0, which are shown as follows.

$$\ddot{x}_0 - \frac{(\ddot{z}_0 + g_z)}{z_0} x_0 = -\frac{(\ddot{z}_0 + g_z)x^*}{z_0} - g_x + \frac{\alpha}{m_0 z_0}$$

$$\ddot{y}_0 - \frac{(\ddot{z}_0 + g_z)}{z_0} y_0 = -\frac{(\ddot{z}_0 + g_z)y^*}{z_0} - g_y + \frac{\beta}{m_0 z_0}$$

$$\alpha = \sum_{i=1}^{8} m_i(\ddot{x}_i + g_x)z_i - \sum_{i=1}^{8} m_i(\ddot{z}_i + g_z)(x_i - x^*)$$

$$\beta = \sum_{i=1}^{8} m_i(\ddot{y}_i + g_y)z_i - \sum_{i=1}^{8} m_i(\ddot{z}_i + g_z)(y_i - y^*)$$

(3)

where x_0 and y_0 are the sagittal and lateral motions of a balancing mass w.r.t the world coordinate, respectively. x^* and y^* are the desired ZMPs. Balancing motion compensates the moment caused by the movement of leg and body, but also raises the spin moment which is derived in equation (2).   The spin moment can vary walking direction if it is lager than the friction between the suproting foot and the ground. Since IWR-III has no spin-moment compensating joints, the third equation in equation (2) is only used as a change index of walking direction. By rearranging equation (3), the information of the actual ZMP can be obtained. The stable ZMP should be inside the supporting area that circumscribes the one or two soles in support.

3   Trajectory Generation by Genetic Algorithm

3.1   Trajectory by Conventional Method

Leg trajectory is generated in the 3-dimensional Cartesian space with the consideration of adaptability on various environments and of avoiding capability for obstacles.   Walking trajectory is represented by the fifth order polynomial interpolation for the time. Coefficients of the polynomial can be determined by 6 via-points data. In this case, acceleration is smooth and continuous on each interpolated interval. However, the combined trajectory by time shifting after each interpolation cannot guarantee the continuity on every via-point because the jerk or impact is not considered in the interpolation.

3.2   Genetic Algorithm

Genetic algorithm is a widely used global searching method for an engineering optimization. It uses a population of parameter set, and works not on a local searching scope but on a global one with simultaneity [12]. Genetic algorithm has three unique operators for its optimization. First, reproduction using a biased roulette wheel increases average fitness values, and then crossover at a random cross-site exchanges chromosomes' information. Finally, mutation prevents an early convergence to a local minimum/maximum.

Searching results are mainly controlled by population size, crossover rate and mutation rate. High crossover and mutation rate generate various new children when the generation is going on, which enables the genetic algorithm to explore the searching space of high fitness values in the beginning of the evolution. But, these also cause the slower convergence to the optimal value by lowering exploiting capability. On the other hand, population size can affect the accuracy of an optimal value, required size of memory storage and computing time. After several pre-testing to find some parameters for genetic algorithm, we determined the string length of 10 bits with the binary representation, as shown in Fig. 3.

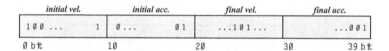

Fig. 3. Coding of chromosome

A chromosome is constructed by serially connecting 4 strings : initial velocity, initial acceleration, final velocity, and final acceleration at the via-points,. To map the searching space into string space, decoding equation is used.

$$ival_{dec} = ival_{bin} \cdot \frac{\left(ival_{MAX} - ival_{MIN}\right)}{2^n - 1} + ival_{MIN} \tag{4}$$

where $ival_{bin}$ and $ival_{dec}$ are the binary and the decoded searching space, respectively. Maximum and minimum values of the searching space are represented as $ival_{MAX}$ and $ival_{MIN}$, respectively. A notation n stands for the string's bit length. The searching range from -5 to 5 is selected to cover peaks of the velocities and accelerations used as strings. Considering computing time, the terminal condition for genetic algorithm is set to 50 as the limitation of generation number, which guarantees the fitness convergence to the best solution in almost cases. Parameters for genetic algorithm and definition of a fitness function considered in this paper are shown in Table 1.

Table 1. Parameters for genetic algorithm

Population	60	Generation	50
Crossover	0.92	Mutation	0.03
String length	10 bit		
Fitness Function	$f = 1 / \sum \left((v_{i+1} - v_i)^2 + (a_{i+1} - a_i)^2 \right)$		

Here, v_i and a_i are the velocity and the acceleration of i-th sampling time, respectively. To have smooth accelerations on the whole interval, acceleration curve should have a parabola shape to minimize the jerk variations. By the above definition of the fitness function, genetic algorithm decreases peak values of the velocity and acceleration for whole interval. The flow chart of trajectory optimization using genetic algorithm is shown in Fig. 4.

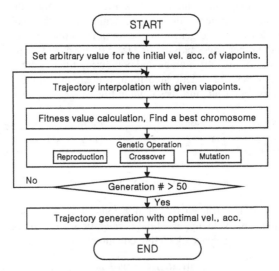

Fig. 4. Flow chart of trajectory optimization

3.3   Trajectory by Genetic Algorithm

Fig. 5 represents the joints velocities and accelerations of a support leg by conventional method, and Fig. 6 depicts those by proposed algorithm. Trajectory by genetic algorithm has a similar effect of interpolation on only one interval as can be seen in Fig. 6. Velocities and accelerations in Fig. 6 have the shapes of the parabola form and are smoother than those of Fig. 5. This result leads to actuators to have a lower maximum speed in the stable operation range with no radical change of speed.

Using a trajectory by genetic algorithm, therefore, smaller maximum torque of a motor can be selected and operated.

Fig. 7 represents a stick diagram of the robot in the sagittal direction. The walking motion by conventional method, left part of the Fig. 7, forms a trapezoidal shape by constraints at mid two via-points. At mid-two positions, the line-representing ankle is concentrated which result from the inadequate or insufficient via-points data. This case has a similar effect of stopping motion at mid-two via-points. The modified walking, however, by genetic algorithm shows a smooth arc shape with the same via-points and optimized velocities, accelerations at the given via-points in Fig. 7. This means the optimal distributions of velocities and accelerations is made during the walking time by reducing the variation of velocities and accelerations.

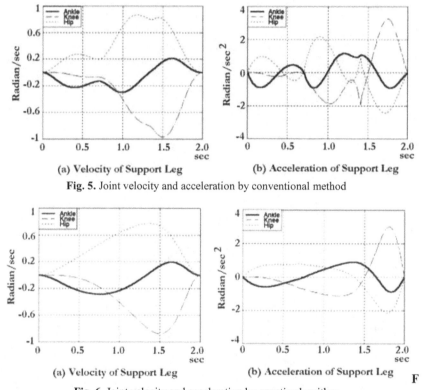

(a) Velocity of Support Leg   (b) Acceleration of Support Leg

Fig. 5. Joint velocity and acceleration by conventional method

(a) Velocity of Support Leg   (b) Acceleration of Support Leg   F

Fig. 6. Joint velocity and acceleration by genetic algorithm

In Fig. 8, the case 1, represented as solid bold line, is the trajectory generation with optimal values by genetic algorithm, and the case 2 by conventional method is the trajectory generation with the average velocities, zero accelerations at mid two via-points. In jerk graph, the right below of Fig. 8, the jerk of the case 1 minimized by genetic algorithm is much smaller than that of the case 2. These peaks of jerk affect the robot system as impacts during the walking. Therefore, it is desirable to reduce the jerk as small as we can such as the case 1 in which the fluctuation of the acceleration curve is very small.

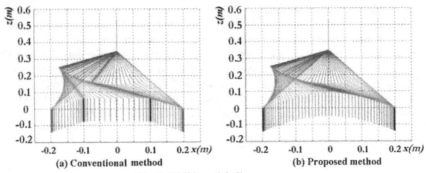

(a) Conventional method   (b) Proposed method

Fig. 7. Walking stick diagram

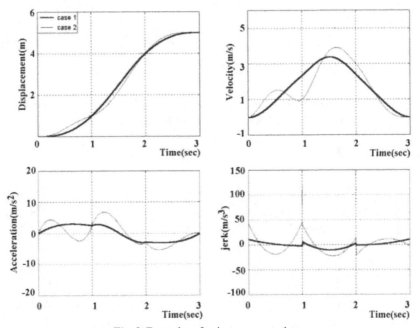

Fig. 8. Examples of trajectory generation

4   Walking Simulation

During the walking, IWR-III has repeated two phases. In single supporting phase, swing leg and the trunk move ahead for the next walk with the compensating motion of balancing joints. In double supporting phase, legs and trunk form a closed chain with no motion, and only balancing mass moves to the next starting position. Continuous walking is performed by an iteration of these phases changing. Fig. 9 shows the structure of numerical simulator for IWR-III and represents the internal data flow.

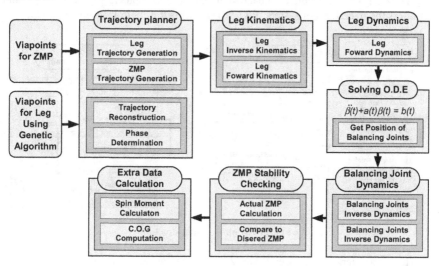

Fig. 9. Simulator structure

Module-type structure of the simulator enables the user to modify the procedure or add one in a simple manner.   A recursive Newton-Euler method are used for dynamics analysis, and the finite difference methods are applied for the solution of system equations. The comparison of spin moments between the conventional and the proposed walking is shown in Fig. 10.

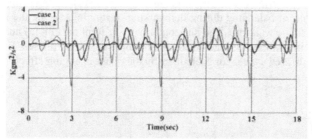

Fig. 10. Spin moments comparison

In the case 2, the spin moment has much larger values at the phase changing time, which can cause the change of walking direction.   The spin moment as a result of applying genetic algorithm, which is represented by case 1, is smaller than that of conventional method.   This implies that robot can go straight with reduced spin moment.

5   Walking Experiments

The system IWR-III has eight AC servomotors and reducers, and is controlled by TMS320C31 DSP controller embedded in the host computer, which analyzes and monitors the overall robot system. Fig. 11 represents the control system structure.

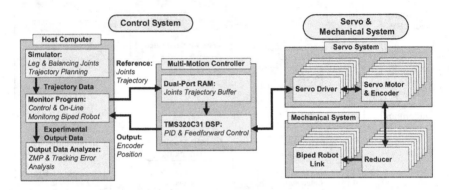

Fig. 11. Control system structure

The controller, which uses TMS320C31 DSP as a main CPU and dual port memory for high-speed communication, is embedded in the PC environment. PID control using memory buffer for 8 servos is done for every 10 ms in the DSP controller, and encoder feedback data is acquired for very 50 ms. Host computer generates reference joints values through the simulation and plots the responses and errors on the multi-view monitoring program coded by Visual C++. Fig. 12 shows the joint position responses of the right, left leg and balancing joints. As unit gaits are repeated except for the start and end of combined gaits, each joint has a periodic form on the whole. Periodic motions of balancing joints, such as the swinging motion of human's arm for balancing during the walking or running, enable the robot to have a steady and stable walking. The horizontal parts of joint data on (a) and (b) in Fig. 12 imply no joints motion of legs during the double supporting phase. The joint data of legs have mirrored shapes for each other by the phase changing effects.

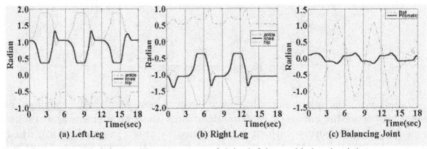

Fig. 12. Joint position responses of right, left leg and balancing joints

Fig. 13 shows the joint position errors of the right, left leg and balancing joints during the walking. It is shown that these errors have little effects on the positions of the robot links, and robot walks well tracking the given command with no tipping in consequence.

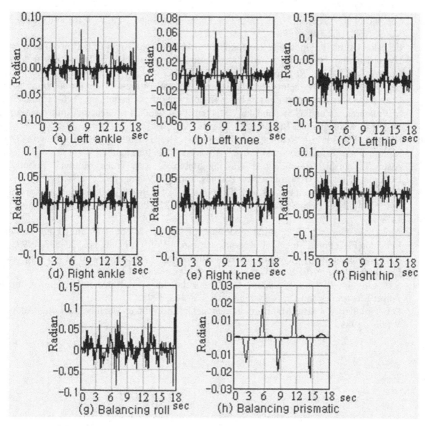

Fig. 13. Joint position errors of left, right leg and balancing joints

6   Conclusion

Genetic algorithm is utilized for the walking trajectory optimization and applied to the real biped robot. Via-points data are used as control parameters determining the trajectory shape. Interpolation is performed with the optimal via-points data that decrease the variation of jerks on the whole trajectory interval. Modified trajectory makes the peak velocities and accelerations smooth and guarantees the accurate ZMP tracking. The experiments on biped robot are made for the verification of improved performance by the proposed method. It is shown that the smooth and stable dynamic walking is achieved through the walking experiment.

References

1. A.Takanishi, M.Ishida, Y.Yamazaki, I.Kato, "The Realization of Dynamic Walking by the Biped Walking Robot WL-10RD," Journal of the Robotics Society of Japan, Vol.3, No,4, pp.325-336, 1985.
2. Furusho et al., "Sensor-Based Control of a Nine-link Biped," Int. J. of Robotics Research, Vol.9, No.2, pp.83-98, 1990.
3. C.Leslie Golliday, Hooshang Hemami, "An Approach to Analyzing Biped Locomotion Dynamics and Designing Robot Locomotion Controls," IEEE Trans. on Automatic Control, Vol.AC-22, No.6, pp.963-972, 1977.
4. C.L.Shih et al., "Trajectory Synthesis and Physical Admissibility for a Biped Robot During the Single Support Phase," IEEE Robotics and Automation, pp.1646-1652, 1990.
5. M.Vukobratovic and J. Stepanenko, "Mathematical Models of General Anthropomorphic Systems," Mathematical Bioscience 17, pp.191-242, 1973.
6. Kenneth J.Waldron, "Realization of Dynamic Biped Walking Stabilized with Trunk Motion Under Known External Force," Advanced Robotics, pp.299-310, 1989.
7. Atsuo Takanishi, "Robot Biped Walking Stabilized with Trunk Motion," Robots and Biological Systems: Towards a New Bionics, Springer-Verlag, pp.271-291, 1989.
8. J.Yamaguchi, A.Takanishi., I.Kato, "Development of a Biped Walking Robot Adapting to a Horizontally Uneven Surface," Proc. of the IEEE/RSJ Int. Conf. on IROS, pp.1156-1163, 1994.
9. Ambarish Goswami, "Postural Stability of Biped Robot and the Foot Rotation Indicator(FRI) Point," Int. J. of Robotics Research, Vol. 18, No. 6, pp. 523-533, 1999.
10. Ching-Long Shih, "Analysis of the Dynamics of a Biped Robot with Seven Degrees of Freedom," Proc. of IEEE Int. Conf. of R & A., pp. 3008-3013, 1996.
11. S.H. Lim, Jin G. Kim, "Adaptive Gait Algorithm for IWR Biped Robot," Int. Conf. on Power Electronics and Drive Systems., pp. 438-443, 1995.
12. D.E. Goldberg, Genetic Algorithm in Search, Optimization & Machine Learning, Addison-Wesley, 1989.

Effect of Fitness for the Evolution of Autonomous Robots in an Open-Environment

Md. Monirul Islam, S. Terao, and K. Murase

Department of Human and Artificial Inteligence Systems,
Fukui University, Fukui 910 8507, Japan,
{monir, terao, murase}@synapse.fuis.fukui-u.ac.jp

Abstract. The choice of a fittness function in artificial evolution has strong consequences on the evolvability of robots, dynamics of the evolutionary process, and ultimately on the outcome of the evolutionary process. In this paper, the effect of fitness functions for the evolution of autonomous robots to navigate in an open-environment by avoiding obstacles is studied. It is found that both the number and description of components of a fitness function affect the convergence of the evolutionary process. However, the performance of evolved robots in an unknown environment is greatly dependent on the description of components of a fitness function.

1 Introduction

Artificial evolution has been widely studied in order to dev elop the control systems of autonomous robots that can perform some useful tasks in partially unknown and unpredictable environments with minimal or without human in tervention. A population of robots, whose components (e.g., con trolsystem, morphology, etc.), is encoded on chromosomes in artificial evolution. According to a pre-defined fitness function, the fittest chromosomes (i.e., individuals) of a generation are selected to create a new population for the next generation [15].

The choice of a fitness function in artificial evolution has strong consequences on the evolvabilit y of the robot, dynamics of the evolutionary process, and ultimately on the outcome of the evolutionary process [7,8]. It is indeed a delicate issue for the evolution of autonomous robots in an *op en-enviornment* wherein each robot in a population may experience different environmental structures and/or conditions during the course of ev olution. Because of the environmental v ariabilit ythe best or worst performing robot at present generation in an *op en-enviornment* evolution may or may not be the best or worst for future generations. Selection pressures therefore may discard or retain some individuals that are good or bad for future generations.

A purely behavioral fitness function is generally used for the ev olution of autonomous robots. It is based only on components that evaluate the effects of behavior. However, an identical behavior of the same individual may get evaluated differently in different environments. This situation may arise for the practical

Y. Liu et al. (Eds.): ICES 2001, LNCS 2210, pp. 171–181, 2001.

application of autonomous robots to real-world problems where environmental conditions are rarely fixed. It has been argued that the design of more fine-grained fitness functions that capture the robot-environment interactions can be a complex task [13]. While few studies consider evolutionary processes under changing-environments (*open-environments*) [3, 17, 18], the effect of fitness functions for the evolution of autonomous robots in such environments is not studied.

In this paper, we study the effect of fitness function for the evolution of autonomous robots to navigate in an *open-environment* by avoiding obstacles. Four behavioral fitness functions are studied for this purpose. One fitness function consists of three components, while others consist of two components. Each fitness function differs from others with respect to number of components and/or description of components. The experimental results show that both the number and description of components affect the convergence of the evolutionary process. However, the performance of evolved robots in an unknown environment is greatly dependent on the description of components.

2   Evolution of the control system for autonomous robots

A large variety of control systems exists in the literature that have been put under evolution [1, 2, 6, 9, 10, 12]. These include feed-forward neural networks, dynamic time recurrent networks, classifier systems, Lisp codes etc. A typical cycle of the evolutionary process to develop the control system for an autonomous robot can be described by the following five steps.

Step i Create an initial population of n individuals where each individual represents a control system.

Step ii Decode each individual to a corresponding control system.

Step iii Allow a robot to perform the evolutionary task as result of the activity generated by the control system for a specified amount of time τ (commonly known as life-time), while its performance is recorded and accumulated according to a pre-designed fitness function. This step is repeated for n individuals.

Step iv Reproduce a number of children for each individual in the current generation based on its fitness value.

Step v Apply evolutionary parameters such as crossover and/or mutation to the children generated above to produce offspring for the next generation.

This cycle is repeated until the desired performance criteria or maximum number of generations has been reached. If any individual of a population during its life-time experiences different environmental conditions than other individuals and/or one generation has different environmental conditions than other generations for the same evolutionary task such evolutionary condition can be considered as *open-environment* condition. The evolutionary procedure such as genetic algorithms (GAs) uses both crossover and mutation operators to produce offspring for the next generation, while a mutation operator is only used by evolutionary programming (EP). However, it has been known that the crossover operator has a greater efficiency for adjusting to a changing environment [4, 5].

3   Fitness function

The evolutionary procedure requires a measure of the utility of its proposed solutions (the chromosomes). This measure is provided by a fitness function - equivalent to the objective function of the traditional optimization techniques. It has been asserted that although fitness cannot be directly measured its distribution in a population can be roughly estimated in any given environmental context on the basis of ecology and functional morphology of the organisms [11].

In artificial evolution, the fitness is defined as the ability to survive and reproduce in a specific environment. That means, it is used as a selection pressure, which is never wholly abated in nature. Indeed, the more normal circumstances are that selection pressures remain generally intense and drive phenotype as close to an optimum as possible. While some phenotypic characteristics are deeply ingrained in the developmental sequence of the organism, a constant pressure of selection, without which they would degenerate due to random genetic drift, maintains other adaptations [5].

A set of fitness functions described in this section forms the basis of experimental studies performed to demonstrate the effect of fitness functions for the evolution of autonomous robots. These functions consist of two or three components that only evaluate the effects of behavior. In other words, they are purely behavioral fitness functions. To formally describe fitness functions, consider an evolutionary task to develop the control system of an autonomous robot for navigating in an *open-environment* by avoiding obstacles. Assume that the robot is equipped with n number of sensors and two motors. The followings are the set of fitness functions.

$$f_1 = \sum_t V(t) \left(1 - \Delta v(t)\right) \left(1 - \sum_{i=1}^{n} s_i(t)\right) \tag{1}$$

$$f_2 = \sum_t \left(1 - \Delta v(t)\right) \left(1 - \sum_{i=1}^{n} s_i(t)\right) \tag{2}$$

$$f_3 = \sum_t V(t) \left(1 - \sum_{i=1}^{n} s_i(t)\right) \tag{3}$$

$$f_4 = \sum_t V(t) \left(1 - \Delta v(t)\right) \tag{4}$$

Here V is the average rotation speed of two motors and is used for rewarding fast controllers. Δv is the absolute value of the algebraic difference between the signed speed values of the motors (one direction is positive and the other is negative) and is used for rewarding straight-locomotion. s_i is the proximity measure of the ith sensor and is used for punishing the robot each time it sensed obstacles. All these values are normalized between 0 to 1.

The fitness function f_1 has been most widely used for avoiding obstacles in *close-environment* wherein each robot in a population can experience same environmental structures and conditions during the course of evolution [6, 7, 12, 16]. In order to understand the effect of fitness functions in the evolutionary process, three fitness functions (i.e., f_2, f_3 and f_4) are made by taking two components from f_1. We think that the performance comparison of these fitness functions will give a clear idea about the effect of fitness functions in the evolutionary process.

4   Experimental setup

A real mobile robot Khepera is used in this study. The structure and function of the robot has been well described elsewhere [14]. In short, the robot is equipped with eight infrared (IR) sensors (six in the front side and two in the rare side) and two dc motors. These sensors work as proximity sensors by detecting objects, emitting infrared light and measuring its reflection. They can also measure the ambient infrared light, which in normal conditions is a rough measure of the local ambient light intensity.

The environment used to accomplish the evolutionary processes is consists of a square area with few obstacles (Fig. 1). The size of the square area is approximately 60 × 60 cm. The walls and obstacles are made of wood and are covered by off-white paper. The floor is made of gray paperboard. Four lights with different intensities illuminate different places in the square environment (Fig. 1). The robot can sense the walls and obstacles within 1 to 5 cm by IR proximity sensors depending on the intensity of lights.

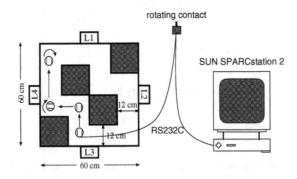

Fig. 1. Experimental step up used for evolution. L1, L2, L3, and L4 are four lights with different intensities illuminate different places in the square environment

The task given to the robot is to navigate by avoiding obstacles in the environment. The task is non trivial because the robot should behave differently at four different light conditions. For example, the robot should move very carefully in the less lighted area because it perceives the obstacles only in their close proximity. In contrast, if the robot lives in the lighted area it can perceive obstacles far way.

A simple two-layered feed-forward neural network is used as a control system to produce control signals for the robot (Fig. 2). In order to keep the system complexity at a manageable level, we use a fixed network architecture in which only connection weights are put for evolution. A direct coding scheme is used to encode the connection weights on to the chromosome [19]. Each chromosome in the population has the same constant length corresponding to the number of connections and biases of the neural network.

Each weight is encoded on a gene of five bits where the first bit determines the sign of the weight and the remaining four bits its strength. Since the neural network has a total of 18 connections and two biases, the length of chromosome became 100 bits. The network produced two output signals to motors by summing up all the values from eight sensors with evolved connection strength. That is, each output was generated by

$$S_p = S_b + G \sum w_i x_i \qquad (5)$$

where, S_p, S_b, G, w_i and x_i represent the output value to the motor, the base navigation speed of the motor, global gain, connection strength and sensor signals value, respectively. The values for S_b and G are set to 5 and $1/1600$, respectively. The global gain determines the sensitivity to the modulation signal from sensors. The values from proximity sensors are 0-1023.

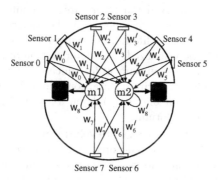

Fig. 2. A two-layered neural network to produce control signals for a real mobile robot Khepera. Note that bias connections are not shown

5   Experiments

Four sets of control experiments, using four different fitness functions described in section 3, are carried out. Each set of experiments consists of three runs using different initial seed, crossover and mutation probabilities. All results presented in this section are the average over three runs. In all control experiments, the initial population is created randomly. The life-time of each individual in a population is chosen short (10 sec) by which a robot can travel only a portion of the evolved-environment (Fig. 1). Thus, different robots in a population may experience different environmental conditions. In this sense, the evolution with such a short life-time can be considered as *open-environment* evolution. We think that the test environment with such a short life-time of each individual in a population can be a model of an *open-environment*. The other parameters used for evolution are given in table 1.

Table 1. GA parameters used for evolution

Population size	30
Number of generations	75
Chromosome length	100
Crossover probability	0.5-0.7
Mutation probability	0.002-0.004
Elite preservation	0.3-0.5
Initial weight range	±0.5
Final weight range	Not bound

The fitness results reported in Fig. 3 show the evolutionary processes of control experiments. It is observed that there are no significant improvements of fitness values during the course of evolution rather deteriorate (in some cases). This may be due to the short life-time of individuals in which any significant improvement may not be visualized. The deterioration of fitness values may be due to the variability of the environment for which the same robot may get evaluated differently in different environmental conditions. Because the performance of an autonomous robot not only depends on its control network but also the environment where it is performing.

In order to observe the progress of evolutionary processes and the performance of evolved robots, all evolved-populations of control experiments are tested in the evolved-environment for a period of 50 sec (life-time), by which a robot can travel the whole environment. The progress of evolutionary process is measured by the fitness value (Fig. 4(a)). However, the performance is measured by sensor values that indicate the distance between a robot and obstacles (Fig. 4(b)). During navigation the values of eight proximity sensors of the robot were sampled every 0.1 sec and were accumulated for a period of 50 sec.

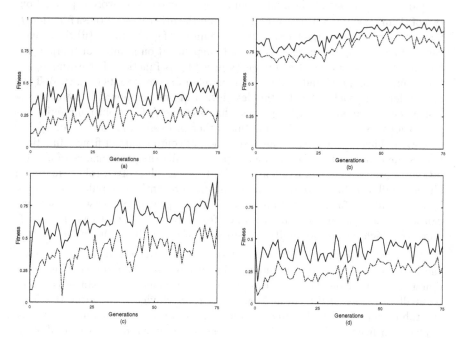

Fig. 3. Population average fitness and best individual fitness at each generation of four evolutionary processes. Figures (a), (b), (c) and (d) use f_1, f_2, f_3 and f_4 fitness functions, respectively, for evolution. Solid and dashed lines represent the best individual and population average fitness values

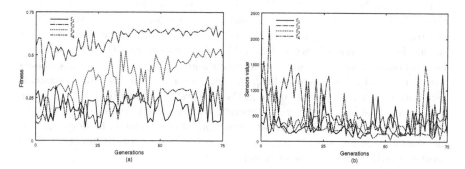

Fig. 4. Testing of four populations, evolved by using f_1, f_2, f_3 and f_4 fitness functions, for a period of 50 sec in the evolved-environment. (a) Population average fitness; (b) Population average sensors value

Three observations can be made from Fig. 4: (i) among four evolved-populations two of them, one uses a fitness function f_2 and the other uses a fitness function f_3 for evolution, are converged for an allotted number of generations; (ii) the convergence of a evolutionary process is not only dependent on number of components but also the description of components in a fitness function. For example, the fitness functions f_2, f_3 and f_4 have same number of components (i.e., two). They only differ from each other by the description of one component. However, the convergence of an evolutionary process that uses the f_2 for evolution require less generations to reach the best value than other processes that use f_3 and f_4 for evolution. This indicates that the description of components in fitness functions affects the convergence of evolutionary process; (iii) the performance of all evolutionary processes is almost identical. That means, all evolved robots maintain almost similar distance from obstacles. Most importantly, a population, evolved by using a fitness function f_4 that does not have a proximity measure of sensors, can also maintain almost same distance from obstacles like others (i.e., f_1, f_2 and f_3), which have a proximity measure of sensors in fitness functions.

We have performed another set of testing experiments in order to assess the performance of evolved-population in an unknown environment, known as dynamic environment (Fig. 5). In dynamic environment, an obstacle is placed manually for 5 sec at 6-8 cm in front of the robot navigating. During this 5 sec the robot can move freely after avoiding the obstacle. Then again an obstacle is placed in front of the robot. This process is continued for the whole life-time of the robot. All evolved-individuals in the final generation of four evolved-populations are allowed to navigate in the dynamic environment. Like other experiments, the values of eight proximity sensors were sampled every 0.1 sec and were accumulated for a period of 250 sec (life-time).

Table 2 shows the population average sensor value of each evolved-population in the dynamic environment. It is seen that the performance of an evolved-population that uses a fitness function f_2 is worst (i.e., highest sensors value) than that of other populations. This is because the use of a fitness function f_2 in the evolutionary process encourages only a straight-locomotion in the evolved-environment, which may or may not be suitable for other environments. Due to the short-life time of each individual in a population, robots with a straight-locomotion thus may dominate in the evolutionary process. That means, robots with other kind behaviors may disappear from a population. By using a straight-locomotion, a robot can only move in the forward or backward direction. When a robot, evolved by using a fitness function f_2, faces obstacles in front of him, he avoids them by moving in backward direction. However, when he does not find any obstacles in his back, he again starts moving in the forward direction. Thus most of the time his front sensors were active.

In contrast, other fitness functions allow a robot to develop various kinds of locomotion, such as turning left and moving forward, during the course of evolution. The development of various kinds of locomotion may be suitable for a robot to avoid obstacles in the unknown environment. The results of these experiments indicate that the description of components in a fitness function is more important than number of components. The use of explicit descriptions for components (e.g., move straight only) in a fitness function indicates that one knows in advance what makes the evolved solution suitable for future. In this paper, we argue that the components of a fitness function should be described implicitly (e.g., move) rather than explicitly (e.g., move straight only). Because if an important aspect of behaviors is absent during the evolution of autonomous robots, they will most likely fail to produce appropriate behaviors in changing environments.

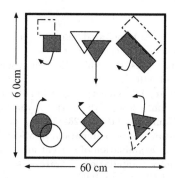

Fig. 5. A Dynamic environment

Table 2. Performance comparison of four populations, evolved by using f_1, f_2, f_3 and f_4 fitness functions, in a dynamic environment. The performance is measured by the sensors value of the robot. The highest value represents the worst performance

Environment	Performance of a population evolved by using a fitness function			
	f_1	f_2	f_3	f_4
Dynamic	1234.5	3054.7	1112.3	1498.3

6 Conclusions

We study the effect of fitness functions for the evolution of autonomous robots in an *open-environment*. Four behavioral fitness functions are used for the evolution of autonomous robots in an *open-environment*. Each fitness function differs from others with respect to number and/or description of components in fitness functions. The experimental results show that both the number and description of components of a fitness function affect the evolutionary process. Testing of evolved-robots in an unknown environment reveals that the performance of evolved-robots is mostly dependent on the description of components of a fitness function. It is, therefore, better to define components of a fitness function implicitly rather than explicitly. The implicit description of components in a fitness function may help to develop various kinds of behaviors during the course of evolution. The development of various kinds of behaviors may be useful for a robot to adapt in changing environments.

References

1. Beer, R.D., Gallagher, J.C.: Evolving dynamical neural networks for adaptive behavior. Adaptive Behavior **1**(1) (1992) 91-122
2. Cliff, D., Harvey, I., Husbands, P.: Explorations in evolutionary robotics. Adaptive Behavior **2**(1) (1993) 73–110
3. Cobb, H.G., Grefenstette, J.J.: Genetic algorithms for tracking changing environments. Genetic Algorithms: Proc. of the Fifth International Conference (ICGA93), San Mateo: Morgan Kaufmann (1993) 523-530
4. Crow, J.F.: The importance of recombination. In Michod, R.E., Levin, B.R. (eds.): The evolution of sex: an examination of current ideas (1998) 56-73
5. Fogel, D.B.: Evolutionary computation: Toward a new philosophy of machine intelligence. New York: IEEE Press (1995)
6. Floreano, D., Mondada, F.: Evolution of homing navigation in a real mobile robot. IEEE Transactions on Systems, Man, and Cybernetics–Part B: Cybernetics **26**(3) (1996) 396-407
7. Floreano, D., Mondada, F.: Evolutionary neurocontrollers for autonomous mobile robots. Neural Networks **11** (1998) 1461-1478
8. Floreano, D., Urzelai, J.: Evolutionary Robots with on-line self-organization and behavioral fitness. Neural Networks **13** (2000) 431-443
9. Harvey, I., Husbands, P., Cliff, D.: Issues in evolutionary robotics. In: Meyer, J.-A., Roitblat, H., Wilson, S. (eds.): From Animals to Animats 2: Proc. of the Second International Conference on Simulation of Adaptive Behavior (SAB92) (1992) 364-373
10. Harvey, I., Husbands, P., Cliff, D.: Seeing the light: artificial evolution, real vision. In: Cliff, D., Husbands, P., Meyer, J.-A., Wilson, S. (eds): From Animals to Animats 3, Proc. of the Third International Conference on Simulation of Adaptive Behavior (SAB'94) (1994) 392-401
11. Hoffman, A.: Arguments on evolution: A paleontologist's perspective. New York: Oxford University Press (1989)
12. Kodjabachian, J., Meyer, J.A.: Evolution and development of neural controllers for locomotion, gradient-Following, and obstacle-Avoidance in artificial insects. IEEE Transactions on Neural Networks **9** (1998) 796-812
13. Mataric, M., Cliff, D.: Challenges in evolving controllers for physical robots. Robotics and Autonomous Systems **19**(1) (1996) 67–83
14. Mondada, F., Franzi, E., Ienne, P.: Mobile robot miniaturisation: A tool for investigation in control algorithm. Proc. of third international symposium on experimental robotics, Kyoto, Japan (1993) 501-513
15. Nolfi, S., Floreano, D.: Evolutionary Robotics: The Biology, Intelligence, and Technology of Self-Organizing Machines. Cambridge, MA: MIT Press (2000)

16. Odagiri, R., Monirul Islam, Md., Okura, M., Asai, T., Murase, K.: Deterministic chaos in sensory information of real mobile robot Khepera. In: Loffer, A., Mondada, F., Ruckert, U. (eds.): The 1st International Khepera Workshop (1999) 49-56
17. Pettit, E., Swigger, K.M.: An analysis of genetic-based pattern tracking and cognitive-based component models of adaptation. Proc. of National Conference on AI (1983) 327-332
18. Sasaki, T., Tokoro, M.: Evolving learnable neural networks under changing environments with various rates of inheritance of acquired characters: Comparison between Darwinian and Lamarckian evolution. Artificial Life 5(1999) 203-223
19. Yao, X: A review of evolutionary artificial neural networks. International Journal of Intelligent Systems 8 (1993) 539-567

Incremental Evolution of Autonomous Robots for a Complex Task

Md. Monirul Islam, S. Terao, and K. Murase

Department of Human and Artificial Inteligence Systems,
Fukui University, Fukui 910 8507, Japan,
{monir, terao, murase}@synapse.fuis.fukui-u.ac.jp

Abstract. An incremental approach is used to develop the control system of an autonomous robot to approach toward the target object by avoiding obstacles in an environment. The approach consists of two-stage evolution. In the first-stage, controllers are evolved to avoid obstacles in an environment. The final population of the first-stage evolution is then used as the initial population for the second-stage evolution. Controllers are evolved in the second-stage to approach toward the target object by avoiding obstacles in the environment. We compare the performance of the incremental approach with that of conventional approach, an one-stage evolutionary approach. It is found that the performance of incremental approach is better than that of conventional approach.

1 Introduction

Artificial evolution has been widely studied in order to develop the control systems of autonomous robots that can perform some useful tasks in partially unknown and unpredictable environments with minimal or without human intervention. A population of robots, whose components (e.g., controlsystem, morphology, etc.), is encoded on chromosomes in artificial evolution. According to a pre-defined fitness function, the fittest chromosomes (i.e., individuals) of a generation are selected to create a new population for the next generation [16].

A number of works have successfully employed artificial evolution to develop the control systems for simulated and real robots [1, 2, 5, 13, 17]. However, most of the control systems are developed for simple tasks, such as avoiding obstacles and reaching a target area. It has been known that difficulties encounter when control systems are sought for complex tasks [8, 19]. Recently, the so-called incremental approach has been used to develop the control systems of autonomous robots for complex tasks. In incremental approach, control systems are developed through successive stages in which good solutions to a simpler version of a given problem are used iteratively to seed the initial population of solutions likely to solve a harder version of the same problem.

Despite some initial theoretical work about incremental approach [9, 11], there are few attempts to apply the approach for the evolution of autonomous robots for complex tasks. Harvey et al. reported that evolving a neural network controller to visually guide a robot toward a small target in an environment

Y. Liu et al. (Eds.): ICES 2001, LNCS 2210, pp. 182–191, 2001.

took less total computational effort if the controllers were first evolved using a larger target [10, 12]. Recently, Floreano and Mondada evolved a controller for a large robot from a controller of a small robot to avoid obstacles in an environment [6]. The use of incremental approach for such simple tasks, which can easily be accomplished by conventional evolutionary approach, does not help much to understand the necessity and utility of incremental approach. Further understanding of incremental approach for its application to complex tasks is necessary.

This paper uses an incremental approach to evolve the control system of an autonomous robot for a complex task, approach toward the observed target by avoiding obstacles in the environment. Animal psychology literature has intensively investigated the task in several species of mammals [14, 18]. The difficulty of the task is that it is contradictory, especially in the early stage of evolution, to generate both behaviors, approaching toward the observed target and losing the target from sight to avoid obstacles. In this study, the control system is incrementally evolved in two-stage. In the first-stage, controllers are evolved to avoid obstacles in the environment by using sensory information of the robot. Controllers are evolved in the second-stage to approach toward the observed target by utilizing acquired behaviors of the first-stage evolution and visual information of the robot.

Fig. 1. The basic structure of a real mobile robot Khepera (*left*). A vision module K213 is plugged in directly on the top of the Khepera robot (*right*)

2   Experimental setup

A real mobile robot Khepera is used in this study. The structure and function of the robot has been well described elsewhere [15]. In short, the robot is equipped with eight infrared (IR) sensors (six in the front side and two in the rare side) and two dc motors (Fig. 1, left). These sensors work as proximity sensors by detecting object, emitting infrared light and measuring its reflection. They can

also measure the ambient infrared light which in normal conditions is a rough
measure of the local ambient light intensity. Several complete modules such as
a vision module and a gripper module can be added to the basic structure.

A vision module K213, which can be supported and managed by the control
protocol available on the basic structure of the robot, is plugged in directly on
top of the basic structure for visual observation (Fig. 1, right). It consists of
a 1D-array of 64 photo-receptors that provide a linear image composed of 64
pixels of 256 gray-levels each. The view angle of the vision module is of about
36° (Fig. 2). However, the vision module also allows detection of the position
in the image corresponding to the pixel with maximal intensity. We used this
facility by dividing the visual field into four sectors of 9° each (Fig. 2).

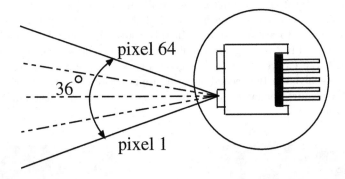

Fig. 2. The vision module K213 of the Khepera robot. The visual angle of the module
is divided into four sectors

The environment used to accomplish the evolutionary processes is consists
of a square area with three obstacles (square boxes) and a target object (circle)
(Fig. 3). The size of the area is approximately 60×60 cm. The walls and obstacles
are made of wood and are covered by off-white paper. The target object is also
made of wood but covered by black-white paper. The height of obstacles is made
of short so that only sensors of the robot can detect them. In contrast, the target
object is made of tall so that the vision module of the robot can see it (within
visual angle). The environment is always illuminated from above by a 60-watt
light bulb.

A simple two-layered feed-forward neural network is used as a control system
to produce control signals for the robot. The connection weights of the neural
networks are put for evolution. A direct coding scheme is used to encode the
connection weights on to the chromosome [20]. Each chromosome in the population has a different length corresponding to the number of connections and
biases of the neural network.

Each weight is encoded on a gene of five bits where the first bit determines the
sign of the weight and the remaining four bits its strength. The network produced
two control signals to motors by summing up the values from IR sensors and

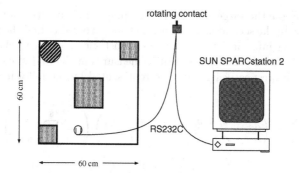

Fig. 3. The environment used for evolution. Square and circular objects in the environment represent obstacle and target object, respectively

vision fields. Here it is worth mentioning that only sensors values are used in the first-stage evolution, while both sensors and visions values are utilized in the second-stage evolution. Each motor output is generated by

$$S_p = S_b + G \sum w_i x_i \qquad (1)$$

where S_p, S_b, G, w_i and x_i represent the output value to the motor, the base navigation speed of the motor, global gain, connection strength and sensor or vision signals value, respectively. The value of S_b and G are set to 5 and 1/1600 (for first-stage evolution) or 1/2200 (for second-stage evolution), respectively. The global gain G determines the sensitivity to the modulation signal from sensors and vision fields. The values from sensors and vision fields are 0-1023 and 0-254, respectively.

3   Evolutionary method

The species adaptation genetic algorithm (SAGA) [9, 11] is used to develop the control system of an autonomous robot to solve successively two problems of increasing difficulty. While most genetic algorithms are essentially performing optimization in a fixed parameter space, SAGA allows for the dimensionality of the parameter space to be under evolutionary control, by employing variable-length genotypes. In terms of neural networks, this means we are able to start with a population of robots each of which has a minimum number of units: extra units may be introduced by mutation and will be retained if they increase the evolutionary success of the mutated robot. This procedure for generating control networks is really incremental [3].

The evolutionary process is incrementally carried out in two-stage by using SAGA. In the first-stage, controllers are evolved to acquire obstacle-avoidance behaviors by using sensory information of the robot. A vision module, by which the robot can see the target object, is then plugged in directly on the top of the robot's body. In the second-stage, controllers are evolved to find the target object by using vision information and acquired obstacle-avoidance behaviors of the first-stage. F_1 and F_2 are used as fitness measures in the first-stage and second-stage evolution, respectively. They are represented by the following equations.

$$F_1 = \left(0.5 + \frac{v_l + v_r}{4.v_{max}}\right)\left(1 - \frac{|v_l - v_r|}{2.v_{max}}\right)\left(1 - \sum_{i=1}^{8} \frac{s_i}{s_{max}}\right) \tag{2}$$

$$F_2 = \frac{1}{4}\sum_{i=1}^{4}\sum_{j=1}^{16} \frac{|p_i(j) - p_i(j+1)|}{p_{max}} \tag{3}$$

Here v_l and v_r are the velocity of the left and right motor. v_{max} is the maximum velocity. s_i is the proximity measure of the ith IR sensor and s_{max} is its maximum value. The first factor of equ. (2) rewards the fast controllers, the second factor encourages straight locomotion, and the third factor punishes the robot each time it sensed obstacles. $p_i(j)$ represents the visual response of the jth pixel of the ith vision sector and p_{max} is its maximum value. When the robot is closer to the target-object, the magnitude of visual responses is larger. Because the image is taken at a time when the vision turret itself emits infrared light. Therefore, we took the sum of absolute changes between neighboring pixels as the measure of distance to the object. The factor of equ. (3) rewards the robot for its closeness to the target object.

4   Experiments

A set of control experiments, consisting of three runs, is performed by using incremental approach. Two populations, each consists of 30 individuals, are e-volved for 75 generations in each experiment. One population is initially generated randomly and is used for the first-stage evolution of incremental approach. The other population uses the final population of the first-stage evolution as an initial population and is used for the second-stage evolution. The life times of each individual of the first-stage and second-stage evolutions are set to 5 sec and 3 sec, respectively. The mutation rate, and the elite preservation rate in the roulette selection used in the SAGA operation are 0.002-0.004, and 0.3-0.5, respectively. All results presented in this section represent the average over three runs.

The fitness results reported in Fig. 4 show the evolutionary processes of the first-stage evolution (graph on the left) and the second-stage evolution (graph on the right) of the incremental approach. Both the obstacle-avoidance and approaching toward the target object abilities that robots acquired in the first-stage evolution and the second-stage evolution, respectively, are improved along the course of evolution. However, it is not clear from Fig. 4 whether the acquired obstacle-avoidance ability is utilized or lost by the second-stage evolution. Thus, two final populations, one evolved by the first-stage evolution and the other by the second-stage evolution, are tested in the evolved-environment (Fig. 3) without a target object and in an unknown environment. The unknown environment is created by rearranging the obstacles of the evolved-environment. The task given to the robot in testing experiments is to avoid obstacles.

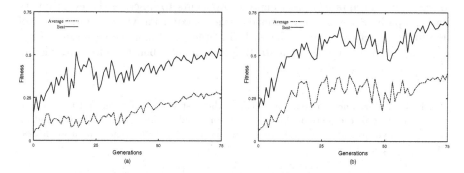

Fig. 4. Evolutionary processes of the incremental approach: (a) the evolutionary process of the first-stage evolution of the incremental approach; (b) the evolutionary process of the second-stage evolution of the incremental approach

The performance of testing experiments is measured according to F_1 (i.e., the fitness function of first-stage evolution). Table 1 shows average population fitness of testing experiments. It is seen that the first-stage evolution reports slightly higher fitness value than that of the second-stage evolution. The reason of a lower fitness value by the second-stage evolution is that the task given in the second-stage evolution and testing experiments is not same. In addition, the second-stage evolution may slightly change the acquired obstacle-avoidance behaviors of the first-stage evolution for finding the target object (Fig. 5). The navigation of the robot evolved by first-stage evolution was smooth (Fig. 5(a)), while it was some sorts of zig-zag for the second-stage evolution (Fig. 5(b)). The zig-zag movement of a robot evolved by the second-stage evolution might be for searching the target object that was the goal of the second-stage evolution.

In order to assess the benefit of incremental approach, an additional set of control experiments by using a conventional approach, one-stage evolution, is carried out. In conventional approach, controllers are evolved to acquire obstacle-avoidance behaviors and to find the target object simultaneously. The fitness function used for evolution by the conventional approach is F_3, where $F_3 = \frac{F_1+F_2}{2}$. That means, the fitness function of the conventional approach is equal to the average of two fitness functions used in the incremental approach. The lifetime of each individual of the conventional approach is set to 8 sec for making fair comparison. Note that the life times of each individual of the first-stage and second-stage evolution of the incremental approach are 5 sec and 3 sec, respectively.

Table 2 compares the performance of the conventional approach and the second-stage evolution of the incremental approach with respect to F_3. The entry of second column in the table represents the average population fitness of the final generation. It is seen that the performance of the conventional approach is much worse than that of incremental approach. There might be two reasons for the worse performance of the conventional approach. First, the fitness function of conventional approach uses many components to describe the evolutionary task properly. Second, one component (i.e., maximizing vision information) of the fitness function is conflicting with the other (i.e., minimizing sensory information). It has been found that starting from small is beneficial for learning and development of neural networks [4].

Table 1. Performance comparison between the first-stage evolution and the second-stage evolution of the incremental approach for avoiding-obstacles in a known and an unknown environments. The performance is measured by the average population fitness of the final generation

Evolution	Performance to avoid obstacles in the	
	known environment	unknown environment
First-stage	0.27130	0.2251
Second-stage	0.23743	0.2037

Table 2. Performance comparison between the second-stage evolution of incremental approach and conventional approach evolution to reach the target object by avoiding-obstacles in an environment. The performance is measured by the average population fitness of the final generation

Evolutionary approach	Performance to approach toward the target object by avoiding obstacles in the environment
Conventional	0.1645
Incremental	0.2531

5   Genetic diversity

The genetic diversity refers to the variation of genes within species. In all organisms that reproduce sexually, each individual plant or animal contains a different mix of genes. The variation within species allows a population to adapt to changes in climate and other local environmental conditions. When a population loses diversity, it becomes genetically uniform and far less adaptable in changing conditions.

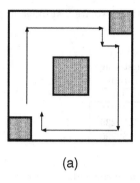
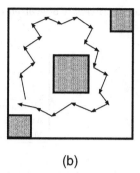

(a)   (b)

Fig. 5. Trajectories of two evolved-robots for avoiding obstacles in an environment: (a) trajectory of a robot evolved by the first-stage evolution of incremental approach; (b) trajectory of a robot evolved by second-stage evolution of incremental approach

The aim of the this section is to determine the effect of genetic diversity on the performance of incremental approach. The genetic diversity is dependent on various factors, such as the diversity of initial population and selection pressure. In the first-stage, the diversity of an initial population is dependent on random initialization of the initial population. In the second-stage, the diversity of an initial population is dependent on the diversity in the evolved-population of the first-stage evolution. Generally, the final evolved-population of the first-stage evolution is used as the initial population for the second-stage evolution. In this study, we are interested to examine the effect of diversity in the initial population of the second-stage evolution for the performance of incremental approach.

To examine the effect of the initial population of the second-stage evolution, the evolved populations at generations 40, 50, and 60 of the first-stage evolution are used as initial populations for the second-stage evolution. Table 3 shows the diversity of the evolved populations of the first-stage evolution at generations 40, 50, and 60. According to [7], the genetic diversity is measured as the average dispersion of individual vectors from the center of mass of the population and further normalized by the string length. Three sets of control experiments using these initial populations are carried out. Each set of experiment consists of three runs. Table 4 shows the fitness performance of the second-stage evolution that uses three different initial populations for evolution. It is seen that the diversity of the initial population does not affect much on the performance of the second-stage evolution. Rather, the performance of the second-stage evolution is greatly dependent on which of the evolved-population of the first-stage evolution is used as an initial population.

6   Conclusion

We have applied incremental approach, a two-stage evolutionary system, to develop the control system of an autonomous robot for a complex task. The exper-

Table 3. Diversity of evolved-populations of the first-stage evolution at different generations

	Diversity of the evolved -population of the first-stage evolution at generation		
	40	50	60
Minimum	0.49130	0.48715	0.48215
Average	0.51078	0.50321	0.50239
Maximum	0.53235	0.52223	0.52230

Table 4. Performance of the second-stage evolution of incremental approach using different initial populations. Evolved-populations of the first-stage evolution at generations 40, 50 and 60 are used as initial populations

	Performance of second-stage evolution using the evolved population of the first-stage evolution at generation		
	40 as an initial population	50 as an initial population	60 as an initial population
Minimum	0.57317	0.48701	0.51130
Average	0.63249	0.56891	0.60245
Maximum	0.87215	0.71230	0.7419

imental results show that the acquired behaviors of the first-stage evolution are slightly modified and utilized by the second-stage evolution. The performance comparison with conventional (one-stage) evolutionary approach shows that the incremental approach outperforms conventional approach. Our experiments have yielded some interesting results, e.g., that the genetic diversity in the initial population of the second-stage evolution does not affect much on the performance of incremental approach. Rather, the performance is greatly dependent on which of the evolved-population of the first-stage evolution is used as an initial population for the second stage evolution. We think that the genetic diversity of a population can be maintained by using learning with evolution together. Thus, more work is necessary in order to determine more precisely the relationship between intermediate evolution and the performance of the incremental approach.

References

1. Beer, R.D., Gallagher, J.C.: Evolving dynamical neural networks for adaptive behavior. Adaptive Behavior **1**(1) (1992) 91-122
2. Cliff, D., Harvey, I., Husbands, P.: Explorations in evolutionary robotics. Adaptive Behavior **2**(1) (1993) 73–110
3. Cliff, D., Harvey, I., Husbands, P.: Incremental evolution of neural network architectures for adaptive behavior. In: Verleysen, M.(ed.): Proc. of the First European Symposium on Artificial Neural Networks (1993) 39-44

4. Elman, J.L.: Learning and development in neural networks: The importance of starting small. Cognition **48** (1993) 71-99
5. Floreano, D., Mondada, F.: Evolution of homing navigation in a real mobile robot. IEEE Transactions on Systems, Man, and Cybernetics–Part B: Cybernetics **26**(3) (1996) 396-407
6. Floreano, D., Mondada, F.: Evolutionary neurocontrollers for autonomous mobile robots. Neural Networks **11** (1998) 1461-1478
7. Floreano, D., Urzelai, J.: Evolution of adaptive-synapse controllers. In: Floreano, D., Nicoud, J-D., Mondada, F. (eds.): Advances in Artificial Life - ECAL99, Berlin: Springer Verlag (1999) 183-194
8. Gomez, F., Miikkulainen, R.: Incremental evolution of complex general behavior. Adaptive Behavior **5** (1997) 317-342
9. Harvey, I.: Species adaptation genetic algorithms: A basis for a continuing SAGA. In: Varela, F.J., Bourgine, P. (eds.): Toward a practice of autonomous systems: Proc. of the First European Conference on Artificial Life (1992) 346-354
10. Harvey, I., Husbands, P., Cliff, D.: Issues in evolutionary robotics. In: Meyer, J.-A., Roitblat, H., Wilson, S. (eds.): From Animals to Animats 2: Proc. of the Second International Conference on Simulation of Adaptive Behavior (SAB92) (1992) 364-373
11. Harvey, I.: Evolutionary robotics and SAGA: the case for hill crawling and tournament selection. In: Langton, C. (ed.): Artificial Life III **XVII** (1994) 299-326
12. Harvey, I., Husbands, P., Cliff, D.: Seeing the light: artificial evolution, real vision. In: Cliff, D., Husbands, P., Meyer, J.-A., Wilson, S. (eds): From Animals to Animats 3, Proc. of the Third International Conference on Simulation of Adaptive Behavior (SAB'94) (1994) 392-401
13. Kodjabachian, J., Meyer, J.A.: Evolution and development of neural controllers for locomotion, gradient-Following, and obstacle-Avoidance in artificial insects. IEEE Transactions on Neural Networks **9** (1998) 796-812
14. Koelher, W.: The mentality of apes. Harcount Brace, New York (1925)
15. Mondada, F., Franzi, E., Ienne, P.: Mobile robot miniaturization: A tool for investigation in control algorithm. Proc. of the Third International Symposium on Experimental Robotics, Kyoto, Japan (1993) 501-513
16. Nolfi, S., Floreano, D.: Evolutionary Robotics: The Biology, Intelligence, and Technology of Self-Organizing Machines. Cambridge, MA: MIT Press (2000)
17. Nolfi, S.: Evolving non-trivial behaviors on real robots: a garbage collecting robot. Robotics and Autonomous System **22** (1997) 187-198
18. Thinus-Blanc, C.: Animal spatial cognition. World Scientific, Singapore (1996)
19. Winkeler, J.F., Manjunath, B.S.: Incremental evolution in genetic programming. In: Koza, J.R., Banzhaf, W., Chellapilla, K., Deb, K., Dorigo, M., Fogel, D.B., Garzon, M.H., Goldberg, D.E., Iba, H., Riolo, R. (eds): Genetic Programming 1998: Proc. of the Third Annual Conference, Morgan Kaufmann (1998) 403-411
20. Yao, X: A review of evolutionary artificial neural networks. International Journal of Intelligent Systems **8** (1993) 539-567

Multi-agent Robot Learning
by Means of Genetic Programming:
Solving an Escape Problem

Kohsuke Yanai and Hitoshi Iba

Dept. of Frontier Informatics, School of Frontier Science,
The University of Tokyo,
{yanai,iba}@miv.t.u-tokyo.ac.jp

Abstract. This paper presents the emergence of the cooperative behavior for multiple robot agents by means of Genetic Programming (GP). For this purpose, we utilize several extended mechanisms of GP, i.e., (1) a co-evolutionary breeding strategy, (2) a controlling strategy of introns, which are non-executed code segments dependent upon the situation, and (3) a subroutine discovery technique. Our experimental domain is an escape problem. We have chosen the actual experimental settings so as to be close to a real world as much as possible. The validness of our approach is discussed with comparative experiments using other methods, i.e., Q-learning and Neural networks, which shows the superiority of GP-based multi-agent learning.

1 Introduction

Recently intelligent agents and multi-agent systems have attracted much interest in Distributed Artificial Intelligence (DAI). GP (Genetic Programming) and its variants have been applied to the multi-agent learning (see [Haynes *et al*.95], [Luke *et al*.96], [Iba96] and [Hara *et al*.99] for example). However, in the multi-agent application, the computational burden is often problematic. This is because the number of GP trees required for the multi-agent task becomes larger with the number of agents. As a result, there are very few researches on the real evolutionary robots by GP.

We have proposed several extended mechanisms of GP for the sake of the multi-agent learning in the following points:

1. The co-evolutionary breeding strategy [Iba96].
   Breeding is performed in the same way like in a distributed GP (see Fig. 1). As generations proceed, some individuals are expected to perform specialized tasks for different agents. We evaluate the fitness of individuals in an agent-type subpopulation as follows: Initially, i.e., at the first generation, the other agents' programs are chosen randomly. At successive generations, we choose, as the other agents' program, the best programs evolved so far

Y. Liu et al. (Eds.): ICES 2001, LNCS 2210, pp. 192–203, 2001.

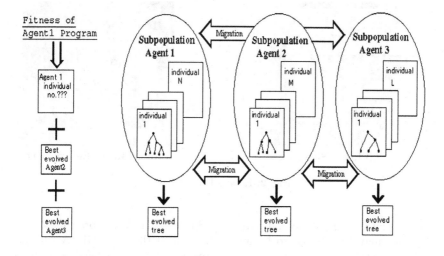

Fig. 1. Co-evolutionary Breeding for Multi-Agent Learning.

in the other agent-type subpopulations. In our previous papers ([Iba96] and [Iba98]), we have empirically shown the superiority of the co-evolutionary breeding over the traditional strategies, such as the homogeneous breeding[1] and the heterogeneous breeding[2].

2. The controlling method of effective introns [Iba and Terao00].
   "Effective introns" are non-functional code segments of a GP tree dependent upon the execution. For sake of identification, we attach an execution counter to each terminal or function symbol in a GP tree. This counter is put equal to zero at the outset of the execution. During the evaluation of the tree, the counter is incremented when its symbol is evaluated. The symbol whose counter remains zero is regarded as an effective intron, which can be removed by an edit operator. We confirmed that (1) the fitness transition is improved during the training phase, (2) the code growth is effectively controlled, and (3) the robustness of acquired programs is increased.

3. The subroutine discovery for re-use [Hondo et al.96a],[Hondo et al.96b].
   The aim is to enhance the robustness of GP by storing a certain number of subroutines of GP subtrees for re-use. The algorithms work in two main parts. The former part is based on ADF version of GP, which generates the main solutions and subroutines. The latter part stores the effective subroutines acquired by this ADF mechanism. These acquired subroutines are

---

[1] In the homogeneous breeding strategy, all agents use the same program evolved by GP.

[2] Each agent uses a distinct program in the heterogeneous strategy.

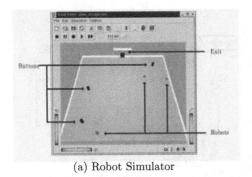
(a) Robot Simulator

(b) Actual Robot Environment

Fig. 2. Escape Problem.

drawn into a library according to a certain criterion. Any subroutine in the library may be referred to by any individual in the main GP. A subroutine is given its fitness value according to the individuals which call it. The subroutine which has the lowest fitness is discarded and a new subroutine, generated by the ADF, is added to the library.

In this paper, we apply the proposed GP system to a multi-agent task with real robots, i.e., Khepara robots. The target task we have chosen is an "escape problem", in which robot agents are supposed to leave a room through a door (or hole) in case of emergency, such as a fire (Fig. 2). However, in order to open the exit door, they have to push all the buttons. This task consists of (1) pushing buttons and (2) escaping from the room through the door. In this paper, we assume robot agents have different speeds and sensors. Thus, the effective cooperation among robots is essential for the efficient solution. The action program can be taken as a program for robots required to "collect valuables in the event of fire and go to the door."

The rest of this paper is structured as follows. Section 2 describes the experimental set-ups for the escape problem with actual robots. Section 3 explains the experimental results and the performance is compared with traditional methods, such as Q-learning and neural networks, which shows the superiority of our approach. Section 5 discusses our approach, followed by some conclusion in Section 6.

2   Application to Actual Robots

2.1   Steps from Simulation to Actual Robots

If simulators are used for learning, the robots can obtain any information and learn smoothly, but this merit becomes a problem in the transition from simulators to actual machine robots.

Although many studies assume that robots recognize their own locations, we try to avoid using information on the coordinates of robot locations. This is because a considerably large-scale device is needed for actual robots to know their own absolute coordinates. In recognizing the external world, only the slight modifications of image outputs of cameras and the values of infrared sensor output are handled. In other words, the above processing is at the level of physical devices such as cameras and infrared sensors. This lowers barriers arising in application to actual robots.

However well designed, a robot system inevitably has to cope with distortions of camera image output, i.e., camera noise. There seem to be two solutions to this.

1. Redesign the function nodes in GP by using the camera output in accordance with actual camera output.
2. Adjust the camera image output of the actual robot close to that of the simulator by putting it through a suitable filter. Use a filter designed to switch in accordance with picture elements.

The first method above can be used only for special GP terminals. The second method is applicable in almost any case, including nodes having direct access to picture elements, but it is difficult to design such filters.

2.2   Design of Entire System

We used Webots of Cyberbotics for the robot experiments. These are simulators for Khepera with several options.

The size of the field is 200cm×200cm in this simulator (the area for actual robots is 100cm×75cm). The initial locations of the robots were fixed and the initial directions of movement were random. The buttons were black and located at three points chosen at random from six. The exit was blue and fixed. If a robot comes within a certain distance from the exit, with all buttons pressed, it is rated as an escaped robot. The limit time is 2000×64ms. A GP individual, i.e., a robot program, was repeatedly executed until the time expired.

In this paper, we assume that there are three robot agents in the environment and that the number of buttons is three. The robot speed was 6×8mm/s for robot A, 8×8mm/s for robot B and 13×8mm/s for robot C.

2.3   Test Environments and Parameters

The function and terminal symbols used in GP are given in Table 1. The simple image processing is done when evaluating the Black, Green, Blue and Wall functions (see Fig. 3). The other parameters are as follows: population size = 512, elite size = 20, maximum generations = 30, mutation probability = 0.3, and maximum tree size = 200.

To calculate the fitness, the following formula is used:

fitness $= a + b$

$\quad\quad$ + (the remaining time when all buttons have been pressed)

$\quad\quad$ + (the remaining time when all robots have escaped), where a is the +5 bonus upon pressing of one button and +10 bonus upon escape of one robot. b is the value depending on the distance between the exit door and each robot at the end of the limit time (maximum if the robot has escaped; the nearer the robot is to the door, the larger the value is). The larger the fitness, the better.

Remember that we have three robot agents in the environment. In this case, the fitness value is expected to be about 120 when the task has been achieved, i.e., when all robots have escaped.

Twenty different training maps with random initial directions are used for the GP learning. This learning is conducted within the Webot simulator. After the GP learning, the acquired robot program is tested against different testing maps both with the simulator and with actual robots.

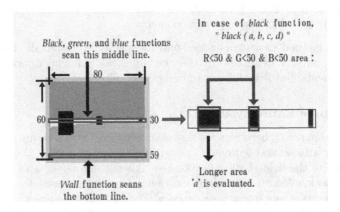

Fig. 3. Image Processing Functions.

3   Experimental Results

3.1   Simulation Results

The experimental result of a typical GP run is shown in Fig. 4, which plots the fitness values with generations. Remember that the fitness value of about 120 is considered as the task complete level. Fig. 5(a) shows the layout of the buttons and the robots in the learning environment. First, a robot presses the nearest button. Robot A, the slowest of the three, goes straight toward the exit, while the other two move to press the other buttons. The pressed buttons disappear from the field and appear outside (bottom of the window). Finally, the three robots escape.

Fig. 5(b) shows the actions for a different testing layout from the learning environment. Robot A is the slowest, and is unable to see the exit obstructed by a button. It stands still, then as the exit comes into sight from behind the disappearing button, it starts for the exit. The other two press buttons chosen by each. Finally the three robots escape.

Table 1. GP Terminals and Functions.

Name	#Args.	Description
Prog2	2	Evaluate two arguments in sequence. Return the result of the last argument.
Black	4	Evaluate one of the four arguments dependent upon the recognition of a black object in the view area.
Green	4	Evaluate one of the four arguments dependent upon the recognition of a green object in the view area.
Blue	4	Evaluate one of the four arguments dependent upon the recognition of a blue object in the view area.
Wall	4	Evaluate one of the four arguments dependent upon the recognition of a white wall near the agent.
Avoid	1	Avoid some collision based upon the output of the ultra-red sensor, and evaluate the argument.
Time_check	4	Evaluate the argument dependent upon the remaining time steps.
Turn_left	0	Turn left at 15 degrees.
Turn_right	0	Turn right at 15 degrees.
Go_forward	0	Move forward for 320 msec.
Go_backward	0	Move backward for 320 msec.

3.2   Use of Actual Robots

In order to apply the acquired GP program to actual robots, the gap of the camera image output was adjusted. Because of the nature of the function nodes in GP, the first method described in section 2.1 was employed. The Black, Green, Blue, and Wall functions were rewritten so that each would work with actual robots. In practice, only the parameters in the parts for discriminating each node color were modified. The allowable ranges of color discrimination were widened a little, which improved robot actions.

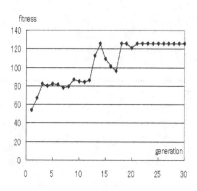

Fig. 4. Fitness Transition.

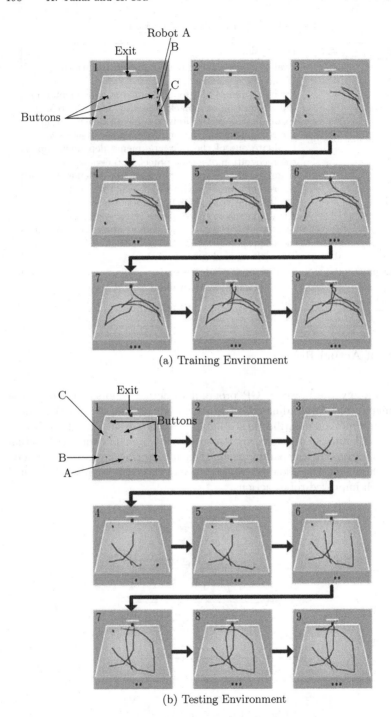

(a) Training Environment

(b) Testing Environment

Fig. 5. Acquired Robot Behaviors in Simulator.

Actual robots also worked well with the acquired GP programs. The actions of the actual robots are shown in Fig. 6[3]. Even when the environment was changed from the training simulation, the robots acted properly. Thus, we can confirm the robustness of the acquired GP program.

4   Comparison with Other Methods

This section compares four methods of approaching the escape problem: heuristic method (human programming), reinforcement learning (Q-learning), neural networks using an error back-propagating method, and GP. The field is an 8×8 square two-dimensional lattice integer coordinate space[4]. Agents A and B move at travel speed of one (1 square at a step) and agent C at travel speed of three. Each agent can recognize only the three squares around it. The direction of travel at one step was obtained under each of the following methods and the task-achieving ratios were compared.

1. Human programming by heuristics
   The algorithm: "each button is pressed by the agent nearest to it" was implemented. This is expected to solve the task heuristically. To any agent not allotted any button as a target, the exit was allotted as the goal. If any agent sees nothing in the three squares around, it can move at random.
2. Reinforcement learning
   We made a Q-table using the states (such as the speed of the agent, the distance to the goal, the vector to the goal, the number of pushed buttons and so on) and nine patterns of actions (moving to one of nine squares around it). Actions are determined by probability in accordance with Boltzmann's distribution. Another step taken was to change the search strategy from exploration to exploitation by reducing the temperature in the course of learning.
3. Neural networks
   Learning with teaching signals was carried out by means of an error back-propagating method. The input signals used were the number of the buttons pressed by the specific agent, the number of the buttons pressed by all agents, the number of the other agents in sight, the direction of the nearest one of the other agents, the number of the buttons in sight, the number of the buttons not pressed in sight, the location of the nearest button not pressed, whether the exit is open or not, whether the previous move was a failure or not, and the direction of the exit.
4. GP
   GP was run by a tournament strategy (size 7) with 200 individuals and 100 generations. The used function and terminal symbols were almost the same

---

[3] The MPEG demonstration of the acquired robot behaviors can be downloaded from http://www.miv.t.u-tokyo.ac.jp/ibalab/. Follow the download instruction.

[4] For the sake of comparison, we chose a lattice field in this experiment. This is because the number of states should be reduced for the sake of Q-learning.

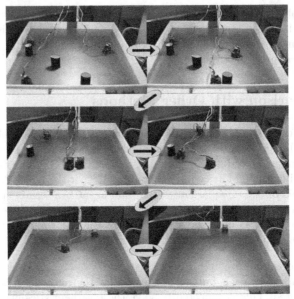

(a) Testing Environment #1

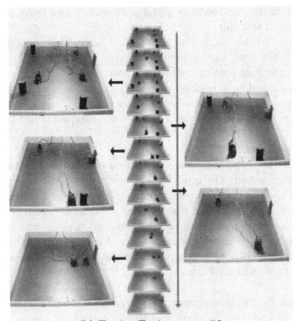

(b) Testing Environment #2

Fig. 6. Actual Robot Behaviors.

Table 2. Task achievement rates by different methods.

Method	Human Programming	Q-learning	Neural Network	GP
First stage	0.88	0.64	0.18	0.85
Second stage	0.35	0.04	0.00	0.40

as the previous ones (Table 1), except that they were modified to use the discrete vector information.

Table 2 shows the results given by various methods in terms of task-achieving ratio, i.e., the success records attained by the agents in case. In the first stage, the subgoal is to push all the buttons so as to open the door, whereas in the second stage, it is for all robots to escape from the door. The data are averaged over 100 runs. Although the settings differ slightly with techniques, the senses of all agents were limited (in practice the visible distance was limited to three). For the fair comparison, the maximum steps of the simulation were uniformly set to be 100 for all methods. Also, those agents that fail to see the goal are allowed to move at random. In Q-learning, it was decided not to take over the matrix of the state and action.

Because Q-learning and neural networks we used are so simple that they may achieve better results if improved. Thus, the comparison in the table does not definitely represent the qualitative ranking of the methods. However, it is observed that the results above show that if the senses of the agents were limited, GP was effective for the category of escape problems.

The performance comparison can be summarized as follows:

1. Human programming by heuristics
   This proved to be an optimal solving method if the senses of the agents are not limited, i.e., the mean time steps taken to carry out the task are relatively few. However, if the senses of the agents are limited, the results rapidly deteriorate, i.e., the performances become worse than the other methods.
2. Reinforcement learning
   If a long time is taken for trials, better results can be attained than with heuristic solving methods, irrespective of limitations on senses. On the other hand, one trial takes longer than under the heuristic method. In an attempt to deal with a more complicated environment, the size of the field needed for the matrix of states and actions will explode.
3. Neural networks
   A network suitable for the first step of the task is successfully attained with some probability. However, it cannot meet the second step of the task. The way how to give the teaching signals in the error back-propagating method needs further discussion.
4. GP
   In examples of training, better results can be obtained compared to the results with other methods. Good results are also obtained in testing environments and the programs acquired are adequately robust.

In this section, when discussing the comparison of techniques, the number of objects in the field of an escape problem, such as buttons and exits, and the size of the field were fixed. If these parameters are changed, different results can be obtained under several methods. For example, if the number of objects is increased, the size of the matrix of states and actions in Q-learning might explode, preventing us from attaining a solution in Q-learning.

It should also be noted that in the present experiments, the time needed for the actual simulation was not taken into account. In other words, no mention was made of the computational costs of algorithms in terms of time under any method. However, we set the actual computation time to be almost of the same order for all four methods.

5   Discussion

The program obtained from the simulation learning of the escape problem can be evaluated as being fairly robust, because the task was accomplished in a different situation from the basic learning environment. However, although in the same situation as the learning environment the task is accomplished 100%, in a different situation the achieving ratio is less than 100%. For example, if robot A (the slowest, ready to start for the exit) is located so that it sees the exit to the right, it tends to take more time to come to the exit. This is because the relative positions of the goal and the robots were fixed in such a way that robot A often sees the goal to the left; so, a lazy program probably evolved. This difficulty can possibly be eliminated by changing the environment; yet there remains a big problem: what changes should be made.

In summary, the actual robots acted much better than expected. This suggests that a program resistant to camera noise and errors was acquired by our method. Thus, we can conclude that GP is one of effective learning methods for application to actual robots.

6   Conclusion

This paper presented the emergence of the cooperative behavior for multiple robot agents by means of Genetic Programming (GP). We showed the following points empirically:

1. GP was successfully extended for the sake of multi-agent robot learning.
2. The escape task was effectively solved with not only simulated but also real evolutionary robots.
3. The performance was compared with traditional methods, such as Q-learning and neural networks, which shows the superiority of our approach.

Our future topics concern the study of this problem in other robot tasks, such as a robot navigation problem [Iba and Terao00] We also plan to conduct an experiment in the difficult situations when the workspace is gradually changing with generations.

References

Hara *et al.*99. Hara,A., and Nagao,T.,   Emergence of Cooperative Behavior using ADG; Automatically Defined Groups, in *Proc. of the Genetic and Evolutionary Computation Conference (GECCO99)*, Morgan Kaufmann, 1999

Haynes *et al.*95. Haynes, T., Wainwright,R., and Sen,S., Evolving a Team, in *Working Notes of the AAAI-95 Fall Symposium on Genetic Programming*, AAAI Press, 1995

Hondo *et al.*96a. Hondo,N., Iba,H., Kakazu,Y., Sharing and Refinement for Reusable Subroutines of Genetic Programming, in *Proc. 1996 IEEE International Conference on Evolutionary Computation (ICEC96)*, pp.565–570, 1996

Hondo *et al.*96b. Hondo,N., Iba,H., Kakazu,Y., Robust GP in Robot Learning, Parallel Problem Solving from Nature 4 (PPSN IV), Springer-Verlag, pp.751–760, 1996

Iba96. Iba,H., Emergent Cooperation for Multiple Agents using Genetic Programming, in *Parallel Problem Solving form Nature IV (PPSN96)*, 1996

Iba98. Iba,H., Evolutionary Learning of Communicating Agents, *Information Sciences*, 108(1-4), 1998

Iba and Terao00. Iba,H. and Terao,M., Controlling Effective Introns for Multi-Agent Learning by Genetic Programming, in *Proc. of the Genetic and Evolutionary Computation Conference (GECCO2000)*, pp.419–426, 2000

Ito *et al.*96. Ito,T., Iba,H. and Kimura,M., Robot Programs Generated by Genetic Programming, Japan Advanced Institute of Science and Technology, IS-RR-96-0001I, in *Genetic Programming 96*, 1996

Koza 94. Koza, J., Genetic Programming II, Automatic Discovery of Reusable Subprograms, MIT Press, 1994

Luke *et al.*96. Luke,S. and Spector,L., Evolving Teamwork and Coordination with Genetic Programming, in Genetic Programming 96, MIT Press, 1996

Soule *et al.*96. Soule,T., Foster,J.A., and Dickinson,J., Code Growth in Genetic Programming, in *Genetic Programming 96*, 1996

Placing and Routing Circuits on FPGAs by Means of Parallel and Distributed Genetic Programming

F. Fernández[1], J.M. Sánchez[1], and M. Tomassini[2]

[1] Departamento de Informática. Escuela Politécnica. Universidad de Extremadura. Cáceres.
fcofdez@unex.es, sanperez@unex.es,
[2] Institut d'Informatique. Université de Lausanne.
mtomassi@iissun4.unil.ch

Abstract. We present results on the application of a new methodology based on Parallel and Distributed Genetic Programming (PADGP). The aim for the methodology we present is to automatically perform the placement and routing of circuits on reconfigurable hardware. The system has been successfully applied to some benchmark problems. For each of the problems we have dealt with, the methodology is capable of finding several solutions. The results show the methodology's feasibility for addressing the problem of placement and routing on FPGAs.

1   Introduction

Field Programmable Gate Arrays (FPGAs) are arrays of prefabricated logic blocks and wire segments with user-programmable logic and routing resources. The circuit design cycle on FPGAs is shorter than conventional design cycle because hardware implementation is relatively easy with FPGAs.

FPGAs are used at present days for solving many problems: DNA sequence matching, signal processing, the emulation of microprocessors, cryptographic attacks and so on [DeHon 2000].

The goal for the methodology we are studying in this paper is to program an FPGA. In order to do so, a logic description obtained during logic synthesis must be mapped and converted into the modules and routing resources available in FPGAs. Bearing in mind that both logic blocks and routing resources are predefined in an FPGA chip, circuits must be laid out within it.

There are two types of FPGAs technologies: *row-based FPGAs* and *island-based FPGAs*. In this paper we dealt with the latter. In the following sections we describe the technology involved and the methodology we proposed for solving the problem of placement and routing on FPGAs. Section 2 describes Island-based FPGAs. Section 3 deals with Parallel and Distributed Genetic Programming (PADGP). Section 4 presents the methodology. Section 5 describes some examples and results that we have obtained, and finally section 6 draw our conclusions.

Y. Liu et al. (Eds.): ICES 2001, LNCS 2210, pp. 204-215, 2001.

2   Island-Based FPGAs

In island-based Fpgas routing resources consists of horizontal and vertical channels and their intersecting areas [Xilinx 1999] (see figure 1).

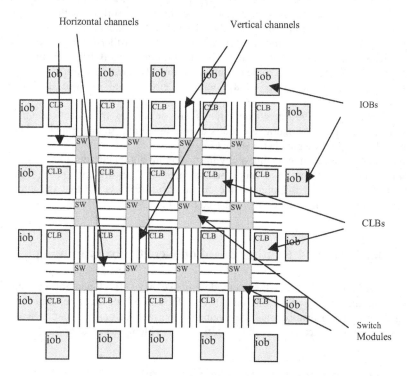

Fig. 1. Island-based FPGAs

Four kind of blocks are available:
- I/O blocks (IOBs): They are used for inputs and outputs.
- Combinational logic blocks (CLBs): They comprise combinational elements such us lookup table-based (TLU) function generators and sequential elements.
- Connection blocks: Comprise a set or routing resources -wire segments for connecting pins of logic cells. They are arranged into tracks
- Switch blocks: They connect the routing resources belonging to four adjacent connection blocks. A static RAM controls each switch block.

The design cycle for FPGA technology includes both the synthesis and mapping of circuits [Maznumder and Rudnick 1999], [Sherwani 2000].

Technology mapping for a TLU-based FPGA architecture describes a circuit which, optimised according to a criteria becomes an interconnection of CLBs. Each of the CLBs implements a logic function.

2.1   Evolutionary Methods for Circuits Design

In this section we briefly describe research work which has applied evolutionary algorithms to circuit design and implementation.

A given circuit can be considered as a graph. Actually, several papers have dealt with the problem of encoding graphs, i.e. circuits, when working with GA and GP [Miller et al 2000], [Thompson et al 1999], [Lohn and Colombano 1998].

For instance, *Cartesian Genetic Programming (CGP)* [Miller 1999] was developed for representing graphs, and is based on Genetic Programming (GP). The aim of CGP is to find complete circuits capable of implementing a Boolean function.

In [Lohn and Colombano 1998] analogue circuits are evolved using linear representation within GA.

In [Koza et al 1999] analogue circuits are developed by means of GP. In [Thompson et al 1999] analogue circuits are designed and implemented in FPGAs. The methodology presented is based on GA.

Nevertheless we want a methodology for managing digital circuits. Our global goal is not to find the design of a circuit, which is capable of performing a given function, but we want to find a mapping of the circuit inside FPGAs, so that the circuit is physically build. Our methodology is based on GP.

3   Placement and Routing Based on Genetic Programming

In this section we describe the methodology we have developed for making the placement and routing of circuits on FPGAs. First of all we describe some Genetic Programming (GP) features that are deemed important for our application and then we describe how GP can successfully address the problem we are describing.

3.1   Parallel and Distributed Evolutionary Algorithms

The basis for evolutionary methodologies were founded 3 decades ago. From Genetic Algorithms [Holland 1975], to the relatively new Genetic Programming [Koza 1992] all of them were thought of as sequential algorithms. Usually, a set of individuals – coded in a different way depending upon the methodology- is breed along generations with the aim of solving an optimisation problem. Although usually results have been encouraging, the time required to find solutions is sometimes prohibitive. This has led researchers toward new studies concerning the way of accelerating the convergence process that populations exhibit when evolution acts. One such ways is the use of parallel or distributed machines working on the same problem.

There are several levels at which an evolutionary algorithm can be parallelised: population, individuals or the fitness evaluation level. The differences in the implementation of the parallel algorithm when using GA with respect to GP stem from how individuals are represented. Basically, authors distribute the set of individuals into several populations. Each of the populations run the same evolutionary algorithm and exchange individuals at a certain rate. This model has been studied within the Genetic Algorithms framework [Cantú-Paz and Goldberg

1997] and has also been applied to Genetic Programming [Andre and Koza 1996],[Punch 1998]. [Fernández et al 2001] gives a description of the different aproaches when using GP. We also focussed on the study of Parallel and Distributed Genetic Programming (PADGP) [Fernández et al 2000a]. We selected some of the important parameters – number of individuals and subpopulations, topology, migration rate- and obtained interesting results on benchmark problems.

As long as the time required for solving the problem of placement and routing on FPGAs is large, we decided to use PADGP instead of classic GP.

Figure 2 graphically depicts the model we use with PADGP. A master process is in charge of managing communication buffers. Populations can send and receive individuals to and from the rest of the populations.

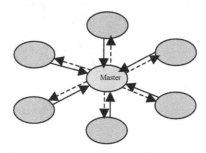

Fig. 2. PADGP model

3.2   Methodology

The aim for the methodology we are proposing is to apply PADGP for solving the problem of placement and routing on FPGAs. We are thus dealing with circuits.

The first step when applying an evolutionary algorithm to a given problem is to decide the way of codifying the problem. Since we use GP it is customary to codify the problem by means of trees. Each circuit to be placed and routed should thus be coded as a tree.

For the sake of simplicity we can say that a circuit is a graph where nodes are occupied by components (see figure 3-a).

If we substitute each of the circuit's components by a box, we can obtain a simplified circuit as depicted in figure 3-b. If each of the components contained in a given boxes is simple enough, we can implement it in any of the CLBs available in the FPGA we are using.

The simplified circuit is therefore useful because it hides components but expresses connections among them. These connections are the only elements we need for performing the placement and routing on FPGAs, given that components on circuits we are dealing with are very simple and can be implemented with any of the available CLBs.

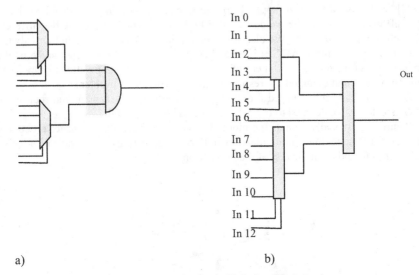

a)   b)

Fig. 3. a) A simple circuit. b) A simplified circuit

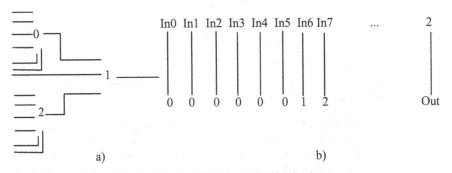

a)   b)

Fig. 4. a) A simplified circuit. b) Circuit wires.

3.2.1   Encoding Individuals

After simplifying circuits, the next step is coding them into individual trees. We know that circuits are graphs, so we must establish a correspondence between graphs and trees.

Let's suppose that the circuit we are dealing with is that depicted in figure 3. We can suppress boxes from the graph and then we just have wires (see figure 4-a).

This procedure can be applied regardless of the circuit complexity. We can now assign labels to each of the wire's extremes see (figure 4-b). Labels will indicate the box to which each of the extremes is connected. In this way, we can build again the whole circuit by placing as many boxes as there are different labels and connecting wire's extremes according to labels.

At this stage we don't need to know extra information relative to CLBs. Just by looking at wire labels, we know the number of CLBs required and how to connect them. We can now represent wires by means of trees by connecting each of the wires as a branch of the tree and keeping them all together in the same tree. By labeling both extremes of branches, we will have all the information required to reconstruct the circuit (see figure 5).

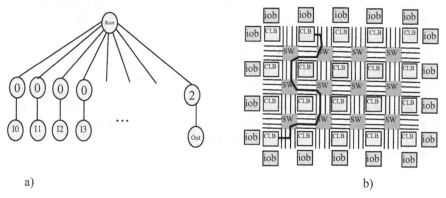

a)   b)

Fig. 5. A) Wires as branches b) Connecting two CLBs

3.2.2   Encoding Physical Connections

The next step is to codify the physical routing of wires into FPGA inside each branch. As described in section 2, island-based FPGAs contains CLBs, switch blocks and wires. Each wire can connect adjacent blocks, both CLBs and switch blocks. Several wires must be connected through switch blocks when joining two CLBs' pins according to a given circuit description.

The main question now is: how can any wire be codified by means of branch of trees so that genetic operations can be applied without problems? And which is the best way of codifying circuits by means of Genetic Programming? The second question is difficult to answer. Let's first see how connections on FPGAs can be expressed by means of branches within trees, and how these branches correspond to circuit's wires.

Every wire in an FPGA is made up of two ends - these can connect to CLB or IOB. On the other hand, as said above, a given number of switch connections may conform the path of the wire. In the representation we have used for codifying wires into tree branches, CLB and IOB connections are described as each of the two end nodes which make up a branch. In order to represent switch connections, we will add as many new internal nodes to the branch as switch blocks are traversed by wires. Each internal node describes a given switch connection.

For the examples in figure 5-b switch blocks are passed through; 6 new nodes must thus be added to the branch. Figure 6 shows the branch representing the wire.

Every node requires additional information: if the node corresponds to a CLB we need to know information about the position of the CLB in the FPGA, the number of pin to which one of the ends of the wire is connected, and which of the wires of the wire block we are using; if the node represents a switch connection, we need information about that connection. Supposing that each wire arriving at a switch block can continue to any of the 4 sides of that block, it is enough to store in that node the direction: north, south, west or east (See figure 6-b).

Once the way of expressing connections by means of nodes has been established we must decide how all different branches make up the whole tree, i.e. the circuit.

After several experiments, we saw that a good way of describing the complete circuit is by means of binary trees. Each of the circuit's wire is placed on a left branch of a different level of the tree (see figure 6-a). If a wire connects to an I/O, no switch blocks will be required for connecting it in the FPGA, so the length of the branch representing that wire will be always short. Wires connecting two boxes will require switch blocks in the FPGA, so the length is not know a-priori.

The length of trees is an important parameter in GP. If we are using the previous way of building trees, branches in the last levels of the trees will always be shorter than branches at the beginning. Therefore, we decided to leave the last branches for I/O wires and the remaining for wires connecting boxes, which actually will require a larger number of nodes.

4   Experiments

In this section we present some experiments we have performed for studying the feasibility of the methodology. Firstly we describe the GP parameters and then we describe two examples of different complexity.

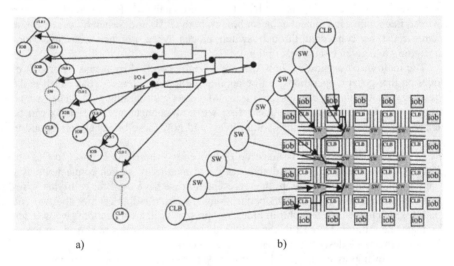

a)   b)

Fig. 6. A) A branch expressing connections. B) Describing circuits by means of binary trees.

4.1 GP Sets

According to the way circuits are encoded into trees, the GP sets we are using are the following:

$F=\{CLB\}$
$T=\{PIN\}$

The CLB function plays a different role according to the position it occupies inside a branch:

- If it appears at the beginning of the branch, it specifies a CLB'S pin connection.
- Otherwise, when appearing as an internal node, it acts as a switch connection.

In the case of the terminal PIN, we can also give it two different meanings. Let's suppose that in a given problem we have n input/output connections and N wires just connecting CLBs:

- If the terminal PIN appears in a branch occupying one of the latest n positions in the tree, then the nodes acts as an IOB connection.
- Otherwise, when it appears among the first N branches of the tree, the node behaves as a CLB connection.

4.2 The Fitness Function

In the context of the problem we are addressing, the fitness function is in charge of two different things: it maps each tree into a simulated circuit and also into a simulated FPGA; secondly, the fitness function must evaluate how good the circuit is, and assign a fitness value, which will later be used by genetic operators.

The fitness function thus represents how well a circuit - described by a tree- correspond to a proposed circuit. Trees are mapped by visiting each of the useful nodes of the tree and making connections on the virtual FPGA. Figure 11 shows the order how individuals are evaluated. We always begin with I/O wires and ends with wires connecting boxes. This is because I/O wires establish some of the boxes that last wires will use.

The fitness value is computed by comparing actual connections and required connections and represents the total error. A simple way of expressing this function is the following:

$$P(i) = d\,(\,pos\,(i)_0\,,CLB\,(i)_0\,) + d\,(\,pos\,(i)_f\,,CLB\,(i))_f$$
$$F \;=\; \sum_{i=1}^{n}\,P\,(\,i\,); \qquad\qquad (1)$$

pos(i)o : position of initial end of wire i in the FPGA.
pos(i)f: position of final end of wire i in the FPGA, i.e.
CLB(i)o: Estimated position for CLB connected to initial end of wire i.
CLB(i)f: Estimated position for CLB connected to final end of wire i.
n: Number of wires for the problem.

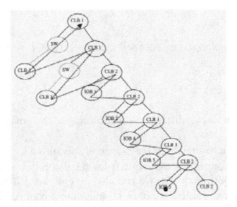

Fig. 7. Describing circuits by means of trees

4.3 Providing Circuits to the GP Algorithm

Describing a circuit is the first step in the process of making the placement and routing by means of GP.

First of all, each box in the circuit is assigned a number. Circuits are then described by a two-dimensional vector. Each coordinate indicates the boxes to which each wire is connected. If a wire connects an input or output to just one box, the pair of numbers making up the coordinate is identical and corresponds to the number of the box. The interpreter used by the fitness function will recognize that both numbers in the coordinate are the same and will be aware that the wire is actually connected to an IOB (see figure 8).

Each time a branch is examined, the fitness function checks which of the CLBs are being connected. A data structure will store the information about the position of the selected CLB that the branch contains. If the CLB has been previously established, the data structure contains its coordinates in the FPGA, and can be used to begin the connection or to compare if the position expressed by the branch corresponds to a position previously calculated by other branch.

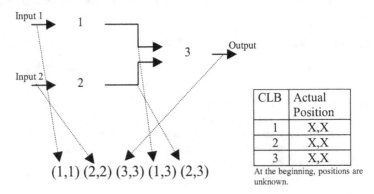

Fig. 8. Describing circuits by means of trees

4.4 Benchmark Problem

In [Fenández et al 2000b] we solved several simple benchmark problems by means of the methodology we describe in this paper. We apply it now to a more difficult benchmark circuit which is depicted in figure 9. Although this circuits doesn't have any specific function, it is useful to study how the methodology can be scaled.

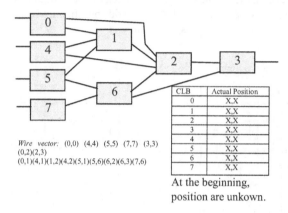

CLB	Actual Position
0	X,X
1	X,X
2	X,X
3	X,X
4	X,X
5	X,X
6	X,X
7	X,X

Wire vector: (0,0) (4,4) (5,5) (7,7) (3,3) (0,2)(2,3) (0,1)(4,1)(1,2)(4,2)(5,1)(5,6)(6,2)(6,3)(7,6)

At the beginning, position are unkown.

Fig. 9. The benchmark problem

5   Results

The main GP parameters employed for the complex circuit were the following:
- Population size=200
- Maximum depth= 30
- Steady State.
- Tournament size=10
- Crossover probabiity=98%.
- Mutation probability=2%.

Three different executions of the experiment were carried out. Each of them found a solution, always different (see figure 10). About 4 hours were required each time to solve the problem.

We have finally applied PADGP to this problem in order to see if the convergence process can be speeded u. Figure 11 shows results obtained with different number of populations and individuals per population, but using always the same total amount of individuals. We send the best individual from each population to another randomly chosen one, each generation. We notice that using 2 or 5 subpopulations is better than using only 1, when using 500 individuals.

6   Conclusions and Future Work

The experiments we performed show the feasibility of GP for solving this problem. The methodology finds different ways of performing the placement and routing of circuits on FPGAs, a hard optimisation problem.   Each of the solutions can be applied to specific situations in which some restrictions must be taken into account.

As in other benchmark problems PADGP speeds up the finding of solutions to real-life problems.   In the future, we will add partitioning techniques to the methodology thus allowing to address larger circuits.

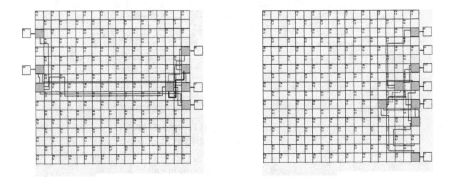

Fig. 10. Some of the solutions that were found.

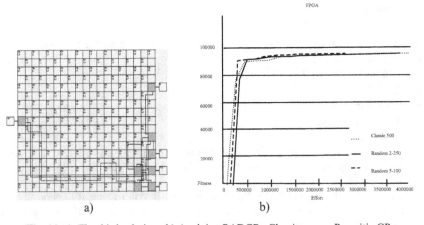

a)   b)

Fig. 11. a) The third solution. b) Applying PADGP. Classic means Panmitic GP.

References

[Andre and Koza 1996] D. Andre and J. R. Koza: Parallel Genetic Programming: A Scalable Implementation Using the Transputer Network Architecture. Piter J.; and Kinnear, Kenneth, Jr. (eds) Advances in Genetic Programming 2. Cambridge, MA: MIT Press. pp 317-337.

[Cantú-Paz, Goldberg 1997] E. Cantú-Paz and D. Goldberg: Predicting Speedups of Ideal Bounding Cases of Parallel Genetic Algorithms. Proceedings of the Seventh International Conference on Genetic Algorithms. Morgan Kaufmann.

[DeHon 2000] A. DeHon. The Density Advantage of Configurable Computing. Computer. April 2000, pp. 41-49.

[Fernández et al 2000a] F. Fernández, M. Tomassini, L. Vaneschi, L. Baucher: A Distributed Computing Environment for Genetic Programming using   MPI. LNCS 1908: Recent Advance in Parallel Virtual Machine and Message Passing Interface. Proceedings 7th Euro PVM/MPI Conference. Pp 322- 329. Sep-2000.

[Fernández et al 2000b] F. Fernández, J.M. Sánchez, M. Tomassini: "Feasibility study of Genetic Programming for solving the problem of Placement and Routing on FPGAs". Proceedings XV Conference on Design of Circuits and Integrated Systems 2000. Nov-2000.

[Fernández et al 2001] F. Fernández, M. Tomassini, L. Vanneschi: Studying the Influence of Communication Topology and Migration on Distributed Genetic Programming. EuroGP 2001. To appear.

[Holland 1975] J.H. Holland: Adaptation in Natural and Artificial Systems, University of Michigan Press, Ann Arbor, 1975 [Koza 1992] J. R. Koza: Genetic Programming. On the programming of computers by means of natural selection. Cambridge MA: The MIT Press.

[Koza 1997] J. R. Koza, F.H. Bennet, D. Andre, M.A. Keane and F. Dunlap: Automated Synthesis of analog electrical circuits by means of genetic programming. IEEE Transactions on Evolutionary Computation, VOL. 1, NO. 2. 1997. Pp 109-128.

[Lohn and Colombano 1999] J. D. Lohn and S. P. Colombano A Circuit Representation Technique for Automated Circuit Design. IEEE Transactions on Evolutionary Computation, VOL. 3, NO. 3, September 1999. Pp 205-219.

[Maznumder and Rudnick 1999] P. Mazumder, E.M. Rudnick, Genetic Algorithms for VLSI Design, Layout & Test Automation. Prentice Hall. (1999).

[Miller 1999] J. F. Miller: An empirical study of the efficiency of learning boolean functions using a Cartesian Genetic Programming Approach, in Proceedings of the 1st Genetic and Evolutionary Coputation Conference, W. Banzhaf, J. Daida, A. E. Eiben, M. Garzon, V. Honavar, M. Jakiela, and R. E. Smith (eds), Morgan Kaufmann, San Francisco, Ca, 1999, vol. 2, pp. 927-936.

[Miller et al 2000] J. F. Miller, D. Job, V. K. Vassilev. Principles in the Evolutionary Design of Digital Circuits – Part I. Genetic Programming and Evolvable Machines, 1, 7-35(2000). Kluwer Academic Publishers. Netherland.

[Punch 1998] W.F. Punch: How effective are multiple populations in Genetic Programming. Genetic Programming 1998: Proceedings of the Third Annual Conference, J. R. Koza, W. Banzhaf, K. Chellapilla, K. Deb, M. Dorigo, D. B. Fogel, M. Garzon, D. Goldberg, H. Iba and R. L. Riolo (Eds),Morgan Kaufmann, San Francisco, CA, 308-313, 1998.

[Shwerwani 1995] N. Sherwani. Algorithms for VLSI Physical Design Automation. Kluwer Academic Publishers, 2nd Edition. 1995.

[Thompson et al 1999] A. Thompson, P. Layzell and R. S. Zebulum: Explorations in Design Space: Unconventional Electronics Design Through Artificial Evolution. IEEE Transactions on Evolutionary Computation, VOL. 3, NO. 3, September 1999. Pp 167-196.

[Xilinx 1999] Xilinx. XC4000E and XC4000X Series Field Programmable Gate Arrays. May 14, 1999.

A Massively Parallel Architecture
for Linear Machine Code Genetic Programming

Sven E. Eklund

Dalarna University, Sweden
sven.eklund@ieee.org

Abstract. Over the last decades Genetic Algorithms (GA) and Genetic Programming (GP) have proven to be efficient tools for a wide range of applications. However, in order to solve human-competitive problems they require large amounts of computational power, particularly during fitness calculations.

In this paper I propose the implementation of a massively parallel model in hardware in order to speed up GP. This fine-grained diffusion architecture has the advantage over the popular Island model of being VLSI-friendly and is therefore small and portable, without sacrificing scalability and effectiveness. The diffusion architecture consists of a large amount of independent processing nodes, connected through an X-net topology, that evolve a large number of small, overlapping sub-populations. Every node has its own embedded CPU, which executes a linear machine code representation of the individuals. Preliminary simulation results (low-level VHDL simulation) indicate a performance of 10.000 generations per second (depending on the application). One node requires 10-20.000 gates including the CPU (also application dependent), which makes it possible to fit up to 2.000 individuals in one FPGA (Virtex XC2V10000).

1   Background

Genetic Algorithms (GA) are a group of stochastic search algorithms which were discovered during the 1960's, inspired by evolutionary biology. Over the past decades they have proven themselves to work well on a variety of problems with little a-priori information about the search space [11].

The search space is sampled by a population of points (individuals) that iteratively move towards an optimal solution. Every iteration (generation) of this population goes through three major steps; an evaluation of the fitness of the individuals, a selection of the most fitted individuals and the application of genetic operations on the individuals, as shown in figure 1.

1.1   Genetic Programming

Genetic Programming (GP) is based on the genetic algorithm (the three steps in figure 1). However, the main characteristic of GP is that the individuals are "programs" or at

Y. Liu et al. (Eds.): ICES 2001, LNCS 2210, pp. 216-224, 2001.

least possible to interpret in a syntactical context. The target system can for instance be a CPU, a compiler or a simulator, which all expect instructions or commands according to a well-defined syntax. These "programs" produce solutions to the problem.

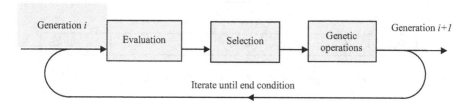

Fig. 1. The three basic steps of GA

GP often, but not always, represent individuals in the form of a tree structure as shown in figure 2. Any function set can be represented in the nodes of the tree, for instance the reserved words of a programming language. It is also the syntax of the function set, which determine the shape of the tree. During evaluation this representation is interpreted, compiled or calculated in order to get a fitness value of the individual.

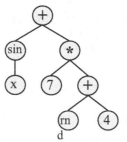

Fig. 2. A tree representation of GP

It has become easier to address harder, human-competitive problems with the genetic algorithm by using the powerful representation of GP. This however, has also increased the demand for efficient computation, since the search space grows rapidly with an increased function set [12].

1.2   Linear Machine Code GP

One of the most effective ways of implementing GP is to use binary machine code as the function set [3]. This eliminates the need for compilation and interpretation of an abstract representation. Instead, the representation is directly executed on a standard CPU in order to get a fitness value.

Since machine code by its nature is linear, a linear representation has been suggested, as shown in figure 3. Genetic operations can be performed directly on this linear structure, sometimes with syntax preserving protection of blocks or with (predefined) blocks of instructions.

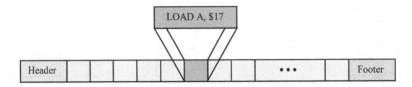

Fig. 3. Linear machine code representation of one individual

2   Parallel GA

It is a well-known fact that the genetic algorithm is inherently parallel, a fact that could be used to speed up the calculations of GP. The basic algorithm by Holland [9] is very parallel, but also has a frequent need for communication and is based on centralized control, which is not desirable in a parallel implementation. This problem has to be addressed in order to make a parallel implementation efficient.

An efficient architecture for GP should be optimized for the calculations and communication involved in the algorithm (the three steps in figure 1). However, it has also to be flexible enough to work efficiently with a variety of applications, which have different function sets. Also, the architecture should be scalable so that larger and harder problems can be addressed with more computer hardware.

2.1   Parallel Models

By distributing independent parts of the genetic algorithm to several processing elements which work in parallel, it is possible to speed-up the calculations. Traditionally, the parallel models have been categorized by the method by which the population is handled. The choice between a global and a distributed population is basically a decision on selection pressure, since smaller populations result in lower selection pressure and faster (sometimes premature) convergence. However, the choice also has a major effect on the communication need of the algorithm.

Bethke [5] made one of the first investigations of parallel implementations of GA in 1976. He described a global population with a partial exchange of individuals in successive generations. In 1981 Grefenstette described four different parallel implementations of GA, with both distributed and global populations [8].

One of the first real implementations of parallel GA was made by Tanese in 1987. She conducted studies of different topologies and migration rates on a distributed population model on a 64-NCUBE system. In some experiments she reported super-linear speed-up compared to sequential GA [14].

The Farming Model. In the model with a global population the algorithm has direct access to all the individuals in the population, either by a global memory or by some type of communication topology, which connects several distributed memories. This model have been reported to scale badly when the number of processing elements grow, due to the communication overhead of the algorithm [1], [2]. This is however heavily dependent on the ratio between communication time and computation time and on the size of the population [6].

The parallel model with a global population is often referred to as the farmer-model or the master-slave-model [6]. Its structure is shown in figure 4. A central unit, a farmer or master, controls the selection of individuals from the global population and is assisted by workers or slaves to perform the evaluation of the individuals.

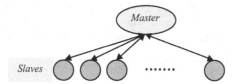

Fig. 4. The farming model

By dividing the population into more independent sub-populations, two alternative parallel models can be identified. Based on the size and number of sub-populations, they are referred to as coarse-grained or fine-grained distributed population models. When dealing with very large populations, which are common in hard, human-competitive problems, these models are better suited since their performance scale better with growing size.

The Island Model. The coarse-grained model, also known as the island model consists of a number of sub-populations (demes) that evolve rather independently of each other. With some migration frequency, however, they exchange individuals between each other over a communication topology. An example of this topology is shown in figure 5.

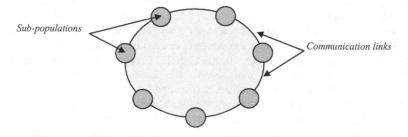

Fig. 5. The island model with seven sub-populations in a ring topology

The island model is a very popular parallel model, mainly because it is very easy to implement on a local network with standard workstations (cluster). A major drawback of the island model is that it modifies the basic genetic algorithm and introduces new parameters, for instance the migration policy and the network topology. Today, there exist little or no theory on how to adjust those parameters [7].

The Diffusion Model. The fine-grained model, often referred to as the diffusion model, cellular GA or massively parallel GA, distributes its individuals evenly over a topology of processing elements. It can be interpreted as a global population laid out on a structure of processing elements, where the spatial distribution of individuals defines the sub-populations. As shown in figure 6 the sub-populations overlap, which makes the communication implicit and continuous and enables fit individuals to "diffuse" throughout the population in contrast to the explicit migration in the Island model. Selection and genetic operations are only performed within these local neighborhoods [4].

The diffusion model is well suited for VLSI implementation since the nodes are simple, regular and mainly use local communication. Further, the nodes operate synchronously in a SIMD-like manner and have small, distributed memories.

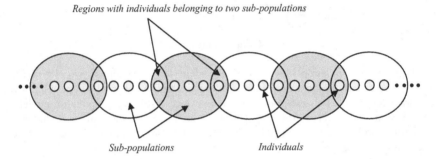

Fig. 6. The diffusion model with a linear topology

3   Architecture

It is my belief that a high performance GP architecture, especially when targeted for portable applications, should be implemented in hardware and be based on the fine-grained diffusion model. Implemented in an ASIC or FPGA it would not only be highly efficient, but also scalable and compact enough to be portable and possible to integrate, for instance, in mobile applications. In [13] several other advantages of the diffusion model are concluded, most notably the absence of a migration parameter and a potentially higher parallelism than in other parallel models.

Machine code representation of the individuals, being one of the fastest ways to evaluate fitness, will further increase the performance. The CPU:s, which execute the machine code, can be integrated on the same chip as the diffusion nodes, making it possible to easily balance the performance of the CPU with the size of the CPU. Integrating a reconfigurable CPU within a FPGA also makes it possible to target the function set to specific applications. In the rest of this chapter I will describe such a system, i.e. a diffusion architecture that evolve individuals represented by linear machine code using reconfigurable, embedded CPU:s.

3.1   Topology

The inter-node communication in the system is based on the X-net topology, a two-dimensional toroidal with extra diagonal links. In [10] this topology has shown to be efficient during simulation of a number of test applications for GA. The size and shape of the local neighborhood is configurable within the topology. In its first implementation the architecture has a nine-node neighborhood as shown to the left of figure 7.

Fig. 7. Two different neighborhoods laid out on the same X-net topology

3.2   Node Architecture

The diffusion node consists of four major parts, a CPU, a memory, a control block and a simple router. Together they perform the three steps shown in figure 1.

The Memory. The memory holds two individuals, A and B, represented by linear machine code. The size of the memory is application dependent but usually holds 128-256 words of instructions.

The Router. The router connects the node to the local neighborhood enabling the node to read both fitness and machine code from selected neighbors. Please note that every node not only is at the center of it's own neighborhood, but also a part of eight other neighborhoods as their north, north-west, west, etc, neighbor.

CPU Architecture. The CPU executes the machine code of the individuals A and B and returns a raw fitness value to the control block. The CPU is based on the 8-bit PIC16Cxx architecture fitted with an extra integer multiplier and executes all instructions in one cycle. The CPU is easily reconfigured with a different word size, a different register set-up or a new instruction set. It is therefore easy to modify the function set of the genetic algorithm and target the system to a specific application.

The Control Block. The control block controls the three parts described above and performs the genetic algorithm. After initializing the two individuals by random generation of machine code the iteration from figure 1 is started. In the first step the CPU evaluates the code of the two individuals in sequence. The raw fitness is then returned to the control block where it is processed according to the application and the fitness function.

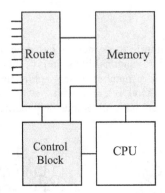

Fig. 8. The four blocks of the diffusion node

During selection, one of the two individuals A and B is chosen on random and its fitness is sent to the neighborhood. The selection is done as a tournament between two randomly chosen individuals amongst the ten individuals received by the router (two local and eight neighborhood individuals). The one-point crossover is done synchronously in a SIMD-like manner by reading two individuals from the router and writing them in the memory at the receiving node. Please note that every node is sending its randomly chosen A/B-individual to every node in the neighborhood, including itself. It is, however, the receiving node that determines which of the ten received individuals that will participate in the crossover operation, based on the result of the tournament.

The mutation is done at instruction level and randomly replaces one instruction with a randomly generated one.

3.3   Design Trade-offs

A crucial trade-off in the system is the size and performance of the CPU. A more powerful CPU will evaluate the individuals faster or support a more powerful function set. Being a major task in the genetic algorithm, this would speed up the whole system. However, it is even more important to fit as many nodes as possible within a single FPGA, since only a limited number of FPGA:s can be used in a portable system. It is therefore important to balance the size and performance of the CPU.

Generally, the size of a node is not dominated by the CPU but of the memory. It would therefore be interesting to evaluate the system performance with two CPU:s per node. Evaluating the two individuals in parallel would almost double the performance of the system but only increase the size of the node by 20-30 %.

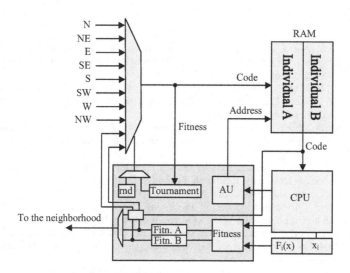

Fig. 9. The diffusion node set up for a regression problem

4   Simulation Results

So far, the design has only been simulated at low level in VHDL to verify its data path and the work of the state machine. However, the simulations have indicated at which frequency the system can operate and the number of gates needed per node.

Given a 8-bit CPU architecture with four general purpose registers, a maximum size of 128 instructions per individual and a 16-point integer regression problem [x, f(x)] as application, the architecture is able to evolve 10.000 generations per second. The gate-count per node is around 20.000 gates making it possible to fit 500 nodes (1.000 individuals) per FPGA (Virtex XC2V10000).

5   Future Work

Being the first architecture of the suggested model, a lot of work remains. To begin with the architecture has to be implemented and verified in the target FPGA. The implemented genetic algorithm must also be evaluated and different applications benchmarked against the execution on standard models. The impact of different neighborhoods and topologies need to be tested and, as mentioned above, different CPU architectures and the impact of two CPU:s per node should be evaluated.

References

1. Abramson, D, & Abela, J, "A Parallel Genetic Algorithm for Solving the School Timetabling Problem", In Proceedings of the Fifteenth Australian Computer Science Conference (ACSC-15), Volume 14, pp 1-11, 1992.
2. Abramson, D, Mills, G, & Perkins, S, "Parallelization of a Genetic Algorithm for the Computation of Efficient Train Schedules", Proceedings of the 1993 Parallel Computing and Transputers Conference, pp 139-149, 1993.
3. Banzhaf, W, Nordin, P, Keller, R, Francone, F, "Genetic Programming – An Introduction", ISBN 1-55860-510-X, pp 330-334, Morgan Kaufmann Publishers Inc, San Francisco and dpunkt Verlag, Heidelberg, 1998.
4. Baluja, S, "A Massively Distributed Parallel Genetic Algorithm (mdpGA)", CMU-CS-92-196R, Carnegie Mellon University, Pittsburgh, Pennsylvania, 1992.
5. Bethke, A. D, "Comparison of Genetic Algorithms and Gradient-Based Optimizers on Parallel Processors : Efficiency of Use of Processing Capacity", Tech. Rep. No. 197, University of Michigan, Logic of Computers Group, Ann Arbor, MI, 1976.
6. Cantú-Paz, E, "Designing Efficient Master-Slave Parallel Genetic Algorithms", IlliGAL Report No. 97004, University of Illinois at Urbana-Champaign, Illinois Genetic Algorithms Laboratory, Urbana, IL, 1997.
7. Cantú-Paz, E "A Survey of Parallel Genetic Algorithms", Department of Computer Science, Illinois Genetic Algorithms Laboratory, University of Illinois at Urbana-Champaign, 1998.
8. Grefenstette, J. J, "Parallel Adaptive Algorithms for Function Optimization", Tech. Rep. No. CS-81-19, Vanderbilt University, Computer Science Department, Nashville, TN, 1981.
9. Holland, J. H, "Adaptation in Natural and Artificial Systems", The University of Michigan Press, Ann Harbor, 1975.
10. Kohlmorgen, U, Schmeck, H, Haase, K, "Experiences with Fine-Grained Parallel Genetic Algorithms", Annals of Operations Research, forthcoming.
11. Koza, J, "Genetic Programming: On the Programming of Computers by Means of Natural Selection", MIT Press, Cambridge, MA, 1992.
12. Koza, J, Bennett III, F, "Building a Parallel Computer System for $18,000 that Performs a Half Peta-Flop per Day", 1998.
13. Schwehm, M, "Parallel Population Models for Genetic Algorithms", Universitat Erlangen-Nürnberg, 1996.
14. Tanese, R, "Distributed Genetic Algorithm", Proc. of 3rd Int. Conf. On Genetic algorithms, pp 434-439, 1989.

Self-Organized Evolutionary Process
in Sets of Interdependent Variables
near the Midpoint of Phase Transition in K-Satisfiability

Michael Korkin

Genobyte, Inc. 1200 Pearl Street, Suite 65, Boulder, Colorado, 80302, USA
korkin@genobyte.com

Abstract. We present a mathematical model of sets of interdependent variables near the midpoint of phase transition in K-Satisfiability, and analyze model's behavior under perturbation. Surprisingly, when noise is introduced into the perturbation pattern, the model reveals a self-organized evolutionary process driven purely by criticality. The self-organized evolution discovers the interdependencies between the variables, which make some partial solutions more persistent in the presence of noise. The model suggests that a higher sensitivity to perturbation near the midpoint of the phase transition is responsible for a higher sophistication in larger sets of interdependent variables, such as organisms with a larger number of genes, larger ecologies and economies, and larger brains throughout the species.

1 Introduction

After a century of neuroscience there is nothing to suggest that a cubic millimeter worth of human cortex is significantly different than that of any other mammalian species. Nature is puzzling in its simplicity: it keeps increasing the numbers of neurons with no apparent changes in neural morphology, neural chemistry, or neural electrophysiology. To make the puzzle even more difficult, evolution preserves the overall brain anatomy as the size increases, except for the late-generated structures getting disproportionately large, especially the cortex [2]. In the process, something seemingly inexplicable happens: although the human cortex is only four times larger than that of the chimpanzee's, there is a striking difference in sophistication. Similarly, the chimpanzee is more sophisticated than the monkey, whose cortex is about one-third of the chimpanzee's, and the monkey is more sophisticated than the rat with one-fifth of the monkey's cortex. What is the enigmatic mechanism of converting more neurons into higher sophistication?

Similarly, organisms with larger sets of genes are more sophisticated, i.e. mammals versus insects, insects versus bacteria. Analogous examples are abundant throughout ecologies and economies.

We turn to the Boolean K-Satisfiability problem [3] as a mathematical tool to model large sets of interdependent variables, and then apply this model to populations of neurons in the brain as an example.

Y. Liu et al. (Eds.): ICES 2001, LNCS 2210, pp. 225–235, 2001.

2   The Satisfiability Problem

The Boolean Satisfiability problem, or SAT, belongs to a class of constraint satisfaction problems. Historically, it was the first member of a notorious group of mathematical problems known as Nondeterministic Polynomial complete, or NP-complete [3]. The NP problems are constraint satisfaction tasks, which require longer than polynomial time (in the number of variables) to solve. The SAT problem is called NP-complete because if a polynomial-time algorithm could be found to solve it, then it could be adapted to all NP problems. In the NP-complete class are the frequently studied Traveling Salesman Problem, and the Graph Coloring problem.

The SAT problem is to find a set of values for Boolean variables $v_1, v_2, v_3 \ldots v_v$, which satisfies a Boolean expression in the conjunctive normal form of the type:

$$(v_1 \text{ OR } v_2 \text{ OR } \sim v_3) \text{ AND } (v_4 \text{ OR } \sim v_5 \text{ OR } v_2) \text{ AND } (\sim v_1 \text{ OR } v_5 \text{ OR } v_3) \qquad \textbf{(1)}$$

Any Boolean expression can be readily converted into a semantically equivalent conjunctive normal form, so there is no loss of generality. The OR expressions in the brackets are called *clauses*. The number of variables in each clause is referred to as parameter K, the total number of clauses as parameter C, and the total number of variables as parameter V. In the example above K=3; C=3; V=5. In most theoretically studied cases [6], the clause size K is kept the same for all clauses in the formula, and the SAT problem is referred to as K-SAT.

The ratio of clauses to variables C/V is a crucial parameter affecting satisfiability. For example, formulas similar to (1) with very few clauses and many variables are easily satisfied (low C/V), since most variables appear rarely, and the conflict between them is unlikely. On the other end of the spectrum there are problems with a large number of clauses and very few variables (high C/V), so that each variable appears in most clauses, and the conflicts are frequent.

2.1   Phase Transition in K-SAT

In theoretical studies of K-SAT [6], it is a common technique to randomly generate multiple formulas similar to (1) with different numbers of clauses and variables, and then run solution-finding algorithms on a computer to accumulate the satisfiability statistics. As the ratio of clauses to variables increases, it gets progressively harder to satisfy the K-SAT expressions.

Surprisingly, the transition from satisfiable to non-satisfiable is not gradual, but sudden and abrupt, after a certain critical C/V ratio is exceeded. The transition statistics, averaged over thousands of runs, is shown in Fig. 1. As the C/V ratio increases, a sudden phase transition takes place.

Satisfiability behavior near the phase transition has very important dynamic properties. As the number of variables increases, while the C/V ratio is kept unchanged, the transition curve gets steeper, approaching a step function at very large number of variables. The midpoint of the phase transition depends on the clause size K (the number of variables in clauses). The critical C/V ratio corresponding to the midpoint of the transition curve is:

$$C/V \approx Ln(2) \times 2^K \qquad (2)$$

The computational cost of solving the K-SAT problems is shown in Fig. 2. For the low C/V ratios the computational cost is low because solutions are abundant. The cost is also low for high C/V ratios, because the solution-finding algorithms quickly discover multiple conflicts between the variables in clauses, and seize searching.

The hard problems are those clustering around the phase transition: they are either almost satisfiable, or almost non-satisfiable, so it takes more computing to find the solution. The larger the number of variables, the harder it is to find the solution near the steep phase transition.

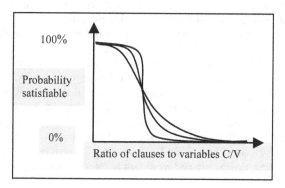

Fig.1. Shape of the phase transition curves from satisfiable to non-satisfiable in K-SAT. Three different curves shown for problems with different numbers of variables.

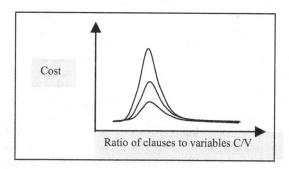

Fig. 2. Computational cost of solving K-SAT for different numbers of variables.

2.2   K-SAT Behavior under Perturbation

Computational cost correlates with the sensitivity to perturbation. We define perturbation as an imposition of partial constrains on a specific subset of variables in a K-SAT expression. These are the constants, not subject to change during the search for solutions.

The K-SAT problems near the phase transition are extremely sensitive to the perturbation. In large problems it may be sufficient to impose just a few constraints to sharply change the satisfiability of the set. On the opposite side of the spectrum there are problems with relatively small number of variables, which are much harder to push through the phase transition by applying a perturbation, because many alternative satisfying solutions may exist.

If a K-SAT expression is such that it resides exactly at the midpoint of the phase transition curve, where the critical C/V ratio equals $Ln(2) \times 2^K$, and the satisfiability is at 50%, the risk of not finding any solutions is equal to the chance of finding many alternative solutions.   Thus, solutions do exist, but not too many of them.   A perturbation may change the satisfiability in either direction.

A surprising dynamics is revealed, when noise is introduced into the perturbation pattern of a K-SAT expression residing at the critical point.

We define noise as a continuous random fluctuation of a small number of the partial constraints, contained within the perturbation pattern, taking place over a period of time. In the presence of noise, the satisfiability of the K-SAT expression will randomly fluctuate around the point on the phase transition curve, which corresponds to the satisfiability of the expression under this pattern of perturbation.

As the satisfiability fluctuates, a multitude of similar, but not identical solutions may be found, which satisfy the K-SAT expression. Taken together as a group over a period of time, these can be viewed as a *population* of responses to the perturbation pattern.

When observed over a period of time, certain *partial* solutions in subsets of variables will appear more frequently in the population, than certain other partial solutions. These are the relatively *fitter* partial solutions in the sense of their high reliability of response to the perturbation pattern in the presence of noise.

In other words, the perturbation pattern applies *selective pressure* on the K-SAT system to select for the more reliable partial solutions.

If these relatively fitter partial solutions were randomly recombined, while the less fit partial solutions were discarded, the new population of solutions would achieve an overall higher reliability of responses to the same perturbation pattern under noise.

In terms of K-SAT, the recombination of partial solutions requires that certain clauses, which prevent these partial solutions from being jointly satisfied, are deleted from the K-SAT expression, while certain new clauses, which facilitate the joint satisfiability, are introduced. In other words, the interdependencies in the given set of variables, described by the K-SAT expression, must be rearranged to raise the reliability of responses to the perturbation pattern.

2.3   Self-Organized Evolutionary Process in K-SAT

Suppose, new clauses with random content are blindly introduced into the K-SAT expression, while the fractionally satisfied old clauses, are deleted. The fractionally satisfied clauses are the ones in which the conjunction of the variables in the clause is not satisfied, when the disjunction is.

It must be stressed, that this clause modification mechanism is purely local, operating at the individual clause level, and does not analyze the overall satisfiability of the K-SAT expression in any way, explicitly or implicitly.

Over a period of time, the partial solutions, which are more frequent in the population, will weaken the interdependency with the inconsistent partial solutions, and strengthen the interdependency with the persistent partial solutions.   This will result in recombination of the persistent partial solutions to form a new population of solutions, inheriting the fitter characteristics. Occasionally, a random change in the persistent partial solution may also be introduced, which will make it more or less persistent.

A continuous succession of recombination and mutation in the populations of similar, but not identical, K-SAT solutions reveals *a self-organized evolutionary process.*   Over time, this process improves the reliability of response to the perturbation patterns in the presence of noise. Multiple patterns can be responded to by the same system of interdependent variables.

This evolutionary process in large sets of interdependent variables is driven purely by the self-organized critical phenomena at the midpoint of the phase transition curve. We next apply this model to biological nervous systems.

3   Nervous Systems and K-SAT

Recent discoveries in neuroscience are changing the traditional views of the nervous tissue in three different ways [7]:

First, the traditional view of the postsynaptic membrane potential as a linear sum of synaptic activation is being revised.   It is replaced with a view that postsynaptic activation is highly nonlinear: multiple synapses clustered close together on the dendritic branch must be co-activated simultaneously (within a short time window) to evoke any postsynaptic potential [8].   Dendritic membrane potentials (dendritic spikes) originating in different branches of the dendritic tree can independently trigger neuronal spikes, which then propagate along the axon to multiple target neurons.

Second, the spiking activity of neurons is no longer viewed in terms of firing rates (average number of spikes in a time window), but in terms of the finely tuned temporal structure of individual spikes and the inter-spike intervals within the spike trains [9].

Third, the traditional view of the synaptic weight as a continuously graded, high-resolution parameter is being replaced with a more recent view of a synapse with very small number of stable states, ultimately a binary synapse [8]. Synaptic modification can rapidly occur on a much shorter time scale than it was previously thought.

3.1   Modeling Biological Neurons in K-SAT

We first model the nonlinear function of the dendritic branch as a Boolean AND of multiple clustered binary synapses.   In order to activate the postsynaptic membrane potential, all of the excitatory synapses in the cluster must receive the afferent spikes, and all of the inhibitory synapses in the cluster must receive no afferent spikes within the same (short) time window.

In turn, each of the afferent spikes is a product of the corresponding neuron with its own dendritic branches. Each individual synaptic cluster on a dendritic branch can fire the neuron. Thus, we model neuronal spike generation as an OR function of multiple dendritic branch potentials. The afferent spikes received by a neuron are designated as Boolean variables $v_1, v_2, v_3 \ldots v_v$; a Boolean expression for spike generation in the biological neuron model is of the type:

$$(v_1 \text{ AND } \sim v_2 \text{ AND } v_3) \text{ OR } (v_1 \text{ AND } \sim v_6) \text{ OR } (v_3 \text{ AND } \sim v_5 \text{ AND } \sim v_7) \qquad (3)$$

Each AND clause in the brackets corresponds to one dendritic branch with clustered synapses (variables in clauses); the OR expression encompasses all branches of the neuron. Some variables in the expression are inverted ($\sim$), which corresponds to inhibitory synapses. Note, that the same variable can participate in multiple AND clauses, because the afferent axons can branch to make multiple synaptic contacts in different clusters.

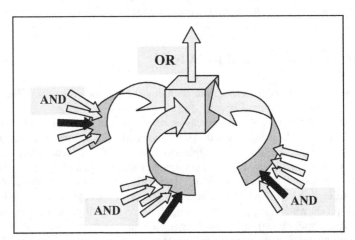

Fig. 3. A model of a biological neuron: three dendritic branches shown with a five-synapse cluster on each branch. Each cluster has four excitatory synapses and one inhibitory.

3.2   Modeling the Biological Brain in K-SAT

We model the biological brain as a system of multiple brain subsystems. Each brain subsystem is a large population of neurons of the type shown in Fig. 3, whose spiking activity is described by the expression of type (3). Each neuron within each brain subsystem is randomly connected to multiple target neurons, and receives connections from multiple afferent neurons.

In our model, each brain subsystem also receives and sends random connections from/to other brain subsystems, however the number of these external connections is small relative to the number of connections within each subsystem.   Some of the subsystems are the peripheral sensory neurons, which respond to changes in the environment.   Other subsystems send activation to muscles and body organs, and receive a reciprocal activation from these muscles and body organs. Most subsystems interconnect exclusively with the other sub-systems within the brain.

Because any cluster in any neuron may receive both excitatory and inhibitory connections, for any neuron to *ever* fire there must exist at least one other firing neuron (including itself). In other words, the probability of firing a neuron is 100%. This condition is expressed as a "long" disjunction (OR) of all the neurons of type (3) in the brain subsystem.

The satisfiability of such an expression is an inverse of the satisfiability of the K-SAT expression in its canonical conjunctive normal form (1).   Although the satisfiability curve in Fig. 1 is flipped vertically, the critical C/V ratio versus K in equation (2) holds.

It is important to stress, that the K-SAT expression for each brain sub-system includes the pattern of spikes arriving externally and independently of the interactions between the neurons within the subsystem. This is a perturbation pattern, as defined earlier.

Depending on the satisfiability of this "long" K-SAT expression, there may be multiple combinations, few combinations, or zero combinations of firing and non-firing neurons capable of satisfying this expression in the presence of a particular perturbation pattern. If multiple solutions do exist, only one combination at a time can be instantiated on the same set of neural "hardware". In other words, the brain sub-system is time-shared between the alternative solutions.

A fundamental question is where in relation to the phase transition midpoint, in terms of C/V ratio and K, did evolution place the biological brain in terms to its neuroanatomy and neuromorphology?

Clause size K corresponds to the size of the synaptic cluster. In the biological neurons this parameter has been estimated both analytically and empirically for pyramidal neurons and other neuronal morphologies in various parts of the brain [8]. It typically varies in the range 8-12 synapses. For K=10, the midpoint of the phase transition in K-SAT corresponding to the critical C/V ratio is: $Ln(2) \times 2^{10} = 709.78$.

In other words, phase transition between non-satisfiability and satisfiability in the interconnected populations of biological neurons occurs when an average neuron has approximately 710 dendritic branches with synaptic clusters. This mathematically derived number is highly consistent with the neuromorphological findings, estimating the number of branches on the order of 10^3 in biological neurons [8]. In other words, in a population of V interconnected neurons with $C=10^3$ synaptic clusters each, the resultant C/V ratio is 10^3.

These numbers suggest that the biological nervous systems may in fact reside near the phase transition boundary in K-SAT. An important question is how would the K-SAT model behave if these parameters were far from critical?

If the synaptic cluster size K were larger (with the same C and V), or the number of branches C was lower (with the same K and V), the satisfiability in K-SAT would be falling to zero. It would be hard to find any combination of neurons to fire together with any other neurons. If, on the other hand, size K of the synaptic cluster was much smaller, or if the number of branches C per neuron was much higher, then the satisfiability would be approaching 100%. In that case, most neurons would fire frequently in multiple combinations in response to almost any activity in the other neurons, without discrimination.

Thus, the brain model in K-SAT requires that parameters C/V and K reside close to the midpoint of the phase transition curve (Fig. 1).

3.3   Neural Dynamics under Perturbation

As follows from the above, biological nervous systems must be near-critical by design, and, as a consequence, highly sensitive to perturbation in terms of K-SAT.

In the brain model, a perturbation is an externally generated pattern of spiking activity arriving to some of the synapses in the brain sub-system. The sub-system reacts to the perturbation by firing a combination of neurons, and not firing some other combination of neurons. This solution must satisfy both the internal constraints within the brain sub-system, and the pattern of perturbation. At the same time, any alternative solution will be necessarily extinguished.

This critical behavior in neural populations is particularly well manifested in the ambiguous images, such as Necker Cube, as well as in the generally narrow attention span [5].

Ideally, the overall satisfiability of the brain sub-system must be placed *exactly* at the midpoint in the phase transition curve at 50%. At this point of instability, the risk of indiscriminate firing in response to a specific perturbation pattern equals the risk of not finding any response.

In our brain model, the pattern of perturbation imposed on any of the brain sub-systems is the afferent activity from the other sub-systems. The peripheral sensory sub-systems are also perturbed by the patterns in the environment, so they receive both external and internal perturbation patterns. In addition, our brain model includes the extracellular neurochemistry and the glia, as the complimentary mechanisms of super-imposing internal global perturbation on top of the external perturbation by varying the excitability of neurons within the brain sub-system.

The neural dynamics of the K-SAT brain model closely matches the experimentally observed biological brain activity, revealed by fMRI imaging techniques [1], and the SQUID array techniques [5]. The model may explain the "binding" phenomenon [1], which takes place within a 100-200 ms time window after a task presentation and encompasses large populations of neurons firing synchronously in various parts throughout the brain. This combination of firing and non-firing neurons is highly specific to each task presented. Different individuals have completely different patterns of activity when presented with the same task. No two brains are exactly alike, including the identical twins [1].

3.4 Evolutionary Processes as Learning in the Brain

As it was shown earlier, in presence of noise in the pattern, the satisfiability of the K-SAT expression will randomly fluctuate around the point on the phase transition curve, which corresponds to the satisfiability of the expression under this pattern of perturbation.

As the satisfiability fluctuates, a multitude of similar, but not identical combinations of firing and non-firing neurons will respond, which satisfy the K-SAT expression. Taken together as a group, over a period of time, these can be viewed as a *population* of neural responses to the perturbation pattern.

When observed over a period of time, certain partial responses in subsets of neurons will appear more frequently in the population, than other partial responses. These are the relatively *fitter* partial responses in a sense of their higher reliability of recognizing the perturbation pattern, when noise is present.

If these relatively fitter partial responses were randomly recombined, while the less fit ones discarded, the new population of solutions would have an overall higher reliability of recognition of the same perturbation pattern under noise, or higher fitness.

In order to improve the fitness of neural responses, a brain sub-system must rearrange the interdependency between its neurons, or rewire.

Our brain model uses a simple rewiring mechanism identical to the earlier described clause modification mechanism to introduce new synaptic clusters with random content, while at the same time deleting the fractionally satisfied clusters. This mechanism requires that all neurons continuously grow new dendritic branches at a certain rate. A new branch contains a cluster of synapses, and the proximal afferent axons randomly connect to them. At the same time, the fractionally activated clusters (with conflicting combinations of excitatory and inhibitory spikes) are depressed and removed, so that the branch completely disappears. Only the branches, which are consistently fully activated, or consistently fully inactivated (no conflicting afferent activity in either case) are retained. This mechanism is highly consistent with the neuroscientific findings [7], [8].

Over a period of time, the persistent partial neural responses will weaken the interdependency with the inconsistent partial responses, and strengthen the interdependency with the other persistent partial responses (in different subsets of neurons). Occasionally, a random change in the persistent partial response may be introduced, which will make it more or less persistent.

A continuous succession of recombination and mutation in the populations of similar, but not identical, neural responses constitutes *a self-organized evolutionary process* responsible for adaptation and learning in the brain. Multiple patterns can be learned by the same brain sub-system of interconnected neurons. Both spatial and temporal patterns will be learned.

This evolutionary dynamics is purely self-organized as a result of model sensitivity to the perturbation in the presence of noise. The time scale of this evolution can be as short as hundreds of milliseconds to seconds, or as long as the lifetime of the brain.

Most importantly, this kind of brain model learns in the presence of noise, i.e. from incomplete or partially corrupted samples, unlike most other models of nervous systems, which require complete and uncorrupted pattern presentation during learning. Neither an external agent, nor an internal "homunculus", is needed to account for any aspects of learning.

3.5   Brain Size and Phase Transition Dynamics

Why evolution favors larger brains as more sophisticated, from mice to men, by simply increasing the total number of neurons V, with no significant modification in C/V, and K?

Since the larger brain has larger parameter V, in order to maintain the critical C/V ratio constant, the length of the K-SAT expression must be longer. For any given neuron to fire, a longer chain of other interdependent neurons must be firing and not firing, than in the smaller brain.

A steeper phase transition curve in the larger brain points to a higher sensitivity to subtle novelties. In a K-SAT model of a smaller brain, a subtle novelty in the perturbation pattern may not result in any change in the current population of responses.

By contrast, an *identical* subtle novelty presented to a larger brain is more likely to change the current population of responses. The changing population will rapidly evolve in the direction of higher reliability of response to this novelty.

Thus, on the one hand, a larger brain is capable of learning larger, more complicated patterns (finding solutions to longer K-SAT expressions), and on the other hand, learning more intricate interdependencies within the same patterns (higher sensitivity to subtlety). The smaller brain remains literally and mathematically "satisfied" in K-SAT with the "simpler" solutions.

4   Towards the K-SAT Artificial Brain

Can an artificial brain based on the presented K-SAT model be constructed? Will it exhibit any of the self-organized properties predicted by the model? How intelligent can it be? Can a very large K-SAT artificial brain exceed the human brain in intelligence?

The answer to these questions, according to the K-SAT brain model, depends purely on the economic feasibility of interconnecting large numbers of neurons while preserving the near-critical C/V and K. An artificial brain, endowed with more than 10^{11} neurons, and more than 10^{15} synapses, all operating in parallel on a time scale of hundreds of milliseconds, will achieve a steeper phase transition curve than the human brain has achieved. According to the presented model, it may be capable of discovering more intricate interdependencies in the perturbation patterns presented to it, as well as learn more complicated patterns.

From the engineering point of view, the real-time computational requirements for the artificial brain (an ability to perform its self-organized processes in response to the real world events) are very severe. On the other hand, the predicted self-organized evolutionary process could be demonstrated in an economically feasible hardware or software model, which the author is currently working on.

5   Self-Organized Evolution in Natural and Artificial Systems

The presented view of the self-organized evolution at the midpoint of the phase transition in K-SAT may be applicable to a large number of natural and artificial systems [4], which reside at the critical C/V ratio versus K, as it was demonstrated for the biological nervous systems.

In some systems the C/V ratio may itself fluctuate over time. For example, in the biological nervous systems the C/V ratio is initially elevated above the critical in the early neonatal period. This is accomplished through a more extensive dendritic branching in the younger neurons to raise the initial satisfiability in order to produce multiple alternative responses to a perturbation at the expense of specificity.

Although not treated here, the Darwinian species evolution can be readily presented in terms of C/V ratio and K. Similarly, certain artificial systems with large sets of interdependent variables, such as manufacturing processes and transportation systems, can be analyzed and even regulated in terms of C/V ratio and K using the presented analysis. Of a particular interest is a conjecture that a self-organized evolution in simple molecules, interdependent through their chemical reactions, may be responsible for the pre-genetic origins of life.

Conclusions

The self-organized evolutionary processes in sets of interdependent variables, driven purely by criticality in the presence of noise, may explain a large number of diverse phenomena in many natural and artificial systems. As long as the system resides near the midpoint of the phase transition curve from satisfiable to non-satisfiable, and it includes a local clause modification mechanism, it will exhibit the self-organized evolutionary process.

References

1.   Edelman G.M., Tononi G. Consciousness, How Matter Becomes Imagination. The Penguin Press, 139-154.
2.   Finlay, B.L., Darlington, R. B. Linked Regularities in the Development and Evolution of Mammalian Brains. Science, 268, (1995) 1578-1584.
3.   Hayes, B. Can't Get No Satisfaction. American Scientist, Volume 85, Number 2, 1997, 108-112.
4.   Kauffman, S. Investigations. Oxford University Press, 2000, 192-194.
5.   Kelso, J.A. Dynamic Patterns. MIT Press, 1995, 274-285.
6.   Kirkpatrick, S., Selman B. Critical Behavior in the Satisfiability of Random Boolean Expressions. Science, 264, (1994) 1297-1301.
7.   Koch C. Biophysics of Computation. Oxford University Press, 1999, 280-306.
8.   Poirazi, P., Mel, B.W. Impact of Active Dendrites and Structural Plasticity on the Memory Capacity of Neural Tissue. Neuron, Vol. 29, 779-796. March, 2001.
9.   Rieke F., Warland D., de Ruyter van Steveninck R., and Bialek W. Spikes: Exploring the Neural Code. MIT Press/Bradford Books, Cambridge, MA, 1997.

Evolutionary Optimization
of Yagi-Uda Antennas

Jason D. Lohn[1], William F. Kraus[1],
Derek S. Linden[2], and Silvano P. Colombano[1]

[1] Computational Sciences Division, NASA Ames Research Center,
Mail Stop 269-1, Moffett Field, CA 94035-1000, USA
{jlohn, bkraus, colomban}@email.arc.nasa.gov
[2] Linden Innovation Research, P.O. Box 1601, Ashburn, VA, 20146, USA
dlinden@lindenir.com

Abstract. Yagi-Uda antennas are known to be difficult to design and optimize due to their sensitivity at high gain, and the inclusion of numerous parasitic elements. We present a genetic algorithm-based automated antenna optimization system that uses a fixed Yagi-Uda topology and a byte-encoded antenna representation. The fitness calculation allows the implicit relationship between power gain and sidelobe/backlobe loss to emerge naturally, a technique that is less complex than previous approaches. The genetic operators used are also simpler. Our results include Yagi-Uda antennas that have excellent bandwidth and gain properties with very good impedance characteristics. Results exceeded previous Yagi-Uda antennas produced via evolutionary algorithms by at least 7.8% in mainlobe gain. We also present encouraging preliminary results where a coevolutionary genetic algorithm is used.

1   Introduction

Automated antenna synthesis via evolutionary design has recently garnered much attention in the research literature [12]. Underlying this enthusiasm is an issue that many designers readily acknowledge - good antenna design requires not only knowledge and intelligence, but experience and artistry. Thus automated design techniques and tools have been lacking. Evolutionary algorithms show promise because, among search algorithms, they are able to effectively search large, unknown design spaces.

The particular antenna we study in this paper is the Yagi-Uda, first proposed in 1926 [14]. We chose this type of antenna because it presents difficult design and optimization challenges, and because it was previously studied with respect to evolutionary design [7]. The Yagi-Uda antenna is comprised of a set of parallel elements with one reflector element, one driven element (driven from its center), and one or more director elements (see Fig. 1). The highest gain can be achieved along the axis and on the side with the directors. The reflector element reflects power forwards and thus acts like a small ground plane. The design parameters consist of element lengths, inter-element spacings, and element diameters.

Y. Liu et al. (Eds.): ICES 2001, LNCS 2210, pp. 236–243, 2001.

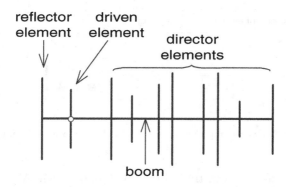

Fig. 1. Typical Yagi-Uda antenna.

The application that we use is taken from [7]. It involves designing a special feed for the Arecibo 305-meter spherical reflector in Puerto Rico [3]. The antenna was to be used to search for primeval hydrogen having a redshift of approximately 5. Neutral hydrogen line emission is at a frequency of 1420 MHz; thus the frequency region of interest was about 235 MHz. Preliminary studies indicated that the band from 219 to 251 MHz was of the greatest interest, particularly from 223 to 243 MHz. The most important design goal was for the feed to have sidelobes/backlobes at least 25 dB down from the mainbeam gain in the region from $70° < \phi < 290°$, due to the interference which came from surrounding radio and TV towers. Of lesser importance was that the E-plane (the plane parallel to the plane of the antenna) and H-plane (perpendicular to the E-plane) beamwidths be about 50°.

Voltage Standing Wave Ratio, or VSWR, is a way to quantify reflected-wave interference, and thus the amount of impedance mismatch at the junction. VSWR is the ratio between the highest voltage and the lowest voltage in the signal envelope along a transmission line [13]. The VSWR was desired to be less than 3 and the gain was to be maximized, limited by the wide beamwidth. The feed would be mounted over a 1.17 meter square ground plane-that is, a ground plane only 0.92λ in size.

2   Antenna Representation and Operators

The representational scheme used is similar to that taken from [7]. As shown in Fig. 2, this scheme is comprised of 14 elements, each one encoding a length and spacing value. Each floating point value was encoded as three bytes, yielding a resolution of $1/2^{24}$ per value. The first pair of values encoded the reflector element, the second pair encoded the driven element, and the remaining 12 pairs encoded the directors. One point crossover was used with cut points allowed between bytes. Mutation was applied on individual bytes.

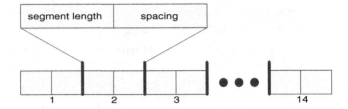

Fig. 2. Genetic representation of a 14-element Yagi-Uda antenna.

Radius values were constrained to 2, 3, 4, 5, or 6 mm. All elements within a given individual were assigned the same radius value. Element lengths were constrained to be symmetric around the x-axis and between 0 and 1.5λ. Elements having zero length were removed from the antenna; as a consequence, a constructed antenna could have less than 14 elements. Spacing between adjacent elements (along the z axis) was constrained to be between 0.05λ and 0.75λ. The wavelength λ was 1.195 meters, the wavelength of 235 MHz.

3   Experimental Setup

Experiments were set up as follows. The NEC2 simulation program [4] was used to evaluate all antenna designs. We used a parallel master/slave generational genetic algorithm with a population size of 6000. One point crossover across byte boundaries was used at a rate of 80%. Mutation was uniform across bytes at a rate of 1%. Runs were executed on a 32-node Beowulf computing cluster [11].

The wire geometry encoded by each individual chromosome was first translated into a NEC input deck, which was subsequently sent to the NEC2 simulator. The segment size for all elements was fixed at 0.1λ, where λ was the wavelength corresponding to 235 MHz. The source element for excitation was specified to be the middle segment of the driven element. The z location of the reflector element was always set to 0. The antenna was analyzed in free space.

The simulator was instructed to sample the radiation pattern of each individual at three different frequency values: 219, 235, and 251 MHz, representing a 13.6% bandwidth. Each radiation pattern was calculated at ϕ set to 0° and θ varying between 0° and 355°, the latter sampled at 5° increments. VSWR values were also calculated for each of the three frequencies.

Fitness was expressed as a cost function to be minimized. The calculation was as follows:

$$F = -G_L + \sum (C * V_i) \qquad (1)$$

where: G_L = lowest gain of all frequencies measured at $\theta = 0°$ and $\phi = 0°$, V_i = VSWR at the ith frequency, and

$$C = \begin{cases} 0.1 \text{ if } V_i \leq 3 \\ 1 \ \ \text{ if } V_i > 3 \end{cases}$$

Lacking from this calculation was a term involving sidelobe/backlobe attenuation. We chose not include such a term because we reasoned that as the mainlobe gain increased, the sidelobes/backlobes would decrease in size.

4   Experimental Results

Thirteen runs were executed under differing random number streams for comparison purposes. Table 1 summarizes the run data for the best antenna found in each run of 100 generations. Fig. 3 shows the radiation pattern from the best antenna found (run 13). It exhibits 10.58 dB and has a VSWR of 2.02 at its center frequency. Its sidelobe/backlobe gain at this frequency is 3.07 dB. Fig. 4 shows a diagram of the antenna's physical structure.

To increase simulation speed, the evolved antennas were produced without the presence of a ground plane – an idealized setting. Adding a ground plane thus simulates more realistic conditions. We removed the reflector element and simulated the best antennas found over a ground plane of 1.17 meters [7]. We found the performance increased – at the center frequency the mainlobe gain was 12.52 dB and the VSWR was 2.39. At 291 MHz, the gain was 11.33 dB, and at 251 MHz, the gain was 11.15 dB. In contrast, the antenna produced in [7] exhibits gains of 10.36, 10.91, 10.34 dB at 219, 235, and 251 dB, respectively. Thus the antenna from run 13 has a minimum performance increase of 7.8% as compared to the previously reported antenna.

Table 1. Results from the best individual after 100 generations for each of the 13 runs (dB is measured at $\phi = 0°$, $\theta = 0°$).

Run	219 MHz		235 MHz		251 MHz	
	dB	VSWR	dB	VSWR	dB	VSWR
1	9.63	2.33	9.64	1.67	10.20	2.99
2	9.49	2.23	9.08	1.85	9.20	1.58
3	9.23	2.89	10.04	1.11	9.62	2.60
4	9.24	2.47	9.23	1.35	9.37	2.83
5	8.73	2.83	8.79	1.51	9.22	2.60
6	9.35	2.87	9.51	1.73	9.28	2.00
7	9.87	2.64	9.82	1.99	9.46	1.98
8	9.04	2.35	9.02	1.64	9.08	2.92
9	9.44	2.96	9.46	1.87	9.51	2.39
10	9.02	1.25	9.12	2.42	9.02	1.41
11	10.01	1.95	9.81	1.97	10.11	1.66
12	9.37	2.55	9.17	1.70	9.41	2.47
13	10.34	2.57	10.58	2.02	10.51	1.70

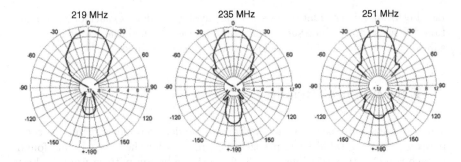

Fig. 3. Radiation pattern of the best evolved antenna without a ground plane, measured at $0° \leq \theta < 360°$, $\phi = 0°$, for 219, 235, and 251 MHz, respectively. (The scale is 2 dB per division. Inner ring is -12 db, outer ring is 12 dB.)

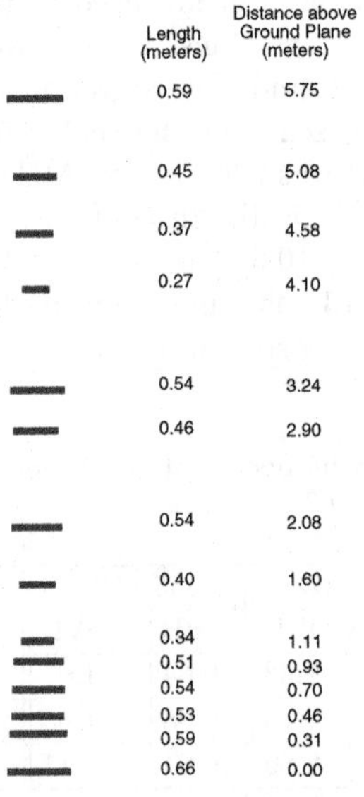

Length (meters)	Distance above Ground Plane (meters)
0.59	5.75
0.45	5.08
0.37	4.58
0.27	4.10
0.54	3.24
0.46	2.90
0.54	2.08
0.40	1.60
0.34	1.11
0.51	0.93
0.54	0.70
0.53	0.46
0.59	0.31
0.66	0.00

Fig. 4. The best Yagi-Uda antenna from run 13. The radius of all elements was 3 mm.

5 Coevolutionary Algorithm – Method and Preliminary Results

A coevolutionary genetic algorithm was also applied to the antenna optimization problem described above. The experiments are ongoing as of this writing, and we briefly mention some encouraging initial results. The algorithm used is similar to that presented in [9]. Two populations are used: one consisting of antenna designs as described above, and one consisting of target vectors. The fundamental idea is that the target vectors encapsulate level-of-difficulty. Then, under the control of the genetic algorithm, the target vectors evolve from easy to difficult based on the level of proficiency of the antenna population.

Each target vector consists of a set of objectives that must be met in order for a target vector to be "solved." A target vector consisting of two values: the mainlobe gain (in dB) and a VSWR value. A target vector was considered to be solved by a given antenna if:

$$G_{\text{target}} < G_L \quad \text{and} \quad V_{\text{target}} > V_L$$

where G_L is lowest gain of all frequencies measured at $\theta = 0°$ and $\phi = 0°$, and V_L is lowest VSWR of all frequencies. For example, an antenna with a G_L value equal to 5 dB and a V_L value equal to 8 would solve the target vector $\langle 2, 12 \rangle$ but not $\langle 7, 12 \rangle$.

Values for target gain ranged between 0 dB (easy) and 12 dB (difficult). Target VSWR values ranged between 12 (easy) and 3 (difficult). Target vectors are represented as a list of floating point values that are mutated individually by randomly adding or subtracting a small amount (5% of the largest legal value). Single point crossover was used, and crossover points were chosen between the values.

The general form of the fitness calculations are from [9]. In summary, antennas are rewarded for solving difficult target vectors. The most difficult target vector is defined to be the target vector that only one antenna can solve. Such a target vector garners the highest fitness score. Target vectors that are unsolvable, or are very easy to solve by the current antenna population, are given low fitness scores.

We ran our coevolutionary algorithm for 200 generations using 1600 individuals in both populations. In the antenna population, crossover and mutation rates were 0.8 and 0.1, respectively. In the target vector population, crossover and mutation rates were 0.8, 0.5, respectively.

The highest-fitness individual came from generation 199. It had mainlobe gains of 8.30, 8.51, and 8.30 dB at 219, 235, and 251 MHz, respectively. While performance is less than the runs from above, it was achieved with a much smaller population, and it is currently our single data point.

Fig. 5 shows a plot of how the highest fitness target vectors varied during the run. Such plots can give insight regarding the difficulty of achieving one objective at the expense of another. In the plot, we see that difficult VSWR levels (near 3.0) are attainable early on and remain so throughout the run. The algorithm focuses on gain, presumably the more difficult objective to meet. We see sudden jumps in gain near generations 13 and 190, accompanied by relaxations in the VSWR.

6 Discussion

Small improvements in antenna performance can be significant in many applications. Because of their numerous design variables, complex behavior, and sensitivity to parameters, Yagi-Uda antennas are notoriously difficult to optimize. Our experiments produced several excellent antennas in a relatively small

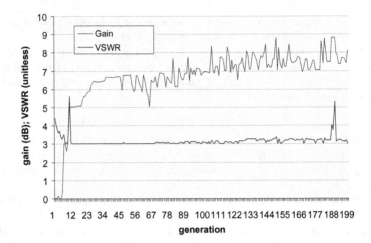

Fig. 5. Coevolution run: plot of gain and VSWR for the best target vector over 200 generations.

number of generations. When simulated over a finite ground plane, the highest performance antenna found exhibiting a mainlobe gain that was 7.8% higher than a previously-reported antenna.

Previous work has explicitly included a sidelobe/backlobe term in the fitness function in order to minimize radiation outside of the desired direction [7]. We did not include an explicit sidelobe/backlobe term but rather relied on the fact that the radiation pattern of an antenna is a zero sum quantity - increasing the intensity in one direction will implicitly reduce the amount of radiation in other directions.

Finally, we are encouraged by our preliminary results produced using coevolutionary optimization. There we saw an antenna generated that had very good properties while requiring less evaluations than the standard GA approach.

Acknowledgments

The research described in this paper was performed at NASA Ames Research Center, and was sponsored by the NASA Intelligent Systems Program. Radiation pattern plots were made using NEC-Win Plus software from Nittany Scientific.

References

1. E.E. Altshuler and D.S. Linden. "Design of a Loaded Monopole Having Hemispherical Coverage Using a Genetic Algorithm." IEEE Trans. on Antennas and Propagation., Vol. 45, No. 1, January 1997.
2. E.E. Altshuler and D.S. Linden. "Wire Antenna Designs using a Genetic Algorithm." IEEE Antenna & Propagation Society Mag., Vol. 39, pp. 33-43, April 1997.

3. I.M. Avruch, et al., "A Spectroscopic Search for Protoclusters at High Redshift." Bulletin of the American Astron. Society, Vol. 27, No. 4, 1995.

4. G.J. Burke and A.J. Poggio. "Numerical Electromagnetics Code (NEC)-Method of moments." Rep. UCID18834, Lawrence Livermore Laboratory, Jan. 1981.

5. C.D. Chapman, K. Saitou, M.J. Jakiela. "Genetic Algorithms as an Approach to Configuration and Topology Design." J. Mechanical Des.,Vol. 116, December 1994.

6. J.H. Holland, *Adaptation in Natural and Artificial Systems*, Univ. of Michigan Press, Ann Arbor, 1975.

7. D.S. Linden, "Automated Design and Optimization of Wire Antennas using Genetic Algorithms." Ph.D. Thesis, MIT, September 1997.

8. D.S. Linden and E.E. Altshuler. "Automating Wire Antenna Design using Genetic Algorithms." Microwave Journal, Vol. 39, No. 3, March 1996.

9. J.D. Lohn, G.L. Haith, S.P. Colombano, D. Stassinopoulos, "A Comparison of Dynamic Fitness Schedules for Evolutionary Design of Amplifiers," Proc. of the First NASA/DoD Workshop on Evolvable Hardware, Pasadena, CA, IEEE Computer Society Press, 1999, pp. 87-92.

10. J.D. Lohn, S.P. Colombano, "A Circuit Representation Technique for Automated Circuit Design," IEEE Transactions on Evolutionary Computation, vol. 3, no. 3, 1999, pp. 205-219.

11. J.D. Lohn, S.P. Colombano, G.L. Haith, D. Stassinopoulos, "A Parallel Genetic Algorithm for Automated Electronic Circuit Design," Proc. of the Computational Aerosciences Workshop, NASA Ames Research Center, Feb. 2000.

12. Electromagnetic Optimization by Genetic Algorithms. Y. Rahmat-Samii and E. Michielssen, eds., Wiley, 1999.

13. D. H. Staelin, et al. 6.014: Electromagnetic Waves (course notes). MIT, May, 1992.

14. S. Uda and Y. Mushiake, Yagi-Uda Antenna, Maruzden, Tokyo, 1954.

Extraction of Design Patterns from Evolutionary Algorithms Using Case-Based Reasoning

E. Islas Pérez[1], C.A. Coello Coello[2], and A. Hernández Aguirre[3]

[1] Instituto de Investigaciones Eléctricas,
Av. Reforma #113, Col. Palmira, 62490 Temixco Morelos, Mexico,
eislas@iie.org.mx
[2] CINVESTAV-IPN, Depto. de Ingeniería Eléctrica, Sección de Computación,
Av. Instituto Politécnico Nacional No. 2508, Col. San Pedro Zacatenco,
México, D.F. 07300, Mexico,
ccoello@cs.cinvestav.mx
[3] Tulane University, Department of Electrical Engineering and Computer Science,
211 Stanley Thomas Hall, New Orleans, Louisiana 70118, USA
hernanda@eecs.tulane.edu

Abstract. In this paper we show a scheme based on case-based reasoning to extract design patterns from a genetic algorithm used to optimize combinational circuits at the gate-level. The approach is able to rediscover several of the traditional Boolean rules used for circuit simplification and it also finds new simplification rules. Also, we illustrate how the approach can be used to reduce convergence times of a genetic algorithm using previously found solutions as cases to solve similar problems.

1 Introduction

Although it is difficult for an evolutionary algorithm to suggest directly new design principles, it is feasible to infer such principles through a careful study and analysis of its behavior on a set of examples. That is precisely the focus of this paper. We propose that by employing case-based reasoning (CBR) techniques, we can extract and reuse design patterns that emerge from the evolutionary process of a genetic algorithm (GA). In fact, we will see how some of these rules are really the same traditionally used by human designers. However, others are entirely new simplification rules which, in some cases, may not even be intuitive to a human designer.

Another interesting aspect of this work is that we show how CBR can be used to solve more efficiently similar circuits to those previously solved. This idea, although intuitive, is not completely straightforward in practice, since the selection pressure of an evolutionary algorithm may destroy partial solutions to a problem. Our approach is therefore, to use a database of solutions previously found that have some (potentially) useful information. Then, using techniques from case-based reasoning, we retrieve this information when designing similar circuits (similarity has to be defined according to certain criteria in this context) and incorporate it in the population of another evolutionary algorithm, as to reduce convergence times. The system will be illustrated with the design of a full adder.

Y. Liu et al. (Eds.): ICES 2001, LNCS 2210, pp. 244–255, 2001.
© Springer-Verlag Berlin Heidelberg 2001

2   Related Work

Apparently, the first attempt to combine case-based reasoning (CBR) and GAs was done by Louis et al. [6]. In this paper, the authors use CBR-principles to explain solutions found by a GA. This same idea was also discussed in Louis' dissertation [4], where he proposed a system that combined CBR with GAs to improve performance of the GA. These ideas were further developed by Louis & Johnson [5] and by Liu [3]. Although Louis [4] and Louis & Johnson [5] used a few examples from circuit design (mainly parity checkers) to illustrate their principles, they did not focus their work specifically on the design of combinational circuits as in our case. Nevertheless, our current proposal has been influenced by this prior work. Several other researchers have proposed approaches that combine CBR and GAs. See for example [9,10]. However, the emphasis of these papers has been to illustrate the benefits of this sort of hybrid scheme rather than emphasizing a certain application domain like in our case. Also, some researchers in evolvable hardware have pointed out the potential benefits of using GAs as a discovery engine capable of producing novel and even inspirational designs. Miller et al. [7], for example, showed that through the evolution of a hierarchical series of examples, it was possible to rediscover the well-known ripple-carry principle for building adder circuits of any size. However, no CBR is used in this work. Recently, Thomson [11] explored the potential of evolving larger systems more quickly via a method of visualizing the subcomponents of the final solution when they appear. Taking these partially evolved solutions from short runs and feeding them to another GA, the convergence time of the GA can be improved. This work is closer to our own, but unlike our proposal, Thomson does not use CBR in his system.

3   Case-Based Reasoning

CBR is a problem-solving paradigm that in many respects is fundamentally different from other major AI approaches [2]. Instead of relying solely on general knowledge of a problem domain, or making associations along generalized relationships between problem descriptors and conclusions, CBR is able to utilize the specific knowledge of previously experienced, concrete problem situations (cases). Finding a similar past case, and reusing it in the new problem situation helps to solve a new problem. A second important difference is that CBR is also an approach to incremental, sustained learning, since a new experience is retained each time a problem has been solved, making it immediately available for future problems. A CBR system can be divided in the following main stages (see Fig. 1):

1. **Identifying the new problem:** The system receives the input case (new problem) and analyzes its most important attributes and characteristics in order to search amongst the cases that are most similar to the cases in the case base. The attributes used to measure the similarities between the cases are called indexes.

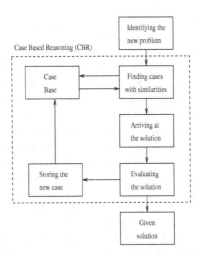

Fig. 1. General structure of a CBR system.

2. **Finding cases with similarities to the new case:** The following step is to find the cases that have more attributes in common with the attributes of the new case using the indexes found in the previous step. Sometimes it is necessary to reduce the subset in order to find the most relevant cases. The algorithm should be fast and efficient and the design is a critical and important aspect when the case base is sufficiently large. The selection of cases from the case base could be considered as analogous to natural selection due to the fact that it is based only on the distance measure (similarity rather than fitness) between the new case and each case in the case base.

3. **Arriving at the Solution:** Once we have the most similar cases, the system starts the adaptation process, which consists of the combination and modification of the most similar cases to form a new solution, and additionally an interpretation or an explanation depending on the application of the system. In most applications it is better if the system explains how it finds the new case.

4. **Evaluating the solution:** The solution obtained in the previous stage is a tentative or potential solution. It is necessary to do an evaluation of the proposed solution before giving it to the final user. This evaluation should show the qualities and weaknesses of the solution for the evaluation of its usefulness.

5. **Assignment and storing of the new case:** Once the solution has been created and evaluated, it is given to the user and then it is possible to create a new case. This new case is formed from the solution found and the original case (problem). Indexes are assigned to the new case and it is stored in the case base.

6. **Explaining, repairing and testing:** If the solution fails, it is important that the system obtains and analyzes the information in order to avoid mak-

ing the same mistakes. If something unusual happens, the system should try to explain it. Subsequently, the system repairs the solution based on the explanation and returns to the evaluation stage.

4   Statement of the Problem

We extract knowledge at two stages of the evolutionary process: at the end of a run and during a run. In the first case, the knowledge to be extracted are the Boolean laws used by the evolutionary algorithm to design a circuit. These laws are obtained after comparing the results produced from two or more runs of the GA (with different parameters) with the solution produced by a human expert. In the second case, the knowledge extracted are the building blocks that the circuit structurally maintains during its evolutionary process. When some individuals arrive at a certain (predefined) threshold in their fitness value during the evolutionary process, it means that these circuits have evolved long enough to contain good building blocks and we can then extract the knowledge that they contain and store it in a case base for further use. Our approach consists of storing solutions that were previously generated by the same GA and use them as a memory of "past experiences". Then, we can use a mechanism to detect cases similar to the one being solved and retrieve from this "memory" some solutions (or past experiences) that can be useful to solve the problem at hand. For the experiments described next, we use a GA with integer representation and matrix encoding (encoded as fixed-length linear chromosomes) that we have adopted in previous work [1].

5   Proposed System

The proposed system that combines a GA with CBR is depicted in Fig. 2.

To understand better the way in which our system works, we will describe in more detail the process of extracting knowledge in the two situations previously mentioned:

1. **At the end of the evolutionary process**: In this case, we perform complete runs of a GA solving a certain circuit. Once a solution is found, a new case is formed with such a solution and the original problem. The original problem will be considered as the attributes in the case base and the solution will be the output of the case. The system will assign other attributes, in order to have indexes that help retrieving the most similar cases in a more efficient way.

2. **During the evolutionary process**: In this case, our work is inspired on the research of Louis [4]. The GA records data for each individual in the population as it is created and evaluated. Such data include a fitness measure, the genotype and chronological data, as well as some information on the individual's parents. This collection of data is the initial case data. Though normally discarded by the time an individual is replaced, all of the case data

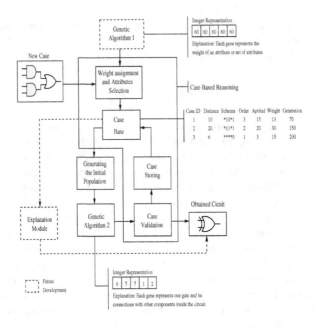

Fig. 2. Proposed system to optimize combinational logic circuits using GAs and CBR.

collected is usually contained in the GA's population at some point and it is easy to extract. When a sufficient number of individuals have been created over a number of generations, the initial case data is sent to a clustering program. A hierarchical clustering program clusters the individuals according to both, the fitness and the alleles of the genotype. This clustering constructs a binary tree in which each leaf includes the data of a specific individual. The binary tree structure provides an index for the initial case base. The numbers at the leaves of the tree correspond to the case number (an identification number) of an individual created by the GA. An abstract case is computed for each internal node based on the information contained in the leaves and nodes beneath.

5.1   Representing Circuits in the Case Base

Depending on the stage at which knowledge is extracted, the representation adopted to store it in the case base can vary:

1. **At the end of the evolutionary process**: The cases will be stored from problems that have been solved previously and they will be used for seeding the initial population of a GA. The attributes contained in this part of the case base are the following[1]: case ID, number of inputs, number of outputs,

---

[1] This scheme presents certain resemblance with the one proposed by Louis [4].

output values, fitness, and genotype. Some examples of these cases are shown in Table 1.

Table 1. Cases for knowledge extraction at the end of the evolutionary process.

Case ID	No. Inputs	No. Outputs	Output Vals	Fitness	Genotype
1	3	2	01101001000010111	39	3230132431232134103231
2	3	2	01101001000010111	38	3230132431232144133204
3	3	2	01101001000010111	39	0200234133241431231130

2. **During the evolutionary process**: The best individuals are recognized during early generations of the evolutionary process. Afterwards they are stored as cases in the case base and retrieved in later generations. Some of the attributes that are contained in this part of the case base are the following: case ID, distance from the root of the tree to the level of the case, schema for the case, scheme order, average fitness, weight (number of leaves or individuals below), and generation info (the earliest and latest leaf occurrence as well as the average in the subtree). Some examples of this sort of cases stored in the case base are shown in Table 2.

Table 2. Cases for knowledge extraction during the evolutionary process.

Case ID	Distance	Schema	Order	Fitness	Weight	Generation
1	5	710*13*2*	6	30	6	50
2	2	**4*50*2*	4	60	8	30
3	8	163*14*41	7	15	4	67

Additionally, we also performed some analysis by hand to try to understand the way in which the GA performs the simplification of a circuit. As we will show in some of the examples presented next, the GA was able to rediscover several of the simplification rules commonly used in Boolean algebra and, furthermore, was able to discover "new" simplification laws that are stored in the case base and can also be used by human designers.

6   An Example

Next, we will provide an example of how is the knowledge extracted both at the end and during the evolutionary process of a GA with integer representation used to design combinational logic circuits at the gate-level.

We want to find the Boolean expression that corresponds to the circuit whose truth table is provided in Table 3. We will start by providing the steps followed to extract knowledge at the end of the evolutionary process. First, we performed

Table 3. Truth table for the circuit of the example (a subtractor).

A	B	C	X	Y
0	0	0	0	0
0	0	1	1	1
0	1	0	1	1
0	1	1	1	0
1	0	0	0	1
1	0	1	0	0
1	1	0	0	0
1	1	1	1	1

10 runs using integer representation and the following parameters: population size=100, maximum number of generations = 2000, crossover rate = 0.5, mutation rate = 0.006. The best solution found from these runs has 7 gates and its corresponding Boolean expression is shown (under "AG Setup 1") in Table 4. This Boolean expression can be contrasted with the best solution found by a Human Designer using Karnaugh maps (this solution has 16 gates). Additionally, we performed 10 more runs using a population size of 700 and a maximum number of generations of 400. The best solution found from these runs has 6 gates and its corresponding Boolean expression is shown (under "AG Setup 2") in Table 4.

Analysis The next step was to analyze (by hand) the solutions produced by our GA with respect to those generated by the human designer: If we take the solution found by the human designer and we factorize C and C' in Y, we have that:

$$Y = A'B'C + A'BC' + AB'C' + ABC = C(A'B' + AB) + C'(A'B + AB') \quad (1)$$

To transform this equation in terms of an XOR gate:
If $S = C$ then it is necessary that $S' = C'$
and if $T = A'B' + AB$ then it is necessary that $T' = A'B + AB'$
If we apply the DeMorgan's theorem to T, we have that:

$$T' = (A'B' + AB)' = (A'B')'(AB)' = (A + B)(A' + B') \quad (2)$$

Applying the distributive law and some basic theorems:

$$T' = (A + B)(A' + B') = AA' + AB' + A'B + BB' = A'B + AB' \quad (3)$$

Verifying that $T' = A'B + AB'$ is the negation of $T = A'B' + AB$. Then we can rewrite the eq. (1) as follows:

$$Y = C(A'B' + AB) + C'(A'B + AB') = C \oplus (A'B + AB') \quad (4)$$

If we apply the operation of a XOR gate to the operand between parentheses, we have:

$$Y = C \oplus (A'B + AB') = C \oplus (A \oplus B) \quad (5)$$

Table 4. Comparison of results between a human designer and two setups of our GA.

Human Designer
$X = A'B + A'C + BC$
$Y = A'B'C + A'BC' + AB'C' + ABC$
16 gates
8 ANDs, 5 ORs, 3 NOTs
AG Setup 1
$X = (B \oplus (A \oplus C)) \oplus (A \oplus AC)$
$Y = (B \oplus (A \oplus C)) \oplus (B(A \oplus C)$
7 gates
2 ANDs, 1 OR, 4 XORs
AG Setup 2
$X = (A'(B \oplus (A \oplus C)) + BC$
$Y = B \oplus (A \oplus C)$
6 gates
2 XORs, 2 ANDs, 1 OR, 1 NOT

Applying the commutative and associative laws, we have:

$$Y = C \oplus (A \oplus B) = B \oplus (A \oplus C) \tag{6}$$

We have the same result obtained from the solution found by the second set of runs of the GA for Y, so we can store this equality in the case base and we will have a reduction in the number of gates. The following can be easily seen from the previous equations:

- In eq. (1) we rediscovered the distributive law
- In eq. (2) we rediscovered the DeMorgan's theorem
- In eq. (3) we rediscovered the distributive law and some basic theorems
- In eq. (4) we rediscovered the operation of an XOR gate
- In eq. (5) we rediscovered the operation of an XOR gate
- In eq. (6) we rediscovered the commutative and associative laws

The cases stored in the case base as a product of the analysis at the end of the evolutionary process are summarized in Table 5.

Using other circuits, we were able to produce several other cases and we were able to rediscover several simplification laws from Boolean algebra and even discover some new ones. For example, we discovered a "new" DeMorgan theorem applied to XOR gates: $(X \oplus Y')' = X \oplus Y = X' \oplus Y'$. This "new" law says that if we apply the DeMorgan's theorem to an XOR gate, then we only need to negate one of the components of the original expression. We also discovered some more complex simplifications, such as $(A + (A \oplus B)) \oplus (A \oplus B) = AB$. All these cases were stored in the case base with the aim of reusing this knowledge with other circuits.

Then, we performed an analysis during the evolutionary process, trying to detect the basic building blocks used by the evolutionary algorithm to generate the best solutions produced. Fig. 3 show several snapshots of the solutions

Table 5. Cases stored in the case base at the end of the evolutionary process for the circuit whose truth table is shown in Table 3.

Original case		Solution	Description	Number of gates eliminated
Case 1	$A'B'C + A'BC'$ $+AB'C' + ABC$	$(A \oplus B) \oplus C$	Comparison between a human designer and the best solution found by the GA	13 - 2 = 11
Case 2	$(A'B' + AB)'$	$A'B + AB'$	Case obtained with the comparison between a human designer and the best solution found by the GA	6 - 5 = 1

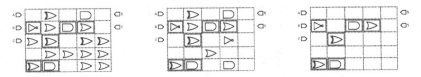

Fig. 3. Solutions obtained by our GA at generations 325 (left), 366 (middle), and 398 (right) for the circuit whose truth table is shown in Table 3.

produced by our GA with the second set of parameters previously described (population size = 700, maximum number of generations = 400). From these pictures, we can see that the circuit has a fitness value of 23 at generation 325 and we were able to recognize the building blocks used by the GA (such building blocks are indicated with a thicker box). At generation 366, the maximum fitness has increased, reaching 29, and we can observe that the building blocks previously mentioned remain in their same position. Finally, when reaching generation 398, we have a fitness of 35 (i.e., a feasible circuit with only 6 gates). Since the building blocks previously mentioned remain in the same position, we proceed to store them in our case base. The building block found will be stored using integers (since our GA used an integer representation), using asterisks (i.e., 'don't care' symbol) for those positions in the circuit different from the building block. This same process was applied to several other circuits, including a parity bit checker, a 2-bit magnitude comparator, a half-adder and a full adder. The details of these experiments are available at [8].

6.1   A Case Study: Use of CBR to Design a 2×2 Bit Adder

To provide an insight into some of the possible applications of our work, we chose a second example in which we want to illustrate how can we use previously acquired knowledge (derived from the design of a half 2×2 adder) to produce a full 2×2 adder. We were interested in analyzing different possibilities regarding the use of CBR to improve the performance of the GA. Therefore, we decided to perform three experiments:

- **First Experiment**: Only previous solutions to the full adder circuit with different fitness values were stored in a case base and some of these individuals were retrieved to seed a percentage of the initial population of a GA before running it. The individuals were taken from different generations with different fitness values in a previous set of runs for the full adder circuit. The initial population was a mixture of previous solutions (10%) and random solutions (90%). This mixture is necessary to avoid an excessive selection pressure that would cause premature convergence. However, the issue of finding the proper number of cases to be injected in the population of a GA is still an open research area [5]. The best known solution to this circuit has a fitness of 36 (i.e., a feasible circuit with 5 gates), and we stored solutions with a fitness value of up to 22. This also intends to reduce selection pressure in the GA.
- **Second Experiment**: Some solutions to different logic circuits including the full adder, the half-adder, the comparator and other circuits were stored in the case base. The most similar cases would then be used to seed a portion of the initial population of a GA before running it. The same mixture of individuals as before was adopted in this case.
- **Third Experiment**: Some solutions to different logic circuits including all the circuits as in the previous experiment, but without including the full adder circuit were stored in the case base. The most similar cases would then seed a part of the initial population of a GA were retrieved before running it. The same mixture of individuals as before was adopted in this case.

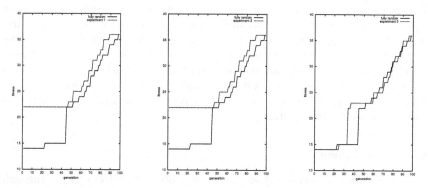

Fig. 4. Comparison of results for the first (left), second (middle) and third (right) experiments. The label "experiment" indicates the runs in which we used cases previously generated by other runs of our GA (i.e., use of the case base).

The results produced from the three experiments are shown in Fig. 4. As we expected, when previous knowledge is used, the GA arrives more rapidly to the best known solution to this circuit. In the first experiment, our GA converges, on average, at generation 87, whereas the GA without knowledge required almost 100 generations to converge (on average). In the second experiment, the GA performed the same as in the first experiment. The interesting issue to analyze here is that our system decided to extract from the case base the previous solution to the full adder (with fitness of 22), instead of the solution to the half adder. This is explained by the fact that the full adder presents a greater resemblance with the circuit being designed. The experiment showed us the capability of our system to discriminate correctly among several circuits. In our third experiment, we can observe that the GA begins to evolve from a fitness value of 14 in generation one, analogously to the GA with its initial population randomly generated. However, the circuit evolves in a completely different way after that, due to the fact that the system retrieves as the most similar case the previous solution found for the half adder circuit. In this case, the GA that uses the case base finds a valid circuit at generation 34, whereas the conventional GA finds a valid circuit at generation 45. This illustrates how the use of case base reasoning can actually help the GA to explore the search space in a more efficient way.

7   Conclusions and Future Work

We have illustrated the potential of combining CBR with a GA to improve performance. The introduction of domain-specific knowledge within a GA is not straightforward, and care must be taken of not biasing the search too strongly as to produce premature convergence. The mixture of individuals proposed in this work (10% of the population were taken from the case base and 90% were randomly generated) seems to be a good choice, at least for the small and medium size circuits used in our experiments [8] (many details were excluded due to space limitations). However, more experimentation in this direction is still necessary. Our approach extends some of the previous efforts to extract design patterns from a GA used to design circuits [7,11], since we show not only how these patterns can be extracted, but also how can they be reused by a GA to design other circuits. More important yet, is the fact that this sort of system can be applied to other domains, and that is precisely one of the future research paths that we would like to explore.

Acknowledgements

The second author acknowledges partial support from CINVESTAV through project JIRA'2001/08, and from CONACyT through NSF-CONACyT project No. 32999-A. The third author states that support for this work was provided in part by grant NAG5-8570 from NASA/Goddard Space Flight Center, and in part by DoD EPSCoR and the Board of Regents of the State of Louisiana under grant F49620-98-1-0351.

References

1. C.A. Coello Coello, A.D. Christiansen, and A. Hernández Aguirre. Use of Evolutionary Techniques to Automate the Design of Combinational Circuits. *International Journal of Smart Engineering System Design*, 2(4):299–314, June 2000.
2. V. Kolodner. *Case Based Reasoning*. Morgan Kaufmann Publishers, San Mateo, California, 1993.
3. X. Liu. Combining Genetic Algorithms and Case-based Reasoning for Structure Design. Master's thesis, Department of Computer Science, University of Nevada, 1996.
4. S.J. Louis. *Genetic Algorithms as a Computational Tool for Design*. PhD thesis, Department of Computer Science, Indiana University, August 1993.
5. S.J. Louis and J. Johnson. Solving Similar Problems using Genetic Algorithms Case-Based Memory. In Thomas Bäck, editor, *Proceedings of the Seventh International Conference on Genetic Algorithms*, pages 283–290, San Francisco, California, 1997. Morgan Kaufmann Publishers.
6. S.J. Louis, G. McGraw, and R. Wyckoff. Case-based reasoning assisted explanation of genetic algorithm results. *Journal of Experimental and Theoretical Artificial Intelligence*, 5:21–37, 1993.
7. J. Miller, T. Kalganova, N. Lipnitskaya, and D. Job. The Genetic Algorithm as a Discovery Engine: Strange Circuits and New Principles. In *Proceedings of the AISB Symposium on Creative Evolutionary Systems (CES'99)*, Edinburgh, UK, 1999.
8. E. Islas Pérez. Development of a Learning Platform using Case Based Reasoning and Genetic Algorithms. Case Study: Optimization of Combinational Logic Circuits. Master's thesis, Maestría en Inteligencia Artificial, Facultad de Física e Inteligencia Artificial, Universidad Veracruzana, November 2000.
9. C. L. Ramsey and J.J. Grefenstette. Case-Based Initialization of Genetic Algorithms. In Stephanie Forrest, editor, *Proceedings of the Fifth International Conference on Genetic Algorithms*, pages 84–91, San Mateo, California, 1993. Morgan Kauffman Publishers.
10. J.W. Sheppard and S.L. Salzberg. Combining Genetic Algorithms with Memory Based Reasoning. In Larry Eshelman, editor, *Proceedings of the Sixth International Conference on Genetic Algorithms*, pages 452–459, San Francisco, California, July 1995. Morgan Kaufmann.
11. P. Thomson. Circuit Evolution and Visualisation. In Julian Miller, Adrian Thompson, Peter Thomson, and Terence C. Fogarty, editors, *Evolvable Systems: From Biology to Hardware*, pages 229–240. Springer-Verlag, Edinburgh, Scotland, April 2000.

Balancing Samples' Contributions on GA Learning

Kwong Sak Leung[1], Kin Hong Lee[1], and Sin Man Cheang[2]

[1] Department of Computer Science and Engineering,
The Chinese University of Hong Kong, Hong Kong.
{ksleung,khlee}@cse.cuhk.edu.hk
[2] Department of Computing,
Hong Kong Institute of Vocational Education (Kwai Chung), Hong Kong.
smcheang@vtc.edu.hk

Abstract. A main branch in Evolutionary Computation is learning a system directly from input/output samples without investigating internal behaviors of the system. Input/output samples captured from a real system are usually incomplete, biased and noisy. In order to evolve a precise system, the sample set should include a complete set of samples. Thus, a large number of samples should be used. Fitness functions being used in Evolutionary Algorithms usually based on the matched ratio of samples. Unfortunately, some of these samples may be exactly or semantically duplicated. These duplicated samples cannot be identified simply because we do not know the internal behavior of the system being evolved. This paper proposes a method to overcome this problem by using a dynamic fitness function that incorporates the contribution of each sample in the evolutionary process. Experiments on evolving Finite State Machines with Genetic Algorithms are presented to demonstrate the effect on improving the successful rate and convergent speed of the proposed method.

1   Introduction

Learning a system directly from input/output samples are widely studied in the area of Artificial Intelligence. Many approaches, e.g. Evolutionary Algorithms (EA) and Artificial Neural Networks (ANN), have been proposed by researchers. These approaches apply sampled inputs to an evolved solution and then compare generated outputs with those sampled outputs. A modification scheme will be applied to modify the internal parameters of the evolved solution in order to reduce the difference of generated outputs and sampled outputs. The advantage of these approaches is learning a system in which internal behaviors are hidden or even evolving a system from a set of incomplete or noisy samples.

Samples captured from a real-world system are usually incomplete, biased and noisy. The distribution of sample points on a sample space affects the quality of the final evolved system. In order to have a complete set of samples, we need to collect large number of samples [3]. In a sample set, some samples may be duplicated, either exactly or semantically. Exactly duplicated samples can be recognized and removed simply by comparing samples. Semantically duplicated or equivalent samples cannot

Y. Liu et al. (Eds.): ICES 2001, LNCS 2210, pp. 256–266, 2001.

be recognized easily because we do not have enough knowledge of the system to be evolving.

For example, the following four input/output samples have the same meaning for a four consecutive 1's detector. If a randomly evolved Finite State Machine (FSM) matches one of the four samples, there is a high probability that all of the other samples are matched. Duplicated samples reduce the convergent speed of a Genetic Algorithm (GA) [5] because of duplicate counting of sample points.

sample 1	sample 2	sample 3	sample 4
011110	0011110	00011110	0000111100
000010	0000010	00000010	0000000100

In this paper, we propose a low overhead fitness evaluation method to balance non-equally distributed samples. A GA-based FSM synthesizer is used to demonstrate the effect of the proposed method. Evolving a state transition table by sampled input/output pairs with GA were studied in the field of machine learning by different researches.[3,7,10,11]

The rest of the paper is organized as follows: section 2 contains a description of FSM; section 3 describes the GA used in this paper; sections 4 and 5 present the experiments and results; finally, section 6 concludes our work.

2   Finite State Machines

Finite State Machines (FSMs) are used to solve various classes of problems, including pattern matching [2], data processing [9], data encoding/decoding [8], machine/robot control [1] and systems testing [6]. There are two popular models of FSMs, namely Mealy machines and Moore machines. Mealy machines and Moore machines are functionally equivalent. Giving a Moore machine, we can construct a Mealy machine and vice versa. In this paper, we focus on discrete-time Mealy machines.

A discrete-time Mealy machine is defined by three sets and two functions. The three sets are the state set S, the input symbol set I and the output symbol set O. The two functions are the state transition function t and the response function r.

$$M = <S, I, O, t, r>$$

$$s_{n+1} = t(s_n, i_n) \tag{1}$$

$$o_{n+1} = r(s_n, i_n) \tag{2}$$

where $s_n, s_{n+1} \in S$, are the current state and the next state

$i_n \in I$, is the current input

$o_n \in O$, is the output at current time instant

In the above equations (1) and (2), the next state and the output of a Mealy machine depends on the current state s_n and the input i_n. The transition function t and the response function r are commonly expressed in a state transition diagram or a state transition table. A state transition table is simply a lookup table that can be realized on a RAM-based hardware. GA operations can be applied to lookup table contents directly.

3   Genetic Algorithms

Genetic Algorithms (GAs) is a search and optimization method inspired by natural evolution. A GA performs a search on the solution space. Individuals of a population are evaluated based on a function. At each generation, genetic operations, e.g. crossover, mutation and selection, are applied to the old population to generate a new population. In the case of FSMs synthesis, a chromosome can encode a state transition table directly. Experiments found that non-equally distributed samples cause premature situation and reduce the convergent speed of the GA. When capturing samples from real-world system, we cannot avoid biased samples. Fig. 1(a) shows a biased sample set and Fig. 1(b) shows an equally distributed sample set on the same sample space. Three chromosomes are used to test both sample sets. In Fig. 1(a), three chromosomes match different number of sample points and their fitness values are different. On the other hand, in Fig. 1(b), three chromosomes match almost same number of sample points and their fitness values are almost the same.

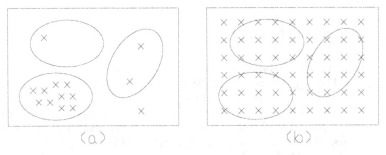

Fig. 1. Sample points distribution on a sample space.

Fig. 2 shows the samples' relative usages in a GA-based Mealy Machines synthesizer. 100 input/output samples were used to train a system. Times of matches for each sample were counted. While finding a complete solution, we calculated the relative percentage usages for all samples and plotted curves in Fig. 2. Most samples were only matched by chromosomes in the last generation. For example, as the curve for a 4-bit reversible counter shown in Fig. 2, there are only 6 out of 100 samples contributed on the evolution process in most generations. This sample set is extremely biased in two areas of the sample space. In order to improve the successful rate and convergent speed of GAs, we are going to propose a new fitness calculation method, namely Dynamic Samples Weighting.

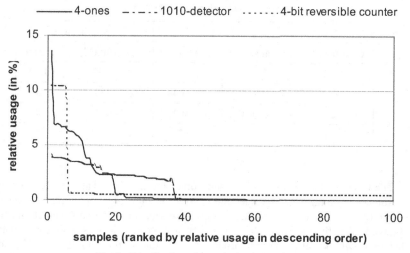

Fig. 2. Distribution of samples' relative usage.

3.1   Dynamic Samples Weighting

For a learning process with a fixed number of samples, N_{sample}, a GA usually adds a constant value, i.e. $1/N_{sample}$, to the fitness value of a chromosome when it matches the sample. Under this arrangement, each sample equally contributes to the fitness function. For example, an individual FSM in a population is tested with 100 input/output samples. By applying the 100 sample inputs to the FSM, if there are 30 exactly matched samples, the fitness is $30\times(1/100)=0.3$. A complete solution is an FSM with a fitness value equal to 1.0. This method can produce good result if sample points are equally distributed in the sample space. However, if duplicated samples exist, fitness values cannot reflect the actual relative fitness of individuals in the population. A simple example is shown in Fig. 3 to illustrate the situation.

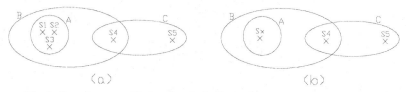

Fig. 3. Sample space with three semantically duplicated samples – S1, S2 and S3.

We are going to evolve a system with five sample points, S1 to S5, as shown in Fig. 3(a). Three chromosomes, A, B and C, are shown as three ellipses. Arbitrary samples are captured from a real-world system. Samples S1, S2, and S3 are semantically duplicated. This means that whenever a chromosome matches S1, it also matches S2 and S3.

Table 1. Effects of duplicated sample points.

chromosome	fitness		actual fitness	
	absolute	relative	absolute	relative
A	3/5=0.6	0.333	1/3=0.33	0.198
B	4/5=0.8	0.444	2/3=0.67	0.401
C	2/5=0.4	0.222	2/3=0.67	0.401
Total	1.8		1.67	

The fitness values of the three chromosomes are shown in Table 1. In the "fitness (absolute)" column, fitness values are simply the ratio of samples that are matched correctly by different chromosomes. As we can see, chromosome A and chromosome B have higher fitness values, 0.6 and 0.8 respectively. There is a higher probability that they will be selected for reproduction. Actually, the samples S1, S2 and S3 should be treated as only one sample point, i.e. Sx, in Fig. 3(b). More precise fitness values for the three chromosomes should be those values shown in the "actual fitness (absolute)" column in Table 1. The chromosome C should have higher probability to be selected for reproduction than the chromosome A because chromosome C matches two sample points, S4 and S5, while chromosome A only matches one sample point, Sx. The misleading fitness comes from the semantically duplicated sample points, S1, S2 and S3.

In order to balance the effect of non-equally distributed sample points, a new fitness calculation method is introduced into the GA, namely the Dynamic Samples Weighting (DSW). Under DSW, all samples are assigned weights. The algorithm counts and records the total number of successful matches for each sample. For a fixed number of generations, say 10, the weight of each sample is recalculated according to the following formulas:

$$E_i = C_i \bigg/ \sum_{i=1}^{N_{sample}} C_i \tag{3}$$

$$W_i = \left(\frac{1}{E_i} \right) \bigg/ \sum_{i=1}^{N_{sample}} \left(\frac{1}{E_i} \right) \tag{4}$$

Where N_{sample} is the number of samples
C_i is the successful matched counts of the ith sample
E_i is the relative contribution of the ith sample on the fitness function
W_i is the new weight of the ith sample

W_i is inverse proportional to E_i. Therefore, low usage samples will have higher weights in the coming generations. The pseudo-code of a GA with DSW is shown in Fig. 4.

```
initialize chromosomes;
for i = 1 to N_sample
    W_i = 1 / N_sample;
    C_i = 1;

loop for generations

    for h = 1 to N_chromosome
        fitness_h ← 0;
        for i = 1 to N_sample
            if evaluate(chromosome_h,input_i) = output_i
                fitness_h ← fitness_h + W_i;
                C_i ← C_i + 1;
        exit on fitness_h = 1;

    perform GA operations;

    if generation no. is divisible by 10
        for i = 1 to N_sample
            E_i ← C_i / Σ(C_i);
        for i = 1 to N_sample
            W_i ← (1/E_i) / Σ (1/E_i);
            C_i ← 1;
```

Fig. 4. Pseudo-code of a GA with DSW.

Table 2. Recalculate the weights of samples on Fig. 3.

Samples	W_i^0	C_i^0	E_i^0	$1/E_i^0$	W_i^1	W_i^*	$|W_i^*-W_i^0|$	$|W_i^*-W_i^1|$
S1	1/5	2	2/9	4.5	1/6	1/9	0.0889	0.0556
S2	1/5	2	2/9	4.5	1/6	1/9	0.0889	0.0556
S3	1/5	2	2/9	4.5	1/6	1/9	0.0889	0.0556
S4	1/5	2	2/9	4.5	1/6	1/3	0.1333	0.1667
S5	1/5	1	1/9	9.0	1/3	1/3	0.1333	0.0000
Total	1	9		27.0	1	1	0.5333	0.3335

Table 3. Effects of duplicated samples and DSW on Fig. 3.

chromosome	fitness (by W_i^0)		actual fitness(by W_i^*)		fitness (by W_i^1)	
	absolute	relative	absolute	relative	absolute	relative
A	0.6	0.333	0.33	0.198	0.50	0.299
B	0.8	0.444	0.67	0.401	0.67	0.401
C	0.4	0.222	0.67	0.401	0.50	0.299
Total	1.8		1.67		1.67	

Table 2 demonstrates the calculation of weights with DSW for samples shown in Fig. 3(a). The W_i^0 column shows the initial samples' weights and W_i^1 shows the new sample weights after one round of recalculation. The weights of samples S1, S2 and S3 decreased from 1/5 to 1/6 and the weight of S5 increased from 1/5 to 1/3. The actual weights in Fig. 3(b) are shown in the W_i^* column. Since S1, S2 and S3 are semantically duplicated, the sum of their weights should be equal to the weight of S4 (or S5). The last two columns show the absolute distance of weights from the actual weights. The total distance of all W_i^0 decreased from 0.5333 to 0.3335.

As shown in Table 3, by using new samples' weights, we can obtain new fitness values in "fitness (by W_i^1)" column. The relative fitness value of chromosome C increased from 0.222 to 0.299 so that chromosome C will have higher probability to be selected for reproduction. The relative fitness values in the "fitness (by W_i^1)" column are closer to the actual fitness values in the "actual fitness (by W_i^*)" column.

4   Experiments

In order to demonstrate the effect of DSW, we use Simple GA (SGA) [4] as the skeleton. Seven sets of samples were generated from seven FSMs as shown in Table 4. In each sample set, there are 100 input/output pairs. For each sample set, we first generated inputs with equal probabilities of 1's and 0's. Then inputs were passed to the targeted FSM to produce outputs. In order to demonstrate the effect of balancing semantically duplicated samples, all exactly duplicated sample pairs were removed.

Table 4. Tested FSMs.

FSMs	no. state	input(bit)	output(bit)
3-ones	3	1	1
4-ones	4	1	1
1010 detector	4	1	1
0101 detector	4	1	1
divide-by-2	4	1	1
modulo-4-counter	4	1	2
reversible 4-counter	4	1	2

For each sample set, both GA without DSW and GA with DSW were tested. The results will be presented in the next section.

4.1   Chromosome Encoding Scheme

The state transition table of a Mealy Machine, $M=<S,I,O,t,r>$, is directly encoded in a chromosome. Each chromosome consists of $n_s \times n_i$ (next_state,output) pairs for all combinations of states and input symbols.

$$[(\delta_{11},\omega_{11}),\cdots,(\delta_{1n_i},\omega_{1n_i}),\cdots,(\delta_{ck},\omega_{ck}),\cdots,(\delta_{n_sn_i},\omega_{n_sn_i})]$$

where n_s is the maximum number of states

n_i is the maximum number of input symbols

c is the current state index and $1 \leq c \leq n_s$

k is the input symbol index and $1 \leq k \leq n_i$

$\delta_{ck}=t(s_c,i_k)$ and δ_{ck}, $s_c \in S$, $i_k \in I$

$\omega_{ck}=r(s_c,i_k)$ and $\omega_{ck} \in O$, $s_c \in S$, $i_k \in I$

For example, Table 5 shows a state transition table of a Mealy machine $M=<S,I,O,t,r>$ with 2 states, $S=\{p,q\}$, 2 input symbols, $I=\{0,1\}$, and 2 output symbols, $O=\{0,1\}$.

Table 5. State transition table for the chromosome [(p,0),(q,0),(q,0),(p,1)]

current state	next state / output	
	input=0	input=1
p	p / 0	q / 0
q	q / 0	p / 1

4.2   GA Operators and Parameters

All experiments used the following GA operators and configurations:

1. **Solution Space:** The maximum number of states for the solution Mealy machine is set to 8. The input is set to 1-bit and the output is set to 2-bit. The size of a chromosome can be expressed as $(b_s+b_o) \times 2^{b_s+b_i}$, where b_s, b_i and b_o are the number of bits for encoding states, input symbols, and output symbols respectively. Thus, the chromosome size is 80 bits for $b_s=3$, $b_i=1$, and $b_o=2$.

2. **Population:** The population size is set to 100. The old population is totally replaced by new offspring.

3. **Selection:** Parents are selected from all individuals of the old population by using roulette wheel weight method.

4. **Crossover:** One point crossover on the boundary of each (next_state,output) pair. The crossover probability equal to 0.4.

5. **Mutation:** Bit mutation rate is set to 0.04.

6. **Sample Weights Recalculation:** For all GA with DSW experiments, samples' weights were recalculated every 10 generations.

7. **Termination of execution:** Each trial is forced to terminate after 2000 generations even a complete solution is not found.

5   Experimental Results

We implemented the described algorithms in C. For each set of samples of the seven FSMs shown in Table 4, we ran 100 trials to collect statistics. Two statistics are shown and discussed in this section.

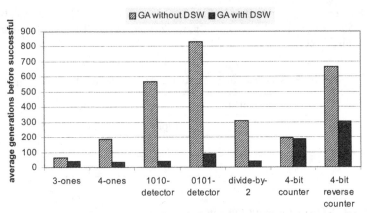

Fig. 5. Average number of generations before obtaining the complete solution for GA without DSW and GA with DSW.

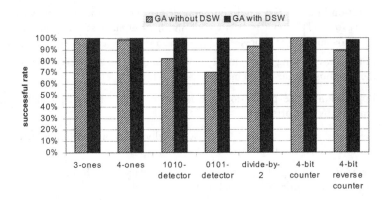

Fig. 6. Successful rate for GA without DSW and GA with DSW in 100 trials.

Fig. 5 shows the average number of generations before producing a complete correct solution. Obviously, GA without DSW needs more generations to evolve a complete correct FSM. With DSW, populations converge faster than without DSW.

For example, the average number of generations to produce a complete solution for the 0101-detector Mealy machine is 832 generations for GA without DSW. Actually, the value should be higher than 832 because 30% of trails, as shown in Fig. 6, did not produce a complete solution within 2000 generations. A great improvement was recorded after introducing DSW to the original GA. It takes only 86 generations to produce a complete solution. The speed up ratio is 832/86≈10. Table 6 summarizes speed up ratios of tested cases. Those values prefixed with ">" indicate that not all trails for GA without DSW produced complete solution. Thus, the actual speed up ratios should be higher than the shown values.

Table 6. Speed up ratio of GA with DSW and GA without DSW.

Tested FSMs	GA without DSW	GA with DSW	Speed Up Ratio
3-ones	64	37	1.73
4-ones	188	32	5.88
1010 detector	567	41	>13.83
0101 detector	832	86	>9.67
divide-by-2	309	37	>8.35
modulo-4-counter	194	186	1.04
Reversible 4-counter	665	304	>2.19

6   Conclusions and Further Work

This paper presented a method to balance samples distribution in a biased sample set. A new GA fitness calculation method, called the Dynamic Samples Weight (DSW) was presented. The DSW method was tested on FSMs evolution. Experimental results show that DSW can improve the successful rate and convergent speed of learning with GA. DSW automatically reduces the effect of duplicated samples by decreasing weights of duplicated samples. This scheme makes the fitness calculation more precise and closer to the actual distribution of sample points.

The effort of calculating weights is much lower than GA operations and fitness evaluation. It is because weights are calculated on every 10 generations and the calculations only involve simple arithmetic operations.

We are going to apply this method to different learning problems. Further studies will be investigated on the frequency of weights recalculation and different weights calculation formulas. We believe that DSW method can be applied to different Genetic Algorithm based learning systems.

References

1. Ashlock D., Freeman, J.: A Pure Finite State Baseline for Tartarus. Proc. of Evolutionary Computation (2000). 1223-1230
2. Bowman, C.F.: Pattern Matching Using Finite State Machines. Dr. Dobb's Journal of Software Tools, Vol. 12, No. 10, (Oct. 1987). 46-7,49-50,52,55,57,92-4,96,106-8
3. Chongstitvatana, P., Aporntewan, C.: Improving Correctness of Finite-State Machine Synthesis from Multiple Partial Input/Output Sequences. Proc. of the 1st NASD/DoD Workshop on Evolvable Hardware (1999). 262-266
4. Goldberg, D.E.: Genetic Algorithm in Search, Optimization and Machine Learning. Addison-Wesley (1989).

5. Holland, J.H.: Adaptation in Natural and Artificial Systems. The University of Michigan Press (1975)
6. Huang, S.H., Kwai, D.M.: A High-Speed Built-in-Self-Test Design for DRAMs. Proc. of Int Symp. on VLSI Technology, Systems, and Applications (1999). 50-53
7. Manovit, C., Aporntewan, C., Chongstitvatana, P.: Synthesis of Synchronous Sequential Logic Circuits from Partial Input/Output Sequences. Proc. of Int Conf. on Evolvable Systems: From Biology to Hardware (1998). 98-105
8. Parhi, K.K.: High-Speed Huffman Decoder Architectures. Proc. of Conf. on Signals, Systems and Computers, Vol. 1, (1991). 64-68
9. Warshauer, M.L.: Conway's Parallel Sorting Algorithm. Journal of Algorithms, Vol. 7, (1986). 270-276
10. Zhou, H., Grefenstette, J.J.: Induction of Finite Automata by Genetic Algorithms. Proc. of Int Conf. on Systems, Man and Cybernetics (1986). 170-174
11. Tanomaru, J.: Evolving Turing Machines from Examples. Proc. of the 3rd European Conf. on Artificial Evolution (1997). 167-180

Solving Partially Observable Problems by Evolution and Learning of Finite State Machines

Eduardo Sanchez[1], Andrés Pérez-Uribe[2], and Bertrand Mesot[1]

[1] Logic Systems Laboratory, Computer Science Department,
Swiss Federal Institute of Technology-Lausanne,
Eduardo.Sanchez@epfl.ch, Bertrand.Mesot@epfl.ch
[2] Parallelism and Artificial Intelligence Group, Department of Informatics,
University of Fribourg, Switzerland,
Andres.PerezUribe@unifr.ch

Abstract. Finite state machines (FSM) have been successfully used to implement the control of an agent to solve particular sequential tasks. Nevertheless, finite state machines must be hand-coded by the engineer, which might be very difficult for complex tasks. Researchers have used evolutionary techniques to evolve finite state machines and find automatic solutions to sequential tasks. Their approach consists on encoding the state-transition table defining a finite state machine in the genome. However, the search space of such approach tends to be innecesarily huge. In this article, we propose an alternative approach for the automatic design of finite state machines using artificial evolution and learning techniques: the *SOS-algorithm*. We have obtained very impresive results on experimental work solving partially observable problems.

1 Introduction

In digital systems we talk about combinational and sequential systems. In the first case, the output of such a system depends uniquely on the input data at the moment, while the behavior of sequential systems depends on the sequence of inputs over a period of time [14]. Thus, in the latter case, a memory is required to save a trace of the sequence of inputs: that is, the *state* of the system. Given the finite character of the memory, the number of states is necessarily finite; therefore, one often speaks of sequential systems as finite state machines (FSM).

One may use a finite state machine to implement the controller of an agent to solve a particular sequential task. However, such a finite state machine must be hand-coded by the user, which may be a tedious work when the tasks are highly complex. Evolutionary techniques may reduce the user's effort to solve such a task. For example, one may use genetic algorithms to evolve finite state machines instead of designing them by hand.

Nevertheless, in some problems, like maze solving, evolutionary techniques may require quite much more computational effort than, for instance, reinforcement learning techniques. Indeed, contrary to reinforcement learning techniques

Y. Liu et al. (Eds.): ICES 2001, LNCS 2210, pp. 267–278, 2001.

that learn by interacting with their environment, evolutionary methods do ignore much of the useful structure of the problem: the states an agent passes through during its lifetime, which actions it selects, etc. [18]

In the reinforcement learning framework, the agent makes its decisions as a function of a signal from the environement, called the environment's *state*. However, it should be noticed that in such framework, the state is any information available to the agent, e.g., a local *observation* of the environment. Thus, in certain reinforcement learning problems, the state does not summarize the past sensations of the agent (i.e., the history of the agent). In the reinforcement learning framework, we speak of Markov states if they succed in retaining all relevant information of the history of the agent, as it is the case of a state in a finite state machine.

Many of the learning techniques developed within the reinforcement learning framework can learn optimal policies if the problem is Markovian; on the other hand, they can only approximate the optimal policy. Partially observable mazes are an example of non-Markovian tasks.

In this paper, we introduce an algorithm for evolving and learning finite state machines, and test its capabilities solving partially observable mazes. In Section 2, we describe the maze learning task and the concept of partial observability. In Sections 3, 4, and 5, we briefly describe finite state machines (FSM), reinforcement learning, and evolutionary techniques. Section 6 describes our *SOS-algorithm* for evolving and learning finite state machines, Section 7 delineates the experimental results, and finally, Section 8 presents some concluding remarks.

2   Maze Learning and Partial Observability

Mazes have long been used as a benchmark for machine learning techniques. Maze learning is presented here as an abstraction of the real navigation problem in mobile autonomous robots: One has an a-priori partition of the sensor space of the robot, a perfect vision, and a discrete world. An autonomous robot is

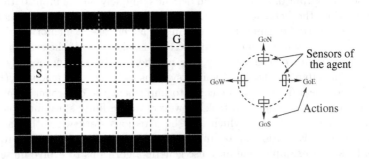

Fig. 1. Example of a maze problem and of an agent. The S location indicates the starting postion, the G location is the goal. The agent is able to detect an obstacle in any of the four directions, and may choose one among four possible actions, or a NOP (no operation) action.

faced to the problem of finding a goal (G) starting from a valid position (S) in a labyrinth-like environment (see figure 1). In some experiments, the robot can only detect an obstacle lying in front of it, and may move one position forward, or turn right or left. In other experiments, the robot can detect an obstacle to the north, south, left and to the right, and may move one position in the same directions. The problem here is to find the correct actions to take, such that the robot avoids obstacles and finds the shortest path to the goal.

In mobile robotics, the agent has only partial information about the current state of the environment, that is, it does not know the state of the whole world from the sensory input alone (i.e., the observations). The agent is said to suffer from *hidden-state* [8] or *perceptual aliasing* [19], and the environment is said to be partially observable.

For an agent to behave in a partially observable environment, a sort of structural abstraction is needed. One technique performs *spatial abstraction*: it consists on aggregating similar observations and treating them to some degree as the same. Spatial abstraction, or function approximation seeks to develop mappings which are more compact than lookup tables. Nearest neighbor, neural networks, fuzzy logic, and other approaches have been widely used [11,15]. Another technique performs *temporal abstraction*, in which certain sequences of actions are treated as higher level actions [12].

In general, the partially observable Markov decision processes (POMDP) provides a formal framework for studying these problems [2,5]. The difficulty of acting and planning in partially observable environments is illustrated by the fact that the optimal policy may need the use of the complete previous history of the system, that is, the whole sequence of observations, actions, and rewards to determine the next action to perform [9].

Researchers attempt to aleviate such difficulty by developing hierarchical reinforcement learning techniques [3], incremental learning methods [13], multi-agent reinforcement learning techniques [16], etc. In this paper, we present an alternative method for solving partially observable mazes by using artificial evolution and learning of finite state machines.

3   Finite State Machines

The behavior of sequential systems depends on the sequence of inputs over a period of time. A memory is required to save a trace of the sequence of inputs: that is, the *state* of the system. Given the finite character of the memory, the number of states is necessarily finite; therefore, one often speaks of sequential systems as finite state machines (FSM) [14].

The state of a sequential system is therefore stored by the state variables of the system; these variables contain all the information concerning the past necessary for calculation of the future behavior.

next state = f(present state, inputs)

The output variables of a sequential system (i.e., the actions of the system) can depend on the present state and present value of inputs, or solely on the present state: one speaks in the first case of Mealy machines and of Moore machines in the second case.

$outputs = \delta(present\ state,\ inputs)$ (Mealy machine)
$outputs = \delta(present\ state)$ (Moore machine)

In order to better visualize the state sequences of the system, the truth tables of the functions f and δ are put together in a single table called *state table* (or state-transition table). The lines in this table correspond to the states, and the columns correspond to combinations of input variables, where for each entry one places the values corresponding to the functions f and δ (value of the next state and outputs).

It is similarly common to give the information of the state tables under another form, more visual: the *state diagram* or state-transition diagram. This is an oriented graph where each state transition is represented by an arrow, linking the present state and the next state. On each arrow, the input values producing the change of state and the ouput values are written.

4   Reinforcement Learning

Reinforcement learning is a computational approach to learning from interaction. This approach offers an attractive paradigm to automating goal-directed learning and decision making [18].

Reinforcement learning tasks are generally treated in discrete time steps: at each time step t, the learning system receives some representation of the environment's *state* s^t, it *tries* an action a, and one step later it is *reinforced* by receiving a scalar evaluation r^t. Finally, it finds itself in a new state s^{t+1}. This is what is called learning by trial-and-error in artificial systems. To solve a reinforcement learning task, the system attempts to maximize the total amount of reward it receives in the long run [18]. To achieve this, the system tries to minimize the so called *temporal-difference error*, computed as the difference between predictions at successive time steps. This is called *temporal difference learning* or *TD*-learning [17].

Reinforcement learning techniques involve the use of a *value function* which can be simply implemented by a look-up table, and is composed of *state values* or *action values*. A state value (denoted as $V(s)$) indicates the expected cumulative reinforcement an agent can receive starting from a given state s; and, an action value (denoted by $Q(s, a)$) indicates the expected cumulative reinforcement an agent can receive taking action a from state s. In the general case, the *policy* of the system (i.e., the way of behaving of an agent) is a simple function of the value function. Therefore, adaptation here is achieved by updating the value function using the scalar evaluation received while interacting with the environment.

5   Artificial Evolution

The idea of applying the biological principle of natural evolution to artificial systems has been used in a wide range of applications. In this work, we use *genetic algorithms*, an iterative procedure that consists of a population of individuals, each one represented by a finite string of symbols (the genome), encoding a possible solution to a given problem [10]. The algorithm starts with an initial population of individuals that is generated at random. At every evolutionary step (a generation), the individuals in the current population are decoded and evaluated according to some predefined quality criterion (the fitness). To form a new population (the next generation), individuals are selected according to their fitness and then modified by using two standard (genetic) operators: mutation (it is applied by flipping one or more random bits in a string with a probability equal to the mutation rate) and crossover (it takes two individuals called parents and produces two new individuals called the offspring by swapping parts of the parents). The genetic algorithm may eventually find an acceptable solution.

Koza [7] and other researchers have used evolutionary techniques to evolve finite state machines. Their approach consists on encoding the state-transition table defining a finite state machine in the genome. Such approach is straighforward, but the size of the genome tends to be innecesarily huge: in a finite state machine of S states (normally selected a-priori), and O combination of inputs (possible observtions), the genome will have a length of $S \times O$ genes, each gene encoding the next state and the action to be executed.

6   Evolution and Learning of Finite State Machines

The resemblance between a FSM state-table and a state-action value function led us to adopt a different strategy to solve partially observable mazes. Instead of considering a table of state-action values $Q(s, a)$, we used a lookup table similar to the FSM state-table. As shown above, a FSM state-table is accesed by a pair (present-state, current-input), and outputs the following state for the FSM. Now, if we consider a Moore machine, the output action of the finite state machine is only a function of the state (i.e., it does not depend on the inputs).

In our approach, to determine the next-state transition for a given pair (state,input), we use a value-state-like function ($V(s)$-like) that will estimate the current best next state for the agent. Thus, to enable the evolution and learning of finite state machines, our lookup table is accessed by the triplet (present-state, current-observation, estimate-value-of-next-state) and outputs the current best next state for the agent. Each state has a unique associated action, which is taken by the agent after its selection of a state transition. Of course, this implies an a-priori association of actions and states. Therefore, we used evolutionary techniques to solve such design problem.

We have denoted the value function by $T(s, o, s')$, where s is the current state of the environment, o is the current observation of the environment made by the agent, and s' is the current best next state. This triplet gives rise to the

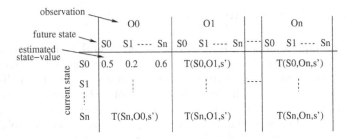

Fig. 2. Lookup table of estimations of the state values $T(s, o, s')$, where s is the current state of the environment, o is the current observation of the environment made by the agent, and s' is the current best next state.

name of our algorithm. Such a function is implemented by a lookup table as shown in Figure 2. The estimates of the value of the states, used by the agent to choose its next state, are updated using a temporal difference formula as shown in Figure 3.

The algorithm starts by using a population of POP individuals whose genomes associate actions to the states of the finite state machine. The length of the genome is equal to the maximal number of states SMAX multiplied by the number of bits (BITACT) needed to code the set of possible actions (notice that the same action can be associated to several different states). Each individual is tested in the maze for a TRIAL number of trials. During each trial, the agent interacts with the maze and uses the learning algorithm of Figure 3 to update its $T(s, o, s')$ value function, until it reaches the goal or completes a number of STEPS agent performed actions (or number of state transitions). The fitness of an individual is computed by a measure of performance depending on the task (e.g., -1 × mean number of steps needed to achieve the goal). The evolutionary process is run for up to G generations. In our experiments, both, mutation and crossover operate at the level of the genes, so, for instance, a mutation is defined as a random change of a gene (i.e., the action associated to a particular state) instead of a random binary change.

7   Experimental Results

7.1   Case 1: The Santa Fe Trail

The Santa Fe trail was proposed by Christopher Langton [7]: an artificial ant is placed on a 32×32 grid of cells, its objective being to collect as many food pellets strewn about as possible in a given number of times steps. The ant starts out facing east, and positioned in the upper left-hand cell of the grid (See figure 4). At each time step it can turn in any direction, walk one cell in that direction, or select the no-operation (NOP) action. The actions are encoded as follows: action 0: NOP, 1: turn left, 2: turn right, and 3: move forward one step. When the agent collects a food pellet, it is reward with positive signal (r=+1). The

1. Initialize the $T(s, o, s')$ value function to a constant value, equal
   to the maximal reward value (optimistic initialization to force
   exploration), where s and o are the current state and observation,
   and s' is the next state,
2. Set the agent to the start position, initialize its state ($s^t = s_0$),
   observe the environment, and take a NOP (no-operation) action,
3. Make $s^{t-1} = s^t$, $o^{t-1} = o^t$
4. Observe the environment and choose as present state, the state with the
   higher estimate $T(s^{t-1}, o^{t-1}, s^t)$. If several states are possible, choose
   the one with the smaller index in table $T(s^{t-1}, o^{t-1}, s^t)$.
5. Take the action a, associated (by the genome) to the current
   state s^t,
6. Update the estimate value $T(s^{t-1}, o^{t-1}, s^t)$ as follows:
   $T(s^{t-1}, o^{t-1}, s^t) = T(s^{t-1}, o^{t-1}, s^t) + \alpha(r + \gamma T(s^t, o^t, s^{t+1}) - T(s^{t-1}, o^{t-1}, s^t))$,
   where α is the step size, r is the reward, and γ is a discount reward factor,
   o^t is the observation at the new state, and s^{t+1} is the next state,
7. Go to step 3 until the state s^t is a terminal state, or a number of
   permitted steps is reached,
8. Repeat steps 2 to 7 for a pre-defined number of trials.

Fig. 3. Learning loop of the *SOS-algorithm*.

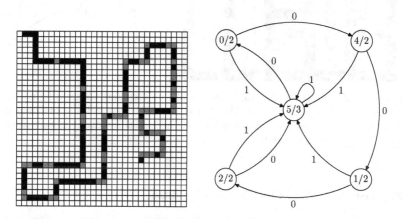

Fig. 4. The Santafe trail and the corresponding finite state machine solution found
by the evolutionary and learning method. The numbers on the arrows indicate the
input value (0: no food, 1: food), the numbers on the circles indicate the state and the
associated action (state/action).

main difficulty faced by the artificial ant lies in the positions of the pellets, which
are not arranged in a simple, consecutive manner. The whole trail contains 89
food pellets.

We have used our *SOS-algorithm* using the parameters shown in the following
table:

G	POP	$SMAX$	$TRIAL$	$STEPS$	α	γ
10	100	10	100	500	0.6	0.9

After runing the algorithm for less than 10 generations (in average), each agent, needing less than 30 trials, learns a finite state machine solution of less than 7 states. Typically, the system is able to learn a 5-state finite state machine without using the evolutionary search after 1 to 25 trials. One such example is presented in Figure 4. As a comparison, Koza [7] describes the use of evolutionary techniques to evolve a finite state machine of up to 32 states, whose complete state-transition table is encoded on the genome, to solve the Santa Fe trail problem. We do not have information of the performance of Koza's method, but instead of our 20-bit genome and the effective use of learning, he used a genome of 65 genes (453 bits). Another solution is given by Angeline et al. [1]; they found a 5-state FSM solution after evolving a population of 100 recurrent neural networks for 2090 generations.

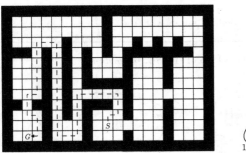 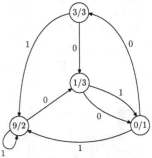

Fig. 5. Complex partially-observable maze and the best state diagram found by the evolutionary and learning method. The numbers on the arrows indicate the input value (0: no obstacle, 1: obstacle), the numbers on the circles indicate the state and the associated action (state/action).

7.2   Case 2: A Partially-Observable Maze

Once our *SOS-algorithm* successfully solved the Santa Fe trail problem, we designed a more complex maze shown in Figure 5. It was motivated by the fact of having multiple paths to reach the goal and a delayed reward signal (i.e., there is a unique reward signal while reaching the goal). In particular, we used our SOS-algorithm and the parameters shown below:

G	POP	$SMAX$	$TRIAL$	$STEPS$	α	γ
10	200	10	500	500	0.6	0.9

After running the algorithm for less than 4 generations (in average), each agent needing less than 250 trials, learns a finite state machine solution of less than 6 states. Typically, the system is able to learn a 9-state finite state machine which guides the robot to the goal in 116 steps, without using the evolutionary search, a 7-state finite state machine which guides the robot to the goal in 96 steps, after the second generation, and a 4-state finite state machine which guides the robot to the goal in 92 steps, after the third generation. This problem was very interesting because of the diverse solutions we obtained: for instance, some solutions needed few states, but needed more steps to reach the goal than other solutions were the FSM was larger but guided the agent through the shortest path. Finally, we succeeded in optimizing both aspects. Our next testbed is a maze with several possible paths to reach the goal we found on the literature.

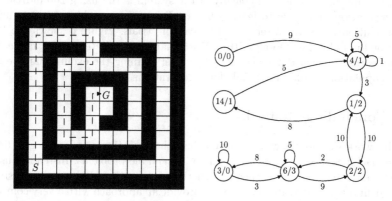

Fig. 6. Wiering-Schmidhuber labyrinth and the best state diagram found by the SOS-algorithm. The numbers on the arrows indicate the input value computed as a funtion of its four sensors (e.g., 0: no obstacle, 1: obstacle to the right, 2: obstacle to the north, 6: obstacle to the north and to the west, etc.), the numbers on the circles indicate the state and the associated action (state/action).

7.3   Case 3: Wiering and Schmidhuber Maze

Wiering and Schmidhuber learns finite sequences of (memoryless) reactive policies using an implicit memory of some of the previous observations [20]. They defined a multi-agent system, where each agent has a Q-table, and an HQ-table that serves as a control transfer unit (except for the last agent which only has a Q-table). The Q-table stores estimates of actual observation/action values and is used to select the next action. The HQ-table stores estimated subgoal values and is used to generate a subgoal once the agent is made active. To test their algorithm, they defined the 12 × 12 maze shown in Figure 6. The system has to discover a path leading from start position S to goal G. There are four actions: go west, go north, go east, go south, and 16 possible observations. Although there are 62 possible agent positions, there are only 9 highly ambiguous inputs. To solve such a maze, they used groups of 3,4,6, 8 and 12 agents using HQ-learning [20]. A reward was only delivered when the goal was reached. Within 20,000 trials all systems almost always found near-optimal solutions. In most cases a 30-step solutions were found, though in 1 out of 8 cases, the optimal 28-step solution was found. The better systems performed using between 3 and 6 agents.

Our evolving and learning approach use the paramters shown below:

G	POP	$SMAX$	$TRIAL$	$STEPS$	α	γ
50	50/200	15	250	1000	0.6	0.9

After runing the algorithm using a population of 50 individuals, in less than 10 generations (in average), each agent, needing less than 250 trials, learns a finite state machine solution of 9 states, which reaches the goal in 28 steps (i.e., the optimal solution). Typically, if the system is run using 200 individuals per generation, it is able to learn a 9-state finite state machine without using the evolutionary search after only 127 learning trials. Figure 6 shows the best solution found by the SOS-algorithm. With that 7-state finite state machine, the agent learns to reach the goal in 28 steps after running the evolutionary search on a population of 100 individuals for 2 generations.

8   Concluding Remarks

In this work we show how to effectively use the evolution and learning of finite state machines to solve complex partially observable problems. Instead of using evolutionary techniques by encoding in the genome the whole state-transition table defining a finite state machine, we provide an approach with a very compact genome, where learning by interaction completes the design of the finite state machine. We compared our results with other approaches in two different tasks, the well-known Santa Fe trail and a 12×12 maze proposed in [20], and provide an intermediate test on a complex maze proposed by one of the authors. The results were very impressive given the compact representation of the genome we used, and the very few learning steps needed by the learning by interaction algorithm, to find optimal solutions.

Our approach (the *SOS-algorithm*) does not solve the curse of dimentionality problem, but provides an efficient algorithm for the design of finite state machines. It is the result of taking into account the normal procedure for designing finite state machines by hand, and the use of artificial evolution and learning techniques. Given that any procedural program can be implemented by a finite state machine, our approach is closely related to genetic programming (GP) [7]. Indeed, a predecesor of GP, evolutionary programming (EP), was proposed in the 1960's by L. Fogel [4], and one of the premises of his work was the automatic development of programs by means of finite state machine evolution [6].

Although evolution was not essential in solving the problems discussed in this paper, in a more general way, evolution enables the solution of the problems using smaller populations. Indeed, without evolution, the learning part of the SOS-algorithm can solve these problems only if the initial population contains all the needed actions, in the sufficient number (i.e., when a maze-solution uses the x action in n different states, the learning part would need an individual with n times the x action in its genome). This condition can only be respected, for any problem, by using big populations.

References

1. P.J. Angeline, G.M. Saunders, and J. Pollack. A evolutionary algorithm that constructs recurrent neural networks. *IEEE Transactions on Neural Networks*, 5(1):54–65, January 1994.
2. A.R. Cassandra. *Exact and Approximate Algorithms for Partially Observable Markov Decision Processes*. PhD thesis, Brown University, 1998.
3. Thomas G. Dietterich. The MAXQ method for hierarchical reinforcement learning. In *Proc. 15th International Conf. on Machine Learning*, pages 118–126. Morgan Kaufmann, San Francisco, CA, 1998.
4. L.J. Fogel and A.J. Owens. *Artificial Intelligence through Simulated Evolution*. John Wiley and sons, New York, 1966.
5. M. Hauskrecht. *Planning and Control in Stochastic Domains with Imperfect Information*. PhD thesis, MIT, Cambridge, MA, 1997.
6. C. Jacob. *Illustrating Evolutionary Computing with Mathematica*. Morgan Kaufmann, San Francisco, 2001.
7. J.R. Koza. *Genetic Programming: On the Programming of Computers by Means of Natural Selection*. The MIT Press, Cambridge, Massachusetts, 1992.
8. J.L. Lin and T.M. Mitchell. Reinforcement learning with hidden states. In J-A. Meyer, H.L. Roitblat, and S.W. Wilson, editors, *From Animals to Animats: Proceedings of the Second International Conference on Simulation of Adaptive Behavior (SAB92)*, pages 281–290, 1992.
9. N. Meuleau, L. Peshkin, K.E. Kim, and L.P. Kaebling. Learning Finite-State Controllers for Partially Observable Environments. In *Proceedings of the Conference on Uncertainty and Artificial Intelligence UAI'99*, Stockholm, Sweden, 1999.
10. Z. Michalewicz. *Genetic Algorithms + Data Structures = Evolution Programs*. Springer-Verlag, 1992.
11. A. Pérez-Uribe. *Structure-Adaptable Digital Neural Networks, Ph.D Thesis 2052*. PhD thesis, Swiss Federal Institute of Technology-Lausanne, 1999.
12. D. Precup, R.S. Sutton, and S. Singh. Theoretical Results on Reinforcement Learning with Temporally Abstract Behaviors. In *Proceedings of the 10th European Conference on Machine Learning ECML'98*, pages 382–393, Chemnitz, Germany, 1998.
13. M.B. Ring. *Continual learning in reinforcement environments*. PhD thesis, The University of Texas at Austin, August 1994.
14. E. Sanchez. An Introduction to Digital Systems. In D. Mange and M. Tomassini, editors, *Bio-Inspired Computing Machines: Toward Novel Computational Machines*, pages 13–48. Presses Polytechniques et Universitaires Romandes, Lausanne, Switzerland, 1998.
15. Satinder P. Singh, Tommi Jaakkola, and Michael I. Jordan. Reinforcement learning with soft state aggregation. In G. Tesauro, D. Touretzky, and T. Leen, editors, *Advances in Neural Information Processing Systems*, volume 7, pages 361–368. The MIT Press, 1995.
16. R. Sun and T. Peterson. Multi-agent reinforcement learning with bidding for segmenting action sequences. In J-A. Meyer, A. Bethoz, D. Floreano, D. Roitblat,

and S. Wilson, editors, *From Animals to Animats: Proceedings of the Sixth International Conference on Simulation of Adaptive Behavior (SAB2000)*, pages 325–332, 2000.

17. R.S. Sutton. Learning to predict by the methods of Temporal Differences. *Machine Learning*, 3:9–44, 1988.

18. R.S. Sutton and A.G. Barto. *Reinforcement Learning: An Introduction.* The MIT Press, 1998.

19. S.D. Whitehead and D.H. Ballard. Learning to Perceive and Act by Trial. *Machine Learning*, 7(1):45–83, 1991.

20. M. Wiering and J. Schmidhuber. HQ-Learning. *Adaptive Behavior*, 6(2), 1997.

GA-Based Learning
of $kDNF_n^s$ Boolean Formulas

Arturo Hernández Aguirre[1], Bill P. Buckles[1], and Carlos Coello Coello[2]

[1] Department of Electrical Engineering and Computer Science, Tulane University,
New Orleans, LA, 70118, USA,
hernanda,buckles@eecs.tulane.edu

[2] CINVESTAV-IPN, Dept. de Ingeniería Eléctrica/Sección de Computación, México,
D.F. 07300, Mexico,
ccoello@xalapa.lania.mx

Abstract. The number of samples needed to learn an instance of the representation class $kDNF_n^s$ of Boolean formulas is predicted using some tolerance parameters by the PAC framework. When the learning machine is a simple genetic algorithm, the initial population is an issue. Using PAC-learning we derive the population size that has at least one individual at some given Hamming distance from the optimum. Then we show that the population does not need to be close to the optimum in order to learn the concept.

1 Introduction

The size of the population is a critical factor for a Genetic Algorithm (GA). Although some theoreticians have favored the use of small populations [17], practitioners and early research indicate the opposite [4] It is normally accepted that a large population will converge at a slower pace but with a higher probability of success. "Small" and "large" are particular estimations derived from the many experimental populations tested on the problem.

In this paper, we seek to provide a characterization of the initial population through the estimation of the distance to the optimum. We apply the PAC-learning (Probably Approximately Correct learning) framework [19] to the estimation of an initial population that guarantees a fitness error between a hypothetical solution and one or more elements of the chromosome set. The metric is based on the Hamming distance formulation. Thus, the fitness function reports the number of bits by which two binary strings differ. A genotype is mapped onto itself to produce a phenotype. Therefore, the Hamming distance in the genotype domain equals that of the phenotype domain. In this way we can provide an environment in which the quality of a population size is estimated by the distance to the solution of the nearest chromosome. We call this size the PAC population.

The PAC framework sets a confidence δ over some predicted error ϵ, hence the terms approximately and correct. The PAC framework is commonly used to predict the generalization error of learning machines, such as neural networks

Y. Liu et al. (Eds.): ICES 2001, LNCS 2210, pp. 279–290, 2001.

[13,22]. A careful reinterpretation of the PAC framework in the GA domain is developed throughout this paper. The class $kDNF_n^s$ (and several other classes) of Boolean formulas has been studied in the PAC framework as well. Thus, expressions for estimating sample complexity are well known and simple to use. We design a code suitable for representing Boolean formulas of the class. We show that the population used to learn a $kDNF_n^s$ concept is quite far from the optimum.

In this paper we develop an application of PAC-learning theory to genetic algorithms. The organization of this paper is as follows: First, we introduce (the simple) PAC-learning model. Then, we state the population size problem using the PAC-learning terminology, and we derive a GA sample complexity expression (population size). At the end, we derive the GA sample complexity for the binomial distribution case, and, we give our conclusions.

2   Definitions and Previous Work

Much of the initial work in PAC learning was devoted to find the sample complexity for learning a Boolean class. The class $kDNF$ is defined as follows (Kearns p.20 [11]):

kDNF: For any constant k, the representation class $kDNF_n$ consists of all Boolean formulae of the form $T_1 + T_2 + \ldots + T_l$, where each T_i is a conjunction of at most k literals over the Boolean variables $x_1, \ldots, x_n$.

The sample complexity for the class $kDNF_n^s$ (Kearns p.97 [11])

For fixed k, let $kDNF_n^s$ be the class of kDNF formulae over $x_1, \ldots, x_n$ with at most s terms, and let A be a learning algorithm for $kDNF_n^s$. Then

$$S_A(\epsilon, \delta) = \Omega(\frac{1}{\epsilon} \ln \frac{1}{\delta} + \frac{s \ln \frac{n}{s}}{\epsilon}) \tag{1}$$

is the smallest sample complexity that upper bounds the error ϵ on future examples with confidence at least $1 - \delta$.

Pagallo [16] uses $S_A(\epsilon, \delta) = \frac{K \log(N)}{\epsilon}$ for learning the smallest DNF expression of the Boolean (target) concept. Iba et. al. [8,9] use Pagallo formula to approach concept learning with four different techniques: neural networks, classifier systems, adaptive logic networks, and evolvable hardware. Based on the failure to learn the target concept in two out of three functions, they indicate ([8] p.14): "we can not conclude that evolvable hardware is superior to the other learning methods". They estimated sample complexity using a rather general formula, and they seem to oversee fundamental aspects of the PAC formalism. Their sample complexity expression assumes that the concept space and the hypothesis space are identical. That is, the classifier machine is assumed to produce only hypotheses in the (target) concept space. But, since their evolvable hardware system produces hypotheses in all the Boolean space (not only DNF expressions), one has to estimate (and work with) the VC-dimension of the hypotheses space spawned by the learning machine (instead of the VC-dimension of the target space). The VC-dimension of such evolvable hardware device would not be easy to estimate, but we can easily predict its size is enormous. The sample complexity would be larger than that reported by Pagallo's formula, thus easier to learn (in probability) [5].

The size of the initial population is the topic of papers by Goldberg [4], Mühlenbein & Schlierkamp-Voosen [14,15], Reeves [17], and others. Cantú-Paz [3] worked out another estimation based on the *gambler's ruin model*. The model predicts the number of correct partitions in a chromosome at the time the convergence is attained.

In this paper we propose the distance to the optimum as a population quality measure (which is rather different than predict the final state of the population from some initial conditions). Thus, between two random populations we prefer the one with the closest element to the solution [6].

3 PAC-Learning and GAs

The problem we address now is that of the population size of a genetic algorithm that contains, with probability greater than $1-\delta$ one or more individuals at a distance no greater than ϵ (of a target hypothesis). By distance we mean the Hamming distance between two binary chromosomes. We use this metric as a measure of the error between two strings. A target hypothesis is anyone of the strings of length l that could encode a solution. PAC-learning uses the error and confidence parameters for deriving the size of the training set that guarantees a particular behavior of a learning machine. The size of this set is called *sample complexity*. Sample complexity is closely related to the population size of a genetic algorithm. In both cases (learning machines and GAs), the concept denotes a set of elements, sampled or generated from the solution space, with some fixed but possibly unknown probability distribution. PAC-learning usually assumes distribution-free data, which, in fact, is a general approach that covers the unknown distribution case.

3.1 PAC-Learning Concepts

PAC-learning is a theory of concept learning through examples [1,20,21]; we proceed as follows: assume a space $X \in \mathcal{R}$. A *concept class* $\mathcal{C} \subseteq X$ is the set of the possible partitions of X; a *hypothesis space* $\mathcal{H}$ is defined in the same way and may or may not be equal to $\mathcal{C}$; a target concept $c \in \mathcal{C}$ is a specific partition of X. Given a finite number of samples of a target concept c, the goal (of a learning machine) is to find the hypothesis $h \in \mathcal{H}$ which is a good approximation to c, with high probability (that is, with high probability, h follows very closely the behavior of c). The "good approximation" of h to c must be measured by some metric (or pseudo-metric at least) that estimates the error ϵ or symmetric difference between the hypothesis and the target concept. The "high probability" is indicated by the confidence parameter δ. More formally, for some small ϵ and δ, the hypothesis h satisfies

$$P_\mu[\ Error(h, c) > \epsilon] \leq \delta$$

for any target concept c and any probability of distribution P_μ. The minimum number of samples needed to attain that goal is called *sample complexity*.

The following example (taken from [7]) illustrates these ideas. We wish to learn the concept "medium build" from a set of examples taken from the space $X \in \mathcal{R}^2$. Every sample $\mathbf{x} = [x_1 x_2] \in X$ is a vector whose first component is "weight", and the second is "height". Let us assume the concept medium build is the area limited by the heights h_{max} and h_{min}, and the weights w_{max} and w_{min}, as shown in Figure 1.

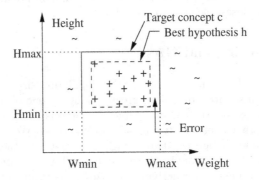

Fig. 1. The concept "medium build"

A set of examples is drawn from the space X and are classified by an *oracle* into either positive (represented by "+") or negative (represented by "$\sim$"). The best hypothesis is the smallest rectangle enclosing all positive examples. We can easily see that larger example sets might derive a better hypothesis, i.e., the closest hypothesis to the target concept.

3.2   Sample Complexity

Now we derive the sample complexity of a finite hypothesis space (the case of a GA). At this point we can estimate the error between the target and best hypothesis. Draw one more example $\mathbf{x} = [x_1 x_2] \in X$, and verify whether or not the sample falls inside the hypothesis rectangle (Figure 1). There is a small chance for an example to fall in the "error area". What is the probability of *not error*? It is at most $(1 - \epsilon)$ because ϵ is the chance of error. Now, if we tested m examples the chance of error becomes $(1 - \epsilon)^m$. We should assume a finite hypothesis space, thus, if all hypotheses have large error the probability that one of them correctly classifies the set X is at most $|\mathcal{H}| \cdot (1-\epsilon)^m$ ($|\mathcal{H}|$ is the cardinality of $\mathcal{H}$). Now, say we want to bound this probability by δ, we have:

$$|\mathcal{H}| \cdot (1-\epsilon)^m < \delta$$

Using the inequality $(1-x) < e^{-x}$, and later logarithms, the number of examples that bound the probability of misclassification by δ is:

$$m > \frac{1}{\epsilon}\left(\ln|\mathcal{H}| + \ln\frac{1}{\delta}\right) \tag{2}$$

With that sample complexity, we can guarantee

$$P_\mu[\ Error(h,c) \le \epsilon] > 1 - \delta \tag{3}$$

will hold for any new example drawn with the same probability of distribution [2]; this is the main PAC-learning equation.

3.3 Interpretation

We have suggested that sample complexity and population size of a GA are concepts of the same kind. A GA population of chromosomes is a population of coded candidate solutions. If the chromosomes are binary coded, and their length is n, the solution is one of the 2^n possible chromosomes; we name the solution chromosome the **target chromosome**.

Similarly, we name the finite number of chromosomes representable by 2^n, the **hypothesis space of chromosomes** $\mathcal{H}$. Therefore, each chromosome is a hypothesis in $\mathcal{H}$. The target chromosome is one such hypothesis which has the additional property of being a solution.

Let us concentrate in the initial population: how different are the initial individuals from the target chromosome? We measure the difference between two chromosomes by their **Hamming distance**. The Hamming distance is just another way to express the error ϵ, and we will find the exact relationship in the next section. Now, **what size must the initial population have to guarantee at least one hypothesis at a distance d from the target chromosome?** Figure 2 exemplifies the question. In an initial population, some individuals lie closer to the target hypothesis than others.

3.4 PAC-Size Population

In the following we modify Equation 2 for estimating the size of the GA initial population that with probability $1 - \delta$ contains at least one individual at a Hamming distance d. For the sake of the discussion, assume the length of the binary strings is 20, $l = 20$. In the sample complexity formula, $m > \frac{1}{\epsilon}\left(\ln|\mathcal{H}| + \ln\frac{1}{\delta}\right)$, we need to determine $\mathcal{H}$, ϵ and δ. The confidence δ is provided by the user and it does not depend on the problem domain.

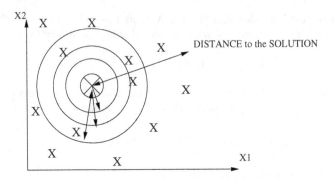

Fig. 2. Distance to the target hypothesis

The size of the hypothesis space of chromosomes is derived as explained in Section 3. The number of hypotheses representable by strings of length l is:

$$|\mathcal{H}| = 2^l$$

Substituting into Equation 2 and using logarithms, the sample complexity formula becomes:

$$m > \frac{1}{\epsilon} \left(l \ln 2 + \ln \frac{1}{\delta} \right) \qquad (4)$$

The error ϵ needs a re-interpretation in the GA context. To any Hamming distance d corresponds an error ϵ. In Figure 1, the "Error" is the area of the target concept minus the area of the best hypothesis. Since the examples are uniformly distributed all over the plane, the area is proportional to the probability of error. The error associated to a Hamming distance $d = 0$, is the probability that $\binom{20}{0}$ out of the 2^{20} hypotheses, match the target hypothesis. This is, for example:

$$- \;\; \epsilon(d = 0) = \frac{\binom{20}{0}}{2^{20}}$$
$$- \;\; \epsilon(d = 1) = \frac{\binom{20}{0}}{2^{20}} + \frac{\binom{20}{1}}{2^{20}}$$
$$- \;\; \epsilon(d = 2) = \frac{\binom{20}{0}}{2^{20}} + \frac{\binom{20}{1}}{2^{20}} + \frac{\binom{20}{2}}{2^{20}}$$

For instance, the error associated with a Hamming distance of 2, $\epsilon(2)$, is the probability that one of $\binom{20}{0} + \binom{20}{1} + \binom{20}{2}$ hypotheses matches the target chromosome.

More formally, for binary strings of length l and Hamming distance d,

$$\epsilon(d) = \frac{\sum_{i=0}^{d} \binom{l}{i}}{2^l} \qquad (5)$$

is the amount of error inherent to the Hamming distance. Equation 5 is part of Equation 4. In the following section we report some population size examples.

3.5   PAC-Population Examples

The computation of a PAC population is simple: we enter the desired Hamming distance into Equation 5 in order to determine the error, then the reported error and our desired level of confidence $(1 - \delta)$ (so, calculate δ) are entered into Equation 4 to determine the sample complexity or population size. The Table 1 shows PAC-population sizes for binary strings of length $l = 20$, for 10 Hamming distances, and 2 levels of confidence.

Table 1. The PAC-population with at least one element ϵ-close to a target chromosome

Error d	Confidence 90%	Confidence 95%
1	807181	841791
2	80336	83781
3	12547	13085
4	2736	2854
5	782	815
6	281	293
7	123	129
8	65	67
9	40	41
10	28	29

We have verified the accuracy of the confidence reported in Table 1 (we generated hundreds of random populations and measured the distance to a target chromosome). The predicted PAC-sized population was always correct. The explanation for this 100% accuracy is the following: we assumed a distribution-free sample complexity, but for the experiments each binary digit is generated with uniform probability. Therefore, the PAC-size is a loose upper bound to the population. In the following section we explain the exact solution to our problem. Since in our experiments we generate binary digits with uniform probability, the probability of having one chromosome in the population is exactly the same as for any other.

3.6   The Binomial Solution

We can find in our hypothesis space $\mathcal{H}$, a binary string of length l, and Hamming distance d from a target chromosome, with probability

$$p = \frac{\binom{l}{d}}{2^l}$$

where $|\mathcal{H}| = 2^l$. For a given confidence, and using the binomial distribution formula, we can say:

$$confidence = 1 - P[X = 0] = 1 - [\binom{m}{0}p^0(1 - p)^m]$$

$$confidence = (1 - p)^m$$

Using logarithms, the "exact" population size is:

$$m = \frac{\ln(1 - confidence)}{\ln(1 - p)} \qquad p = \frac{\binom{l}{d}}{2^l} \qquad (6)$$

This expression is useful when we are aware of the density function. The Table 2 shows exact population sizes for binary strings of length $l = 20$, for 10 Hamming distances, and 2 levels of confidence.

Table 2. The "exact" population that provides at least one element ϵ-close to a target chromosome

Error d	Confidence 90%	Confidence 95%
1	120721	157062
2	12707	16532
3	2117	2754
4	498	647
5	156	202
6	62	80
7	30	39
8	18	24
9	14	18
10	10	16

When we contrast Tables 1 and 2 the increase in population due to the distribution-free assumption is apparent. In any case, we shall apply either formula according to the current conditions of our problem or experiments.

3.7   Non Binary Alphabets

In the previous sections we assumed binary alphabets, thus from there we generalize the results to alphabets of any cardinality. The sample complexity formula 2is not affected but error expression 5 involves the cardinality of the alphabet. Thus,

$$\epsilon(d) = \frac{\sum_{i=0}^{d} \binom{l}{i} * (R - 1)^l}{R^l} \qquad (7)$$

is the estimation of the error at a distance d for chromosomes of length l in alphabets of cardinality R. The Table 3 shows the PAC-population size for different distances and cardinality $R = 3$. Confidence $1 - \delta = 90\%$ in all cases.

The binomial solution for non-binary alphabets is the following:

$$m = \frac{\ln(1 - confidence)}{\ln(1 - p)} \qquad p = \frac{\sum_{i=0}^{d} \binom{l}{i} * (R - 1)^l}{R^l} \qquad (8)$$

The Table 4 shows the "exact" size for different distances and cardinality $R = 3$. Confidence is 90% in all cases.

Table 3. The PAC-population for cardinality $R = 3$ and confidence $\delta = 90\%$

Error d	l=18	Error d	l=40
5	25895	23	321
6	5634	24	201
7	1530	25	136
8	510	26	99
9	206	27	77
10	99	28	64
11	57	29	56
12	38	30	52
13	29	31	49
14	25	32	48

Table 4. The "exact" population for cardinality $R = 3$ and confidence $1 - \delta = 90\%$

Error d	l=18	Error d	l=40	Error d	l=180
5	3253	18	941	107	294
6	750	19	406	108	218
7	218	20	193	109	165
8	79	21	101	110	127
9	35	22	58	111	101
10	19	23	37	112	82
11	13	24	26	113	68

4   GA Learning $kDNF_n^s$ Formulas

We use the GA as the machine to learn concepts of the $kDNF_n^s$ space. In the following three experiments, generalization error $\epsilon = 0.1$ and confidence $1 - \delta = 90\%$. The sample complexity is estimated using Equation 1. Phenotypes are encoded into genotypes through an alphabet of cardinality $R = 3$. A Boolean formula instance of the $kDNF_n^s$ space is: $f(x_1, x_2, x_3, x_4, x_5, x_6) = x_1 x_2 + x_3 x_4 + x_5 x_6$. The parameters are: fixed number of literals $k = 2$, number of variables $n = 6$, and number of terms $s = 3$. The codification is as follows: each loci in the chromosome represents a variable, and the allele values are 0,1, and 2. A negated variable has allele 0, a positive variable allele 1, and allele 2 for a variable not present in the term. Therefore, the formula $f(x_1, x_2, x_3, x_4, x_5, x_6) = x_1 x_2 + x_3 x_4 + x_5 x_6$ is encoded as: 112222221122222211 The string length is determined as $number of terms \times number of variables$

The GA parameters are: binary tournament selection, $P_{crossover} = 0.7$, uniform crossover, $P_{mutation} = 0.2$ per individual.

4.1   Experiment 1

Learn the Boolean formula of 6 variables and 3 terms:

$$f = x_1\bar{x}_2x_3x_6 + \bar{x}_1x_3\bar{x}_5\bar{x}_6 + \bar{x}_3\bar{x}_4x_5x_6$$

The sample complexity is $S_A(\epsilon, \delta) = 44$. A population of 50 chromosomes finds a solution identical to the formula in generation 10 (of 20). Generalization of the Boolean expression was tested with a random population of 50 elements and has confidence greater than 98% (minimum estimated is 90%). The length of these chromosomes is 18, thus, in Table 4 we can find that the (estimated) closest element to the optimum lies at a Hamming distance between 8 and 9.

4.2   Experiment 2

Learn the Boolean formula of 10 variables and 4 terms:

$$f = x_1\bar{x}_2x_3\bar{x}_4x_7x_8\bar{x}_9 + \bar{x}_1\bar{x}_2x_5x_6x_9x_{10} + x_1\bar{x}_2x_4\bar{x}_5x_6x_8x_9\bar{x}_{10} +$$
$$x_1\bar{x}_2x_4\bar{x}_5x_6x_8x_9 + x_2\bar{x}_3\bar{x}_4x_6x_7x_{10}$$

The sample complexity is $S_A(\epsilon, \delta) = 60$. A population of 70 chromosomes finds a solution identical to the formula in generation 12 (of 20). Generalization of the Boolean expression was tested with a random population of 512 elements ($\frac{2^{10}}{2}$) and has confidence greater than 95% (minimum estimated is 90%). The length of these chromosomes is 40. Thus, in Table 4 we can find that the (estimated) closest element to the optimum lies at a Hamming distance between 21 and 22.

4.3   Experiment 3

Learn the Boolean formula of 30 variables and 6 terms:

$$f = x_1\bar{x}_2x_3\bar{x}_4x_7x_8x_9\bar{x}_{11}\bar{x}_{12}x_{15}x_{16}x_{19}x_{20}x_{21}\bar{x}_{22}x_{24}\bar{x}_{25}x_{26}x_{28}x_{29}\bar{x}_{30} +$$
$$x_2\bar{x}_3\bar{x}_4x_5x_6\bar{x}_8x_{10}x_{11}x_{12}\bar{x}_{13}\bar{x}_{16}\bar{x}_{18}x_{19}\bar{x}_{20}x_{21}\bar{x}_{23}x_{24}x_{25}\bar{x}_{26}x_{27}\bar{x}_{29}x_{30} +$$
$$\bar{x}_1x_4\bar{x}_5\bar{x}_6\bar{x}_7x_9\bar{x}_{10}x_{13}x_{14}\bar{x}_{15}x_{17}x_{18}x_{22}x_{23}\bar{x}_{27}x_{28}\bar{x}_{29}x_{30} +$$
$$x_1x_2\bar{x}_3x_7x_8x_{11}x_{15}x_{19}\bar{x}_{20}\bar{x}_{21}x_{22}\bar{x}_{23}x_{24}x_{26}x_{28}x_{30} +$$
$$\bar{x}_1x_4\bar{x}_5\bar{x}_6\bar{x}_7x_9\bar{x}_{10}x_{13}x_{14}\bar{x}_{15}x_{17}x_{18}x_{22}x_{23}\bar{x}_{27}x_{28}\bar{x}_{29}x_{30} +$$
$$x_2x_5x_7x_9x_{10}x_{11}\bar{x}_{12}\bar{x}_{13}\bar{x}_{16}\bar{x}_{17}\bar{x}_{19}x_{20}\bar{x}_{22}\bar{x}_{24}x_{25}x_{26}x_{27} \qquad (9)$$

The sample complexity is $S_A(\epsilon, \delta) = 120$. A population of 120 chromosomes finds a solution with optimum fitness in generation 5 (of 20). This chromosome is different to the original formula in 112 locus (not alleles), and still the generalization is amazingly high. We tested with a random population of 100,000 elements (out of 2^30) to find a confidence greater than 98% (minimum estimated is 90%). The length of these chromosomes is 180. Thus, in Table 4 we can find that the (estimated) closest element to the optimum lies at a Hamming distance between 110 and 111.

5 Final Remarks

We have shown the application of the PAC-learning formalism for estimating the population size of a GA that guarantees a distance requirement. Furthermore, we have made a new interpretation of the concepts to fit the GA paradigm.

It has been reported by several researchers that Evolvable Hardware (at gate-level) needs enormous computational resources [18,12,10]. Perhaps we have been approaching the problem in the incorrect way. In these experiments we can see how easy it is to find complex Boolean formulas. The key is to work in a hypotheses space whose instances are also members of the concept space. We show that in these conditions the GA does not need large populations and, in fact, the predicted best element of the initial population is quite far from the optimum. This is due to the proper representation of phenotypes, and of course, to the correct interpretation of the PAC theory that assumes a learning machine generating only hypotheses in the concept space.

Acknowledgments

This paper describes research done in the Department of Electrical Engineering and Computer Science at Tulane University. Support for this work was provided in part by DoD EPSCoR and the Board of Regents of the State of Louisiana under grant F49620-98-1-0351. The third author acknowledges partial support from CINVESTAV through project JIRA'2001/08 and from the NSF-CONACyT project 32999-A.

References

1. M. Anthony and N. Biggs. *Computational Learning Theory*. Cambridge University Press, Cambridge, England, 1992.
2. Anselm Blumer, Andrzej Ehrenfeucht, David Haussler, and Manfred K. Warmuth. Learnability and the Vapnik-Chervonenkis Dimension. *Journal of the ACM*, 36(4):929–965, October 1989.
3. Erick Cantú-Paz. *Efficient and Accurate Parallel Genetic Algorithms*. Genetic Algorithms and Evolutionary Computation. Kluwer Academic Press, 2000.
4. David E. Goldberg. Sizing Populations for Serial and Parallel Genetic Algorithms. In J. David Schaffer, editor, *Proceedings of the Third International Conference on Genetic Algorithms*, pages 70–79, San Mateo, California, 1989. Morgan Kaufmann Publishers.
5. Arturo Hernández-Aguirre. *Sample Complexity and Generalization in Feedforward Neural Networks*. PhD thesis, Department of Electrical Engineering and Computer Science, Tulane University, 2000.
6. Arturo Hernández-Aguirre, Bill Buckles, and Antonio Martínez-Alcántara. The pac population size of a genetic algorithm. In *Twelfth International Conference on Tools with Artificial Intelligence*, pages 199–202, Vancouver British Columbia, Canada, 13-15 November 2000. IEEE Computer Society.
7. S.B. Holden and P.J.W. Rayner. Generalization and PAC Learning: Some New Results for the Class of Generalized Single-layer Networks. *IEEE Transactions of Neural Networks*, 6(2):368–380, March 1995.

8. Hitoshi Iba, Masaya Iwata, and Tetsuya Higuchi. Machine Learning Approach to Gate-Level Evolvable Hardware. In Tetsuya Higuchi, Masaya Iwata, and Weixin Liu, editors, *Evolvable Systems: From Biology to Hardware. First International Conference (ICES'96)*, pages 327–343, Tsukuba, Japan, October 1996. Springer-Verlag.

9. Hitoshi Iba, Masaya Iwata, and Tetsuya Higuchi. Gate-Level Evolvable Hardware: Empirical Study and Application. In Dipankar Dasgupta and Zbigniew Michalewicz, editors, *Evolutionary Algorithms in Engineering Applications*, pages 259–276. Springer-Verlag, Berlin, 1997.

10. Tatiana G. Kalganova. *Evolvable Hardware Design of Combinational Logic Circuits*. PhD thesis, Napier University, Edinburgh, Scotland, 2000.

11. Michael J. Kearns. *The Computational Complexity of Machine Learning*. MIT Press, Cambridge, Massachusetts, 1990.

12. Julian F. Miller, Dominic Job, and Vesselin K. Vassilev. Principles in the Evolutionary Design of Digital Circuits—Part I. *Genetic Programming and Evolvable Machines*, 1(1/2):7–35, April 2000.

13. Tom Mitchell. *Machine Learning*. McGraw-Hill, Boston, Massachusetts, 1997.

14. Heinz Mühlenbein and Dirk Schlierkamp-Voosen. Predictive Models for the Breeder Genetic Algorithm, I: Continuous Parameter Optimization. *Evolutionary Computation*, 1(1):25–49, Spring 1993.

15. Heinz Mühlenbein and Dirk Schlierkamp-Voosen. The Science of Breeding and Its Application to the Breeder Genetic Algorithm (BGA). *Evolutionary Computation*, 1(4):335–360, Winter 1994.

16. Giulia Pagallo and David Haussler. Boolean Feature Discovery in Empirical Learning. *Machine Learning*, 5:71–99, 1990.

17. Colin R. Reeves. Using Genetic Algorithms with Small Populations. In Stephanie Forrest, editor, *Proceedings of the Fifth International Conference on Genetic Algorithms*, pages 92–99, San Mateo, California, July 1993. University of Illinois at Urbana Champaign, Morgan Kaufmann Publishers.

18. Adrian Thompson, Paul Layzell, and Ricardo Salem Zebulum. Explorations in Design Space: Unconventional Design Through Artificial Evolution. *IEEE Transactions on Evolutionary Computation*, 3(3):167–196, September 1999.

19. Leslie G. Valiant. A Theory of the Learnable. *Communications of the ACM*, 27(11):1134–1142, November 1984.

20. Vladimir Naumovich Vapnik. *The Nature of Statistical Learning Theory*. Springer-Verlag, New York, 1995.

21. Vladimir Naumovich Vapnik. *Statistical Learning Theory*. Wiley, New York, 1996.

22. M. Vidyasagar. *A theory of learning and generalization: with applications to neural networks and control systems*. Springer-Verlag, London, 1997.

Polymorphic Electronics

Adrian Stoica, Ricardo Zebulum, and Didier Keymeulen

Center for Integrated Space Microsystems
Jet Propulsion Laboratory
California Institute of Technology
Pasadena CA 91109, USA

Abstract. This paper introduces the concept of polymorphic electronics (polytronics) –referring to electronics with superimposed built-in functionality. A function change does not require switches/reconfiguration as in traditional approaches. Instead, the change comes from modifications in the characteristics of devices involved in the circuit, in response to controls such as temperature, power supply voltage (VDD), control signals, light, etc. For example, a temperature-controlled polytronic AND/OR gate behaves as AND at 27°C and as OR at 125°C. The paper illustrates polytronic circuits in which the control is done by temperature, morphing signals, and VDD respectively. Polytronic circuits are obtained by evolutionary design/evolvable hardware techniques. These techniques are ideal for the polytronics design, a new area that lacks design guidelines/know-how,- yet the requirements/objectives are easy to specify and test. The circuits are evolved/synthesized in two different modes. The first mode explores an unstructured space, in which transistors can be interconnected freely in any arrangement (in simulations only). The second mode uses a Field Programmable Transistor Array (FPTA) model, and the circuit topology is sought as a mapping onto a programmable architecture (these experiments are performed both in simulations and on FPTA chips). The experiments demonstrate the polytronics concept and the synthesis of polytronic circuits by evolution.

1   Introduction

The classic approach to multifunctional system design is based on switching/multiplexing the output of single-function modules/subsystems, each with its stand-alone independently implemented circuit. When a condition is triggered, either by a command, or by the signal from a sensor/detector, a switching action takes place routing the output of one module instead of another. If N functions are needed, area for implementation of N modules needs to be physically present.

Reconfigurable devices allow the ensemble to collapse possibly within the size of one module, resources being shared, different functions being achieved following a reconfiguration based on switches. One of the consequences is efficient adaptive computation. Circuits can react to environment or context and change functionality as appropriate. A simple example is that of a power aware DAC in a portable device, capable of 16 bits resolution if the battery is loaded, and only 8 bits if battery is low (a

Y. Liu et al. (Eds.): ICES 2001, LNCS 2210, pp. 291-302, 2001.

fine-grained resolution with no resource overhead can be envisioned for graceful degradation). Similarly, a speed/resolution compromise can be imagined; e.g. 16 bits at 100kHz and 8 bits at 1MHZ.

In this paper we introduce another approach to multi-functionality, based on the polymorphic electronics concept first described in [1]. The term *polytronics* is derived from *poly*morphic elec*tronics*, but covers a wider range of polymorphic information processing structures, referring to primitive computational elements with built-in, superimposed multi-functional designs. This contrasts not only with today's digital logic circuits, but, in fact, with all currently used information processing structures (electronic and non-electronic, such as optical), which are based on primitive components designed for single function. The concepts of polytronics can be applied to multi-functional devices (for an example of a multi-functional device see [2] , for evolving devices one can follow a methodology as in [3]), or to multi-functional circuits, which is the focus of this paper. Polytronic circuits have several intrinsically built-in functions, and can have the same output provide different functional response under the control of certain global parameters, such as the supply voltage. In a different embodiment the circuit can provide different desired functional response simultaneously at different probing points. Polytronics could constitute the fabric of a new type of versatile, multi-functional systems. The capacity of storing/hiding "extra" functions provides for watermark/invisible functionality, thus polytronics may find uses in intelligence/security applications. Built-in environment-reactive behavior (e.g. changing function with temperature) may also find uses in a variety of space and military applications.

A simple example of multiple functionality can be considered in the context of a configurable logic block (CLB) that needs to provide, selected as needed, either an OR function or an AND function. A common implementation technique uses a circuit implementing the AND, a circuit implementing the OR, and a selection logic that, based on a control signal, activates the desired circuit and routes its output to the output of the block. In a polytronic implementation a single circuit would be designed. The function of the circuit would change as a result of changes that a control parameter produces in the parametric characteristics of its constituent devices. The control parameter could be voltage, temperature, light, radiation, or any other parameter that changes the characteristic (and operational point) of a device. In a different embodiment, passing data in one direction gives an AND, and passing the data in another direction through the circuit, gives an OR. At extreme, fully reversible circuits passing the same data in opposite directions, or passing different (desired) functions in opposite directions can be conceived.

This paper demonstrates the concept of polytronics, and in particular the use of evolutionary/evolvable hardware techniques to obtain polytronics. The paper is organized as follows: Section 2 discusses the evolutionary approach to polytronic design and detail the two techniques employed in the experiments – free evolution and evolution on the FPTA. Section 3 presents five experiments in which polytronic circuits are evolved. Section 4 presents a discussion on the evolved polytronics as well as plans for follow-on experiments. Section 5 discusses possible applications of polytronics in defense/security/intelligence and space applications. Section 6 presents the conclusions. The SPICE code for circuits discussed in this paper can be obtained from the URL provided as reference [13].

2   Evolutionary Approach to Polymorphic Design

How to design polytronic circuits? Unlike the case of traditional circuits there are no design guidelines or handbooks. The approach relies largely on changes in the device characteristics, usually subtle effects, commonly ignored in a first order approximation by traditional design (e.g. changes with temperature). Evolution however, can do without design rules, as long as the circuit specifications are straightforward, which is the case, and candidate circuits can be evaluated and ranked - thus, this is a problem well suited for evolutionary approach. An automated synthesis system based on evolutionary algorithms is presented with the multiple requirements that the circuit needs to satisfy. A generative process determines candidate solutions that are evaluated against a fitness function incorporating desired criteria and compete against each other, the best candidates being selected for reproduction and the process repeats; in most cases after a number of generations an acceptable (perhaps sub-optimal) solution can be found. For details on different ways of applying evolutionary techniques to design of electronic circuits see for example [4-6].

The resources used in the experiments are of two different kinds. In one case, unconstrained evolution allows the free exploration of the search space, with no topological restrictions – this can lead to new (patentable) designs. The disadvantage is that it must all happen in simulations, since there is no hardware implementation that would support it. Different loads were used in experiments to explore their influence on the convergence of the evolutionary algorithm. The second approach uses the FPTA model introduced in [7] and further detailed along with various evolutionary experiments in [8-10]. This approach has the advantage that its solution can be implemented after evolution, or evolved directly in hardware on a programmable FPTA chip. Moreover, the chip can be reconfigured to map different polymorphic gates as needed. The disadvantage is that the topology has certain restrictions imposed by architectural constraints. Also, the evolved circuits may in certain cases rely/make use of the non-ideal characteristic of the switches (i.e. the non-zero ON resistance and finite OFF resistance), which means that the transistors acting as switches can not be ignored and may lead to a topology that involves more actual components than may be possible if connections were ideal and the topology unconstrained.

2.1   Unconstrained Evolution

The unconstrained/free evolution was described in [11]. The experiments described in this paper use only NMOS and PMOS transistors, which can be interconnected in arbitrary topologies. The width and length of the transistor channel were also parameters for search. The advantage of this representation is the flexibility to map circuits with arbitrary types of interconnections, by establishing a straightforward mapping between the electronic circuit topology and the chromosome. Each functional block of the chromosome, also called gene, states the nature, value, connecting points, width and length of the MOS   (Metal-Oxide-Semiconductor) transistors. However, there is no integrated circuit model that supports the hardware implementation of the evolved solutions.

2.2   Evolution on FPTA

The FPTA is a cellular architecture, with transistor-level reconfigurability. Its flexibility in comparison with other devices was discussed in [10]. The elementary cell has a number of "fixed" transistors interconnected by transistors acting as switches. The number of "fixed" transistors per cell varied in different generations of the FPTA. The experiments presented here used the early version of the cell, with 8 transistors interconnected by 24 switches. [7]. Each switch is associated with a bit in the chromosome describing the cell. A bit being "1" translates to a closed switch, a "0" to an open switch. One can configure candidate circuits by programming the switches with binary string chromosomes produced by the Genetic Algorithm.

3   Evolutionary Experiments

The experiments presented in this paper demonstrate the evolution of polymorphic gates that change logic function under control of a) temperature, b) control signal or c) VDD. The temperature controlled polytronic AND/OR gates are AND for 27°C and become OR at 125°C (in other experiments at 5°C/90°C – one can choose the desired temperature). In a second set of experiments we evolved a AND/OR/XOR polymorphic gate with 10 transistors, which reacts at the change of a control signal Vmorph as follows: the gate is OR if $V_{morph} = 0V$, XOR if $V_{morph} = 1.5V$, AND if $V_{morph} = 3.3V$. In a third set of experiments we evolved polytronics gates changing from AND when power supply was 1.2V to OR when power supply was 3.3V.

The following indicates the evolutionary parameters used in the experiments. The population size was 50; the number of generations ranged between 100 to 200; the mutation rate was 8% and crossover rate was 30%.

3.1   Control by Temperature

The experiments were performed in SPICE simulations as well as on the FPTA chip, with the chip immersed in a temperature chamber. (For more extreme temperature experiments and a study on using evolution to expand the operation domain of electronics at high temperatures see [12]) The simulation experiments performed SPICE analysis at two temperatures of interest, circuit response was evaluated against two different criteria (for the lower and for the higher temperature). The fitness function was based on a combination of the quality of solutions at the two temperatures.

3.1.1   Free/Unconstraint Topology Search

In a first experiment, a AND/OR polymorphic gate was evolved. The gate behaves as an AND gate at 27°C and as an OR gate at 125 °C. Figure 1 depicts the circuit and its response. The circuit receives two inputs, In1 and In2, and it uses a 3.3V voltage supply. The output was a 10MOhms resistive load. This figure depicts the circuit inputs, In1 and In2, as well as the circuit output for 27°C (AND gate) and 125°C (OR gate).

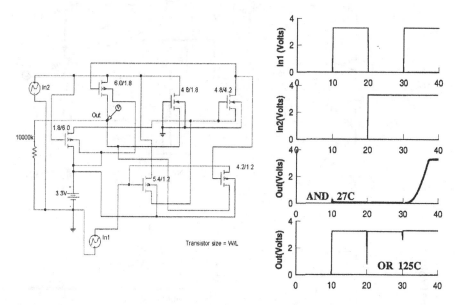

Fig. 1. Schematic of the Polymorphic circuit evolved for different temperatures (left). Circuit inputs and outputs (at 27 and 125°C) in the left. Axis X shows time in miliseconds.

3.1.2   Evolution on FPTA

A similar AND/OR gate was evolved using two FPTA cells. The evolved circuit behaves as an AND gate at 5°C, and as an OR gate at 90°C. Figure 2(A) shows the evolved circuit and Figure 2(B) shows the circuit response. It can be seen that the inputs In1 and In2 are applied to the first FPTA cell, while the output is collected from the second FPTA. Each re-configurable cell consists of 8 transistors interconnected through 24 switches.

3.2   Control by Dedicated Input Signal V_{morph}

Two experiments with unconstrained representation were performed, in which polymorphic gates with respectively two and three different logic functions have been evolved.

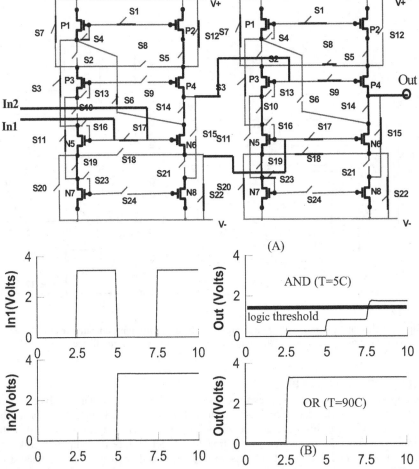

Fig. 2. Schematic of the evolved circuit on the FPTA(A) and its response at different temperatures(B). Axis X shows time in miliseconds.

3.2.1   AND/OR
In the first experiment, a circuit with an AND/OR functionality has been evolved. The circuit performs an AND function for $V_{morph} = 0V$ and an OR function for $V_{morph} = 3.3V$. Figure 3 depicts both the circuit schematic and its response. The fact that the

effect of an external signal compared to the change with that particular temperature range was more powerful was not surprising.

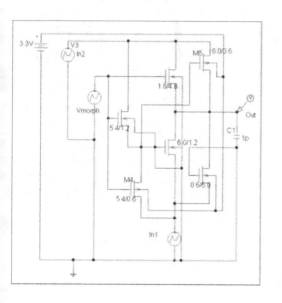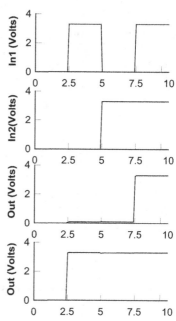

Fig

3.2.2   OR/AND/XOR

More than two functions can be superimposed. Figure 4 depicts a circuit that behaves as an OR for $V_{morph} = 0$, as an XOR for $V_{morph} = 1.5V$ and as an AND for $V_{morph} = 3.3V$. The schematic in Figure 4 provides the information on transistors width and length, parameters that have been evolved together with the circuit topology. Also shown in this figure the response of the evolved polymorphic gate.

3.3   Control by Supply Voltage (VDD)

In this experiment we evolved a circuit that performs different functions depending on the level of the power supply voltage, VDD. When VDD = 3.3V, the gate behaves as an OR gate; when VDD=1.2V it behaves as an AND gate. A possible application would be to endow circuits with built-in different behavior for *active* or *sleeping (power saving)* mode. Figure 5(A) displays the evolved circuit and Figure 5(B) shows the response. Note that the input voltage levels are adjusted with VDD.

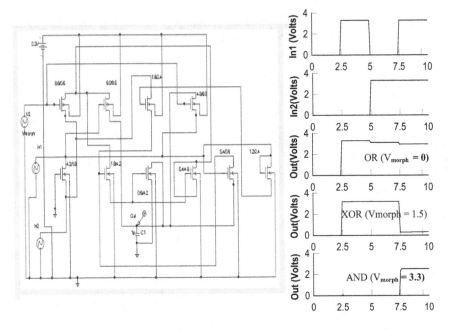

Fig. 4. Evolved Polymorphic circuit exhibiting three different functionalities , OR, XOR and AND.

4   Discussion

The experiments presented in this paper show solutions that satisfied the imposed stopping conditions for evolution. Additional constraints are required to produce circuits closer to practical use. The gate in Figure 1 evolved to a satisfactory solution as far as the logic level, which was the objective of the experiment, yet it is a slow gate, and its response can be definitely improved (perhaps even by a solution with fewer transistors). If we don't ask for it, evolution will not volunteer a solution for what we think (but not specify clearly) should be good. In the second experiment Figure 2 illustrates a compliant response: the level in the last interval, corresponding to the (1, 1) input combination is interpreted as a (1), being above the threshold defined as half-way between VDD and GND. Again, this is to illustrate the concept – more robust circuits further away from the threshold can be obtained, which will become the focus once we find the most useful context (from an application point of view) as far as temperature range and voltage level. Another observation is that the first experiment (without constraints in the possible connections) led to a better solution. The freedom of choosing unconstrained connections appears to have helped. The polymorphic gates with voltage control are not the first ever multi-functional gates that change function at the modification of an input control signal. For example U.S. patent 042335245 describes a 4 transistor circuit that performs XOR/OR/NAND

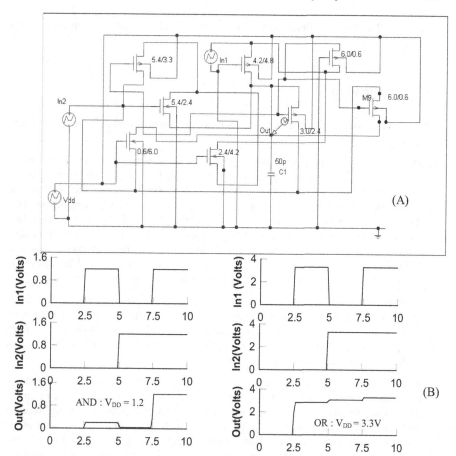

Fig. 5. Schematic of the polymorphic circuit controlled by supply voltages(A). Circuit inputs and response for two cases, V_{DD}=1.2V (left) and V_{DD}=3.3V (right). Axis X of the graphs gives the time in milliseconds.

and a 6 transistor that performs ADD/NOR/OR/NAND/AND. It is Ok for polymorphic circuits to exploit switch-like functionality of transistors – the main difference compared with the classic switch-based approach to multi-functional circuits would be that these switches are not (only) for multiplexing the outputs from constituent stand-alone functions. VDD control can be used in having reactive power-down change of functionality. It is also to observe that with VDD one can quasi-instantly change the function of the entire circuit, no matter its size! In fact all global controls including supply voltage, temperature, or control signal, etc. can be used for quasi-instant control of an entire circuitry. To change the function of a classical reconfigurable circuit all configuration bits need to be loaded and the associated time increases with the size. Fractions of second are needed for million gates components. Using a flash context-switching scheme is rapid but requires extensive extra resources. Polymorphic circuits are fast – circuits could completely change function on a clock edge.

5   Applications

The polytronics concept opens a new domain of commercial and defense applications. For example:

- Polytronics provides a new way to obtain circuits with one or more conceived "extra" functions in addition to the "main" function of the circuit. The "extras" can be activated under certain conditions or can coexist. Possible uses of the "extras": an authentication signature / watermark, extra protection from reverse engineering (the real operational function of the circuit shows up only in special conditions), protection from unauthorized usage by incorporating biometric info part of circuit design, providing an additional communication channel. This technology could be used as a non-traditional technique for insertion of sensors into denied areas and facilities data infiltration from denied areas and facilities, innovative tagging technology, etc.

- Polytronics allows for a built-in reactive behavior surfacing/taking control in specified conditions: for example, smart fuses in which the increased temperature triggers a new functionality of the guidance electronics. It would also enable systems that rapidly morph between functions, without switching overhead. It would also provide more compact multi-functional designs.

Certain applications may require a hidden/secret function, hard to detect and/or understand if reverse engineered. Polytronics could provide this feature. For example, a circuit may for all purposes look and act as a clock generator. In reality, when a control key – such as temperature level or pattern, EM pattern, VDD control etc is applied, it would exhibit a burst that unlocks/resurrects a special encoding scheme. This "extra" function may be a watermark visible only when certain conditions are created. This can be used for tagging, or other ID/verification need. The control may also be a biometric pattern. For example, a circuit can be designed to produce its essential function only if its components receive individual specific biometric signal. More specific, the array of voltages generated after a preprocessed fingerprint scan influences different areas of the circuit "biasing" it variably to create the condition in which the system is ok to operate. This offers a unique "personalized" custom chip with biometric info part of its hardwired design.

6   Conclusion

This paper introduced a new paradigm of circuits with super-imposed multiple functionality. Polytronics (short for polymorphic electronics) circuits are multi-functional circuits in which the functional changes come not from a switch-based routing of outputs of modules designed for individual functions, but more from superimposed functional design and changes from modifications of device characteristics and operating points. The paper demonstrates the approach for several cases of morphing control – using temperature, VDD and control voltage signals. Evolvable hardware appears an ideally suited technology for the design/determination of polytronics, since this is an area without any design know-how, but it is easy to specify requirements in an objective function. Circuits were evolved both with a free/unconstrained topology search, and using a FPTA model. The experiments show the successful evolution of polytronic AND/OR and AND/OR/XOR gates behaving differently at different temperatures (27°C/125°C, 5°C/90°C), VDD (1.2/3.3) or morphing voltage signal (0V/3.3V), (0V/1.5V/3.3V).

Acknowledgements

The research described in this paper was performed at the Center for Integrated Space Microsystems, Jet Propulsion Laboratory, California Institute of Technology and was sponsored by the Defense Advanced Research Projects Agency (DARPA) under the Adaptive Computing Systems Program managed by Dr. Allan O. Steinhardt. A. Stoica wishes to express special thanks to Dr. Jose Munoz, former Program Manager at DARPA, now at DOE, for funding the Evolvable Hardware research at JPL, and for his continuous encouragement and suggestions. The authors are also grateful for the support received from JPL program and line management, in particular from Drs. Leon Alkalai, Benny Toomarian, Anil Thakoor, and Taher Daud.

References

1. A. Stoica, Polymorphic electronics – A novel type of circuits with multiple functionality, NASA New Technology Report NPO-21213, 10/06/2000, Patent pending.
2. Multifunctional device http://www.ele.kth.se/FMI/research/hiep/oeic.htm
3. Stoica, A. Klimeck, G., Salazar-Lazaro, C. Keymeulen, D. and Thakoor, A. (1999) "Evolutionary design of electronic devices and circuits". *Proceedings of the 1999 Congress on Evolutionary Computation*, Washington, D.C. July 6-9, pp. 1271-1278
4. Koza, J., F.H. Bennett, D. Andre, and M.A Keane, "Genetic Programming: Darwinian invention and problem solving", Morgan Kaufmann Publishers, San Francisco, CA, 1999
5. Thompson, A., Layzell, P. and Zebulum, R., Explorations in design space: unconventional electronics design through artificial evolution, IEEE Transactions on Evolutionary Computation, September 1999, V.3, N.3 pp. 167-196
6. Higuchi, T. et al., Real-world applications of analog and digital evolvable hardware, IEEE Transactions on Evolutionary Computation, September 1999, V.3, N.3 pp. 220-235
7. Stoica, A. Towards Evolvable Hardware Chips: Experiments with a Programmable Transistor Array. Proc.7th International Conference on Microelectronics for Neural, Fuzzy and Bio-inspired Systems, *Microneuro'99,*. Granada, Spain, April 7-9, 1999, pp. 156-162
8. Stoica, A., Keymeulen, D., Tawel, R., Salazar-Lazaro, C. and Li, W. "Evolutionary experiments with a fine-grained reconfigurable architecture for analog and digital CMOS circuits. *Proceedings of the First NASA/DOD Workshop on Evolvable Hardware*, Pasadena, CA, July 19-21, IEEE Computer Society Press, pp. 76-84, 1999
9. Zebulum, R. Stoica, A. and Keymeulen, D., A flexible model of CMOS Field Programmable Transistor Array targeted to Evolvable Hardware, ICES2000, pp.274-283.

10. A. Stoica, Ricardo Zebulum, Didier Keymeulen, Raoul Tawel, Taher Daud, and Anil Thakoor.Reconfigurable VLSI Architectures for Evolvable Hardware: from Experimental Field Programmable Transistor Arrays to Evolution-Oriented Chips. IEEE Transactions on VLSI. February 2001

11. R. Zebulum, M.A. Pacheco, M. Vellasco, and H. T. Sinohara, "Evolvable Hardware: Automatic Synthesis of Analog Control Systems". In *IEEE Aerospace Conference,* Big Sky, Montana, March 14-25, 2000. IEEE Press.

12. Stoica, A. Keymeulen, D. and Zebulum, R. "Evolvable Hardware Solutions for Extreme Temperature Electronics", 3rd NASA/DOD Workshop on Evolvable Hardware, July 2001, Long Beach, CA, pp.93-97, IEEE Press.

13. http://cism.jpl.nasa.gov/ehw/public/ices01.

Initial Experiments of Reconfigurable Sensor Adapted by Evolution

D. Keymeulen, R. Zebulum, A. Stoica, and M. Buehler

Center for Integrated Space Microsystems
Jet Propulsion Laboratory
California Institute of Technology
Pasadena CA 91109, USA

Abstract. Missions to planets with unknown environmental condition, have recently been approached with new ideas, such as use of biology-inspired mechanisms for hardware sensor adaptation. In this paper we describe the initial development of efficient mechanisms for smart sensing which will lead to higher quality data. The self-reconfigurable pre-processing analog electronics is based on evolvable hardware.

1   Introduction

Modern sensors provide high data rates with only a small fraction of the data carrying quality information. The current pre-processing electronics is not smart enough to eliminate useless/redundant data and the on-board real-time processing capabilities are limited. These restrictions impose large on-board storage memory and high communications bandwidth. If the electronics could adapt to incoming signals and the context of the measurement, more information could be obtained from the sensor and sent back to earth.

The concept of reconfigurable and adaptive electronics for signal conditioning has led to the design of a series of recent chips that allow programmable adjustment of amplifier gains, memory-based compensation of sensor non-linearity, etc [11]. However, the flexibility of these programmable devices is limited by the high level of reconfiguration granularity, and requires that all compensation data be predetermined through lab experiments and then stored in ROM; also no later changes in sensor characteristics or electronics itself could be considered once the sensor is in operation. A complementary technique, called evolvable hardware (EHW), allows the automatic determination of optimal electronic circuit configurations by evolutionary algorithms [1][2][3]. In particular, a chip designed for EHW experiments at the Jet Propulsion Laboratory (JPL) called a Field Programmable Transistor Array (FPTA) has high flexibility by reconfiguration at transistor level [4][5][14][16]. EHW has also been considered for various application hardware, from antennas to complete evolvable space systems that could adapt to changing experimental environments and, moreover, increase their performance during the mission [15].

Y. Liu et al. (Eds.): ICES 2001, LNCS 2210, pp. 303-313, 2001.
© Springer-Verlag Berlin Heidelberg 2001

In this paper we describe the initial development of efficient mechanisms for smart on-board sensing, adaptively controlling the reconfigurable pre-processing analog electronics using EHW, which will lead to higher quality data. The target is to demonstrate the mechanisms for the MARS'01 MECA (Mars Environmental Compatibility Assessment) Electrometer. We identify one application of the electrometer for which the reduction of the data can be considerable: discrimination task of materials with different triboelectric properties. The discrimination task requires a sophisticated signal conditioning able to analyze multiple responses in order to extract differences in signal and to adapt to ambient environmental conditions. The discrimination task is translated to fitness evaluation metric that is used by an evolutionary algorithm to determine the optimal configuration of the electronics. At this stage of the research, the search for an electronic circuit realization of a desired transfer characteristic is made in software as in *extrinsic* evolution using signals obtained from the electrometer [14]. In the near future we will use *intrinsic* evolution where the hardware actively participates in the circuit evolutionary process and is the support on which candidate solutions are evaluated.

This paper is organized as follows: Section 2 presents a description of the electrometer sensor array. Section 3 presents the adaptive sensor architecture. Section 4 presents the FPTA, the experiments and results obtained for the adaptive electrometer for a discrimination application in a changing environment. Section 5 provides conclusion.

2   Electrometer Sensor Array

The electrometer is a part of the MECA project that has its objective as a better understanding of the hazards related to the human exploration of Mars [10]. The MECA project also has a material patch experiment to determine the effects of dust adhesion, a wet chemistry laboratory to characterize the ionic content of the soil, and microscopy station to determine particle size and hardness. The electrometer was built into the heel of the Mars '01 robot arm scoop and has four sensor types (Fig. 1): (a) triboelectric field, (b) electric-field, (c) ion current, (d) temperature. The triboelectric field sensor array contains five insulating materials to determine material charging effects as the scoop is dragged through the Martian regolith and the insulating materials are rubbed against the Martian soil.

In the rubbing sequence, depicted in Fig. 2, the scoop is first lowered against the Martian soil. During the start of the traverse, the electrometer is calibrated to zero by closing a switch. After reaching the end of its traverse, the scoop is abruptly removed from the soil at which time the triboelectric sensor response is measured. As seen on the left in Fig. 2, charge is generated triboelectrically across capacitor C3 as the insulator is rubbed on the Martian surface. Since the charges are in close proximity across C3, no charge appears across capacitors C1 or C2. As the insulator is removed from the surface, the charges redistribute themselves across C1 and C2 and provide the signal for the amplifier.

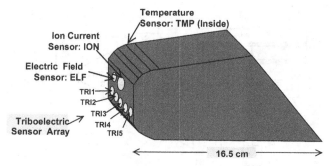

Fig. 1. Electrometer sensor suite mounted in the heel of the Mars'01 scoop

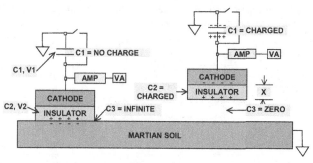

Fig. 2. Operational scenario for the scoop and charge distribution in the electrometer during rubbing (left) and after removal from the surface (right).

This electrometer is an induction field meter [6] operated in a direct current mode, where the operational amplifier input current charges C1. The electrical schematic of the non-adaptive component of the triboelectric sensors is shown in Fig. 3. The design of the electric field sensor follows from the traditional electrometer [7]. The instrument is composed of a capacitive divider where C2 is the field sensing capacitor and C1 is the reference capacitor. The point between the capacitors is connected to the positive terminal of the first stage amplifier (terminal +5 of U3) operated in the follower mode. The sensing electrode is protected by a driven guard that is connected to the negative terminal of the first stage amplifier (terminal -6 of U3). A second operational amplifier (U4) is added to provide additional amplification. At the beginning of the measurements, C1 is discharged using the solid-state switch, S1 which has very low leakage. In the TRI sensor, C2 has an insulator dielectric which acquires charge during rubbing.

Four different insulating materials were loaded into the titanium triboelectric sensor head (Fig. 1). A typical laboratory experiment consists of manually rubbing a wool felt on the triboeletric head at room temperature. The results are shown in Fig. 4. The falling period between 10 and 20 seconds represents the rubbing period. The large negative response is for the Rulon-J which is to be expected for Rulon-J rubbed on wool.

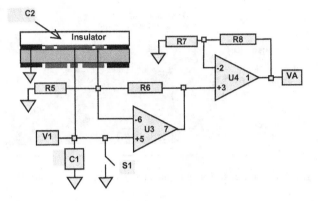

Fig. 3. Schematic circuit representations for the non-adaptive component of the Triboelectric sensor (TRI) fully characterized before field use.

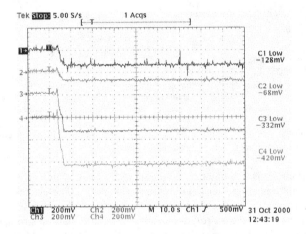

Fig. 4. Response of triboelectric sensor array to white **wool felt** (For all figures: response C1 is ABS (TRI1), response C2 is **Polycarbonate** (TRI2), response C3 is **Teflon** (TRI3) and response C4 is **Rulon-J** (TRI4)).

3   Adaptive Sensor Architecture

The triboelectric sensor array is an example of a hybrid integrated array devices where the sensors are grouped on the same devices but where the signal processing is done on a separate device [8]. This sensor array employs similar sensors (in terms of the measurand) but sensors have subtle differences (i.e. partially correlated outputs) related to the triboelectric properties of materials, known as the triboelectric series and respond to a very wide range of materials. (The triboelectric series orders the materials in a single list and indicates the direction and the amplitude of the charge

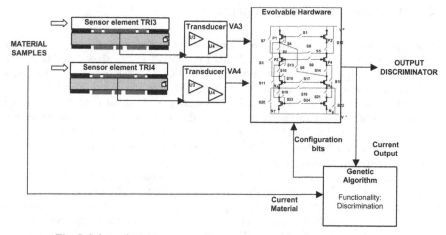

Fig. 5. Schematic arrangement of an adaptive ecltrometer sensor array device

transfer that occurs when two materials are rubbed). The signal processing must therefore carry out a sophisticated analysis of the responses to extract the subtle differences in signals. The approach we have chosen, as shown in Fig. 5, is to use an EHW discrimnator signal conditioner connected to the triboelectric sensor array and that will be able, after evolution, to discriminate the response of different materials with high precision.

Another important reason to use an adaptation mechanism is to be able to do *in-situ self-calibration* of the sensors and its electronics [9]. Indeed the sensors are very sensitive to ambient conditions, such as temperature, humidity, atmospheric and contact pressure, ambient gas, materials. They are also sensitive to the material and surface condition of the sensors. For example, the dust clinging on the insulator surface considerably affect the response of the triboelectric sensor arrays. Finally the sensor array has poor aging characteristic, that is the triboelectric sensing element is slowly corroded and thus changes its response characteristics with time. To remedy this high sensitivity to the ambient conditions and sensor conditions, we performed an *in-situ self-calibration*: calibrate the sensors right at the site with a set of reference materials with known triboelectric properties in the current environmental conditions.

Fig. 5 shows the basic arrangement of an adaptative electrometer sensor array system for discriminating different materials. The triboelectric property of the material is sensed by an array of sensors, each with its response, which is converted to an electrical signal via suitable transduction circuitry. The voltage signal VA_i is then injected to the evolvable hardware optimized for the current environment and a set of reference materials. The prediction of the triboelectric property of the material compared to the one of the reference materials is given in terms of output voltage. In the next section, we describe the EHW developed by JPL, called FPTA, and the experiments.

4   Adaptive Sensor Experiments

Our experiment was performed using the FPTA and an electrometer testbed. The idea
of a FPTA was introduced first by Stoica [5]. The FPTA is a concept design for
hardware reconfigurable at the transistor level. As both analog and digital CMOS
circuits ultimately rely on functions implemented with transistors, the FPTA is a
versatile platform for the synthesis of both analog and digital (and mixed-signal)
circuits. Further, it is considered a more suitable platform for synthesis of analog
circuitry than existing Field Programmable Gate Arrays (FPGAs) or Field
Programmable Analog Arrays (FPAAs), extending the work on evolving simulated
circuits to evolving analog circuits directly on the chip.

The FPTA module is an array of transistors interconnected by programmable
switches (Fig. 9). The status of the switches (ON or OFF) determines a circuit
topology and consequently a specific response. Thus, the topology can be considered
as a function of switch states, and can be represented by a binary sequence, such as
"1011...", where a 1 represents an ON switch and a 0 represents an OFF switch. The
FPTA architecture allows the implementation of bigger circuits by cascading FPTA
modules with external wires. To offer sufficient flexibility the module has all
transistor terminals connected via switches to external terminals (except for power
and ground) [10]. One FPTA module was fabricated through MOSIS, using 0.5-µm
CMOS technology. We built a testbed to acquire the signals of the electrometer and
extended it for future developments with a test board with four chips mounted on it
and connected with the electrometer (Fig. 6).

Fig. 6. Module of the Field Programmable Transistor Array connected to the electrometer
testbed.

The experiment shows that the EHW approach finds a FPTA circuit that is able to
discriminate between the responses of the electrometer to three different materials
chosen as reference. The experiments used three rubbing material samples
(Polystyrene, wool felt and Teflon) and used only two insulating materials of the

electrometer (Teflon and Rulon-J). The materials are ordered in the triboelectric series as follow with the first one becoming positively charged when rubbed on the other materials: Polystyrene, wool felt, Teflon and Rulon-J. The experiments start by an initialization procedure which puts the electrometer in a known state: the five electrometer insulators were cleaned by brushing followed by Am-241 alpha particle deionization. The deionization process was observed by running a trace and noting when the response no longer changed. After cleaning and deionization, the samples were placed in the apparatus as seen in Fig. 6. The data acquisition was started and five points were acquired every second. The first fifty points were baseline points. During the next 200 points, the samples were rubbed by the apparatus from left to right as shown in Fig. 4, Fig. 7 and Fig.8. During the final data points, the rubbing was stopped and the rubbing material was no longer in contact with the electrometer insulating materials. At this stage of the research the circuit was obtained by *extrinsic evolution* using the SPICE simulator and the response of the electrometer.

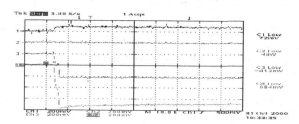

Fig. 7. Response of triboelectric sensor array to **Polystyrene** (C1 is ABS, C2 is Polycarbonate, C3 is Teflon, C4 is Rulon-J).

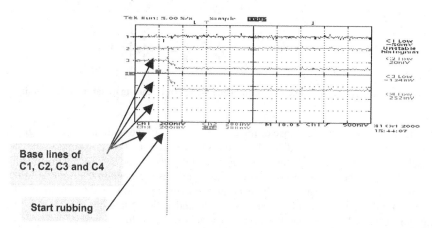

Fig. 8. Response of triboelectric sensor array to **Teflon** (C1 is ABS, C2 is Polycarbonate, C3 is Teflon, C4 is Rulon-J).

The response of the electrometer was obtained in air at a pressure of 970mb, relative humidity of 33 percent and a temperature of 21°C. The evolvable hardware system used one FPTA cell. The circuit had two inputs and one output. At the two inputs, we injected the sensor responses of the insulating material TRI3 (Teflon, response C3) and TRI4 (Rulon-J, response C4) to the three rubbing materials in addition to the baseline as shown in Figs. 4, 7 and 8. The outputs were collected as a voltage signal (Vout on figure 9).

The following Genetic Algorithm parameters were used: Population: 40, Chromosome size: 24 bits for 1 FPTA, Mutation rate: 10%, Crossover rate: 90%, exponential selection, elite strategy: 20% population size. The fitness function seeks to maximize the voltage difference at the output when different materials are used for rubbing. It can be described by the following equation:

$$Fitness = \frac{1}{T} \int_t \sum_{i \neq j} |Vi(t) - Vj(t)|$$

where the indexes i and j sweep the four patterns of the three materials and the baseline and T is the period of time used to evaluate the fitness.

The main task of evolution is to synthesize a circuit able to discriminate among the three materials and the baseline. The solution circuit, shown in Fig. 9, amplifies the voltage differences among the materials measured by the sensor such that the output voltages of the FPTA for each reference materials are distributed between Ground and 2.3 Volt.

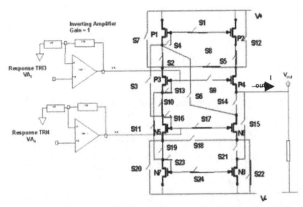

Fig. 9. Evolved circuit able to discriminate among 3 reference materials and the baseline.

Figure 10 shows the response of the evolved circuit. In the bottom part of the graph are the responses (VA3 and VA4) of the electrometer to the 3 materials and the baselines. Before being applied to the FPTA, they pass through a unit gain inverter stage (Fig. 9). The circuit response for the four patterns is shown in the top part of the graph. In the circuit response, there is an average separation of 0.6V between the adjacent materials, except for the wool felt and teflon materials, for which the difference is 1.2V. The overall output range achieved a value around 2.3V, whereas the input range given by the responses of the sensor is around 0.7V (Table 1). The circuit allows for an unknown material rubbed on the electrometer to classify in-situ with high precision its triboelectric property compared to the reference materials and to send only to earth its classification.

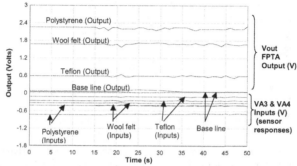

Fig. 10. Response of the evolved circuit for 3 reference materials and the baseline. The time starts when the material sample is rubbed on the isolating materials of the electrometer

Table 1: Inputs/Outputs of the Evolved Circuit

	Teflon	**Wool**	**Polystyrene**
- VA3 (TRI3)	0.124 V	0.332 V	0.412 V
- VA4 (TRI4)	0.252 V	0.420 V	0.684 V
Vout (output FPTA)	0.5 V	1.7 V	2.3 V

To assess the generalisation of the circuit solution we have tested the evolved circuit with sensor responses under slightly different environmental conditions which resulted in a decrease in the response of the sensors. As expected, the difference in response of the evolved circuit was smaller but it still captured the correct order of the patterns corresponding to the triboelectric series [12,13] (Fig. 11).

Although the task in this experiment was easy due to the nearly uniform distribution of the input Voltages (Table 1), the next generation of FPTA, which integrates programmable resistors [20], will be able to tackle discrimination tasks with more sensor inputs and non-uniform distribution of the input Voltages. Moreover the programmable capacitors of the next generation of FPTA will also allow to work with time dependent input signals.

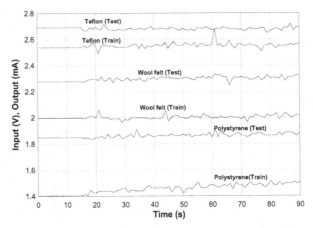

Fig. 11. Response of the evolved circuit for 3 materials for slightly different environmental conditions than for experiment of Fig. 10. The output measures the output current I_{out} at the drain of transistor P4.

5   Conclusion

These initial experiments, although illustrating the power of evolutionary algorithms to design analog circuit for sophisticated analysis of responses of sensor array and to maintain functionality by adapting to changing environments, only prepare the ground for further questions. The long term results of the proposed research would allow sensor electronics to adapt to incoming data and extract higher quality data, making available information otherwise not accessible. It will make sensor systems adaptive and intelligent. It will increase the amount of information available from sensors, while actually decreasing the amount of data needed for downlink in case of space missions.

Acknowledgements

The research described in this paper was performed at the Center for Integrated Space Microsystems, Jet Propulsion Laboratory, California Institute of Technology. The authors are grateful for the support received from Drs. Leon Alkalai and Nikzad Toomarian and the valuable discussions and comments from Drs. Anil Thakoor and Taher Daud.

References

1   Higuchi, T., Niwa, T., Tanaka, T., Iba, H., de Garis, H., and Furuya, T. Evolving hardware with genetic learning: A first step towards building a Darwin machine. In Meyer, Jean-

Arcady, Roitblat, Herbert L. and Wilson, Stewart W. (editors). *From Animals to Animats 2: Proceedings of the Second International Conference on Simulation of Adaptive Behavior.* pp 417 - 424. 1993. Cambridge, MA: The MIT Press.

2   A. Thompson, "An evolved circuit, intrinsic in silicon, entwined in physics". In *International Conference on Evolvable Systems.* Springer-Verlag Lecture Notes in Computer Science, 1996, pp. 390-405.

3   J. R. Koza, F. H. Bennett III,, D. Andre and M. A. Keane, Genetic Programming III – Darwinian Invention and Problem Solving, Morgan Kaufman, San Francisco, 1999

4   Stoica, A. Klimeck, G. Salazar-Lazaro, C. Keymeulen, D. and Thakoor, A. Evolutionary design of electronic devices and circuits, *Proceedings of the 1999 Congress on Evolutionary Computation,* Washington, DC, July 6-9. pp 1271-1278, 1999

5   Stoica, A. Toward evolvable hardware chips: experiments with a programmable transistor array. *Proceedings of 7th International Conference on Microelectronics for Neural, Fuzzy and Bio-Inspired Systems,* Granada, Spain, April 7-9, pp. 156-162. IEEE Comp Sci. Press, 1999.

6   J. A. Cross, *Electrostatics: Principles, Problems and Applications,* Adam Hilger (Bristol, UK).

7   *Electrometer Measurements,* Keithley Instruments (Cleveland OH), 1972

8   Julian W. Gardner, *Microsensors: Principles and Applications.* New York: John Wiley & Sons, 1994.

9   Gert van der Horn and Johan L. Huijsing, *Integrated Smart Sensors: Design and Calibration.* New York: Kluwer Academic Publishers, 1998.

10 M. Buehler et al., MECA Electrometer: Initial Calibration Experiments. In *Proceedings of the 10th Internaltional Conference (Electrostatics 1999).* Institute of physics Conference Series No 163, pp 189-196, Institute of Physics Publishing, Bristol, UK, 1999.

11 B. Schweber, Programmable analog ICs: designer's delight or dilemma ?. In *EDN.* April 2000. Cahners Publication.

12 O.J. McAteer, *Electrostatic Discharge Control.* McGraw-Hill, 1989.

13 N. Jonassen, *Electrostatics,* Chapman & Hall, 1998.

14 A. Stoica et al.. Reconfigurable VLSI Architectures for Evolvable Hardware: from Experimental Field Programmable Transistor Arrays to Evolution-Oriented Chips. In *IEEE Transactions on VLSI.* 9(1):227-232, April 2001. Piscataway, NJ: IEEE Press.

15 A. Stoica, J. Lohn and D. Keymeulen *Proceedings of the First and Second NASA/DoD Worshop on Evolvable Hardware,* IEEE Computer Society Press, 1999 and 2000.

16 Zebulum, R. Stoica, A. and Keymeulen, D., A flexible model of CMOS Field Programmable Transistor Array targeted to Evolvable Hardware, In *International Conference on Evolvable Systems,* pp. 274-283. Springer-Verlag Lecture Notes in Computer Science, 2000.

A Lossless Compression Method for Halftone Images Using Evolvable Hardware

Hidenori Sakanashi, Masaya Iwata, and Tetsuya Higuchi

National Institute of Advanced Industrial Science and Technology (AIST)
AIST Tsukuba Central 2, 1-1-1 Umezono, Tsukuba, Ibaraki 305-8568, Japan
{h.sakanashi, m.iwata, t-higuchi}@aist.go.jp

Abstract. This paper proposes a lossless (reversible) data compression method for bi-level images, particularly printing images. In this method, called *Dispersed Reference Compression (DRC)*, the coding scheme is changed according to the characteristics of the images to be compressed by Evolvable Hardware. Computational simulations demonstrate that DRC provides compression ratios that are up to 30% better than the current international standard for bi-level image compression. This paper also reports on the progress in discussions to incorporate a DRC-based compression method as an improvement to the international standard.

1   Introduction

On-demand publishing (ODP) is starting to revolutionize the publishing world and our concept of *books* [1]. With ODP it will soon be possible to order any book from home using the Internet and have it delivered immediately from the nearest print-shop.

This is one of latest publishing technologies combining the Internet and on-demand printing, where printing presses controlled directly by a PC can output highly precise printed materials. As this does not involve the initial setup costs, such as those for films and plates, involved in conventional printing processes, small runs of books are relatively inexpensive. The Internet provides the means for online marketing.

However, there are a number of problems which ODP must overcome before it becomes a popular way of publishing and marketing books. One serious problem is how to transfer the huge amounts of data required for book images.

Full-color printed materials are produced by sequentially pressing 4 halftone images for each of the 4 colors, Cyan, Magenta, Yellow and blacK (CMYK). These huge-sized bi-level images are created by a Raster Image Processor (RIP) system. For example, the total data size for the 4 halftone images for an A4-sized, full color print at 2400 dpi is about 2.2 GBytes. Thus, it is impractical to send the data for many pages via the Internet, because of the costs and time involved.

Data compression techniques are therefore essential for ODP to overcome this problem. However, no compression methods currently available satisfy all the following requirements simultaneously; (1) lossless (reversible) compression, (2) high compression ratios for halftone images, particularly for digital printing images.

Y. Liu et al. (Eds.): ICES 2001, LNCS 2210, pp. 314–326, 2001.
© Springer-Verlag Berlin Heidelberg 2001

Although the well-known compression methods for the facsimile, G3 [2] and G4 [3], are lossless compression techniques, they do not handle digital printing images well, which consist of many small dots which, in turn, are made up from many pixels, as shown in Fig. 1. The current international standard for bi-level image coding, called JBIG2 [4], also executes lossless compression and gives better compression ratios than G3 and G4. However, this method is also not sufficient to support ODP, because it cannot compress halftone images with large resolutions well, due to limitations that are discussed in more detail below.

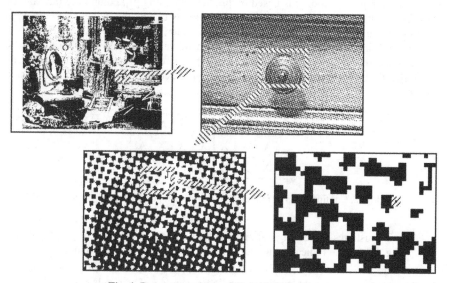

Fig. 1. Data composition of digital printing images

To overcome these kinds of problems, it is advantageous to change the compression scheme using Genetic Algorithm (GA) [5][6][7]. The Dispersed Reference Compression (DRC) method proposed in this paper adopts Evolvable Hardware (EHW) to search for better compression schemes according to the characteristics of the target images to be coded. EHW has reconfigurable circuits and can change its circuit structure by GA [8]. Thus, DRC with EHW can simultaneously provide both fast compression/decompression and high compression ratios.

The rest of this paper is organized as follows: Section 2 describes the features of the traditional lossless compression method for bi-level images. In Section 3, the concepts and mechanisms of DRC are explained in detail. The results of performance tests are presented in Section 4. Section 5 provides an overview of the hardware implementation of DRC. Section 6 addresses the standardization of DRC, and Section 7 concludes this paper.

2   Lossless Compression of Bi-level Images

There are many methods for lossless image compression, including the G3 [2], G4 [3], and JBIG2 (Joint Bi-level Image experts Group, 2) which is the international standard for bi-level image coding. JBIG2 has compression modes for lossless and lossy (irreversible) compression of pictures and texts. In this paper, we focus on the lossless compression mode, called *generic region encoding*, because G3 and G4 were developed for facsimile and because JBIG2 has better compression performance.

Generic region encoding is based on prediction coding [9], combining the Marcov model and arithmetic coding. The basic principle is that as the neighboring pixels in normal images strongly correlate with each other, the value of a pixel can be predicted fairly well by observing the pattern of values for its neighboring pixels. And, since the values of pixels that can be predicted correctly do not need to be recorded in the compressed data, precise pixel-prediction makes the compressed data smaller.

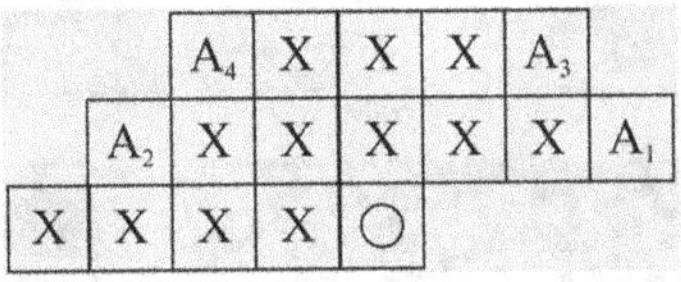

Fig. 2. Configuration of reference pixels used in generic region encoding

Fig. 2 illustrates the configuration of the reference pixels used in generic region encoding. The circle indicates the location of the pixel to be coded, and A_1 to A_4 and X indicate the locations of *reference pixels* that are referred to in prediction. This pattern of reference-pixels locations is called a *template*. The pixels marked A_1 to A_4 are called *AT pixels* and the locations of these can be changed almost anywhere within a field consisting of 256 x 256 locations, as shown in Fig 3. The current pixel, again marked by a circle, is always at the location to the right of the center on the bottom row of this field. This location and those to its right on the bottom row are not possible candidate locations for the AT pixels to move to, because pixel prediction must be executed using the values of previously coded pixels.

The template is used to extract a bit sequence called *context*. Pixel prediction is executed on the basis of the context. Accordingly, the size and the shape of a template will strongly influence the prediction-hit rate and consequently the compression ratio. As the procedure for pixel prediction is explained in the JBIG2 documentation [4], this is not described further here.

JBIG2 is claimed to have compression ratios that are 30% better than the previous version, JBIG [10], because it has 6 more reference pixels [4]. However, it cannot compress halftone images with large resolutions well, because of the following problems;

(1) Difficulties in optimizing the locations of AT pixels, and
(2) Limited number of AT pixels.

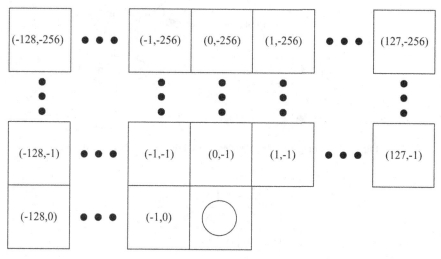

Fig. 3. Field of cancidate AT pixels locations

As there are about 256 x 256 candidate locations for each AT pixel, as shown in Fig. 3, there are about $_{(256 \times 256)}C_4 = 7.68 \times 10^{17}$ possible patterns of AT pixel locations, which makes very difficult to search for the optimal pattern for every target image to be compressed. Moreover, when the number of AT pixels is increased, the search space expands exponentially, making template optimization even more difficult.

3   Dispersed Reference Compression Method

The Dispersed Reference Compression (DRC) method, which is proposed in the following section, overcomes these problems. In this section, the mechanism of DRC is explained in detail.

3.1   Overview

As shown in Fig. 4, DRC consists of 4 parts; (1) the template, (2) the encoder, (3) the evaluator, and (4) the template optimizer. These are explained in turn.

First, as explained in the previous section, the template is used to extract the relevant context for each pixel to be coded. Similar to the JBIG2 method, DRC also adopts 16 reference pixels, however, unlike JBIG2 where only 4 of the 16 reference pixels can move, in DRC it is possible to set the number of fixed and AT pixels that are free to disperse within the template field. Fig. 5 shows the initial configuration of reference pixels in the DRC template. The numbering of the reference pixels in this figure represents the likely effectiveness of a given reference pixel in pixel prediction, where A_1 usually yields better predictions than A_{16}. Thus, when setting which

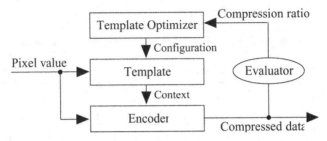

Fig. 4. Block diagram of the Dispersed Reference Compression Method

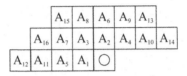

Fig. 5. Initial configuration of AT pixels used in DRC

reference pixels are free to move, it is generally better to allow those pixels to lowest likely effectiveness levels (e.g., A_{15} and A_{16}) to move.

Second, like JBIG2, DRC utilizes an MQ-Coder as the encoder, which works as follows;

(a)   pixel prediction is executed using the context extracted by the template, and

(b)   each current pixel is encoded, according to the result of prediction.

Repeating these steps for all pixels, the compressed data is generated.

Third, the evaluator calculates a fitness value using the compressed data generated by the MQ-Coder. In this paper, the compression ratio is adopted as the fitness value. The compression ratio (CR) is calculated as

$$CR = [\text{original data size}] / [\text{compressed data size}].$$

The objective of DRC is to find good templates, and make the original data as small as possible using the templates. This function yields high CR values when a template gives smaller compressed data.

Finally, the template optimizer searches for the optimal template using a Genetic Algorithm. As shown in Fig. 3, because the field for the AT pixel locations is restricted to 256 x 256 = 2^{16}, the location of each AT pixel can be specified by 16 bits. Thus, a template can be represented as a chromosome for the GA in the form of 16 x n bits, where n is the number of AT pixels, as shown in Fig. 6. The other parameters for the Genetic Algorithm are set as follows; elitist strategy, population size = 30, tournament selection (tournament size = 2), uniform crossover (crossover ratio = 0.8), point mutation (mutation rate = $1/(n \times 16)$), and maximum generation = 50.

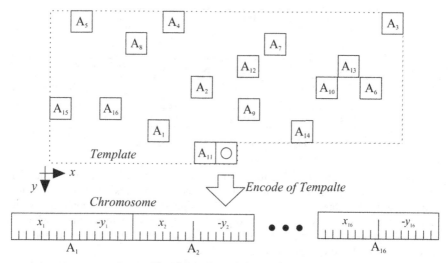

Fig. 6. Structure of chromosome

3.2   Behavior

DRC works as follows: First, the target data to be coded is divided into several sets of lines. These sets are called *stripes*. In this paper, an image is divided into stripes according to the average run length, which is the length of a sequence of repeated pixel values in each line. That is, the average run length is observed for every line, and a new stripe begins when a line has an average run length that is 30% smaller or larger than that of the previous line.

Second, to optimize the template in each stripe, the GA repeats the following steps until the generation exceeds the maximum number of generations:

(1) evaluation (repeating the following steps to all chromosomes),

    (1-1) decode ith chromosome to obtain a template,

    (1-2) the stripe is compressed using this template,

    (1-3) the compression ratio is calculated as the fitness value,

(2) tournament selection,

(3) uniform crossover,

(4) one point mutation.

In this paper, to reduce the costs involved in evaluations at step (1-2), the stripe image is sub-sampled by 10 in both the horizontal and vertical directions.

Third, every stripe image is compressed, using the template discovered by the GA. The compressed stripes are recorded one after another with their corresponding templates in a file. The file format of the compressed data is shown in the Fig. 7. The compressed data is composed of two parts; the header part and the body part. The header part records the size of the original image, the number of reference pixels, the number of AT pixels, and other additional information, such as the resolution of the image.

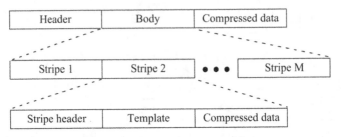

Fig. 7. File format of compressed data

The body parts consist of fractions corresponding to the stripes. Each fraction has three parts; the stripe header, the template part, and the compressed data of the stripe. The stripe header stores information about the stripe, such as the location of the stripe and the number of lines. The template part records the encoded template, which is used to compress the corresponding stripe. The compressed data part is the data stream output from the MQ-Coder.

Finally, when all stripes are processed in this way, compression is completed. Using a Pentium III (800 MHz), a simulation of the DRC requires about 15 minutes to compress an image of 7 MB.

4   Performance

This section reports the results of experiments to evaluate the performance of DRC. We prepared 5 types of images, as detailed in Table 1 and shown in Fig. 8. The image sets #1 to #3 contain 4 images that correspond to the 4 colors of CMYK.

The results of the simulations are given in Table 2 that show the compression ratio for each image. To compare performance, this table also gives the results achieved by other methods, including gzip, G4 (MMR), JBIG, and JBIG2. A performance result for G4 on the image set #5 is not included in the table because the G4 simulator was unable to handle such large images.

This table shows that DRC provided better compression ratios than the others methods. For the yellow and black images in the image sets #1 to #3 and #5, as these were very well compressed by the other methods, the margin of performance improvement with DRC is smaller than for the other images.

Table 1. Specifications of test images

Type	Size		Resolution	lpi	Note
#1	6656 x 9280	(7.72 MB)	800 dpi	100 lpi	CMYK
#2	6656 x 9280	(7.72 MB)	800 dpi	100 lpi	CMYK
#3	6656 x 9280	(7.72 MB)	800 dpi	100 lpi	CMYK
#4	12134 x 15117	(22.93 MB)	1200 dpi	175 lpi	
#5	43434 x 58100	(315 MB)	2400 dpi	175 lpi	CMYK

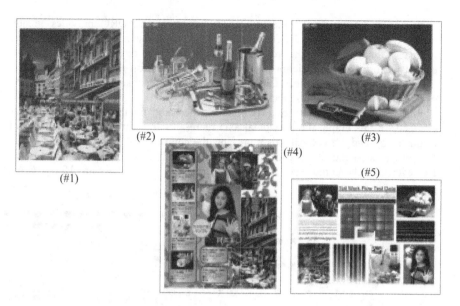

Fig. 8. Images used in simulations

Table 2. Compression rates

		gzip	G4[1]	JBIG[2]	JBIG2[3]	DRC	
#1	C	3.27	1.95	4.61	5.63	6.81	(+21.1%)
	M	3.23	1.75	4.77	5.46	6.47	(+18.5%)
	Y	3.75	1.92	6.08	6.63	7.45	(+12.4%)
	K	4.36	2.61	7.15	7.95	8.08	(+1.7%)
#2	C	5.53	1.97	6.76	12.66	14.82	(+17.1%)
	M	4.83	1.87	6.33	9.39	10.88	(+15.9%)
	Y	6.02	1.99	9.76	12.78	13.75	(+7.6%)
	K	8.42	2.72	13.30	20.20	21.78	(+7.8%)
#3	C	5.49	2.00	7.17	10.36	11.94	(+15.3%)
	M	5.12	1.78	6.60	9.56	11.51	(+20.4%)
	Y	6.00	1.46	8.82	12.16	12.72	(+4.6%)
	K	8.61	2.45	19.08	17.73	17.91	(+1.0%)
#4		2.80	2.37	3.30	4.72	5.85	(+23.9%)
#5	C	6.30	---	11.84	18.61	23.72	(+27.5%)
	M	6.10	---	11.29	17.10	22.18	(+29.7%)
	Y	10.48	---	17.00	18.05	20.07	(+14.7%)
	K	16.37	---	27.70	34.27	37.42	(+9.2%)

†1) JBIG-KIT ver.1.1 <http://www.cl.cam.ac.uk/~mgk25/>
†2) UBC-powerJB2 ver.1.0 <http://spmg.ece.ubc.ca/>
†3) 4 AT pixel version of DR simulator

It is also clear that as the resolution increases, the DRC provides greater improvements compared to JBIG2. As the distance from one halftone dot to the others is larger in images with higher resolutions, templates with converged reference pixels cannot perform pixel prediction well, and the compression ratio becomes smaller. DRC gives better performance than JBIG2, because it has many AT pixels and they can be dispersed by GA optimization.

Fig. 9 shows the results of experiments where the number of AT pixels was changed. In this graph, the horizontal axis represents the learning costs (number of evaluations), and the vertical axis represents the compression ratio. From this graph, the compression ratio seems to improve as the number of AT pixels is increased. However, and the compression ratio drops drastically when there are 15 or 16 AT pixels.

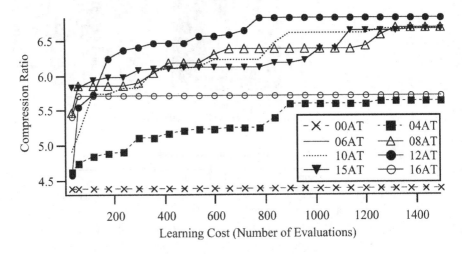

Fig. 9. Transitions in compression ratios

The reason of these phenomena is as follows: First, the search space expands exponentially as the number of AT pixels is increased, and the search ability of GA could not cope with this combinatorial expansion.

Second, in the case of 15 or 16 AT pixels, the reference pixels A_1 and A_2 begin to move. As these strongly influence the compression ratio (the compression ratio achieved with a template with these reference pixels gives better compression ratios than a template without them), the population prematurely converges onto chromosomes representing templates including these pixels in early generations.

In the future, these problems will be solved by,

    adaptive sub-sampling and striping of the image in evaluation,

    autonomous determination mechanism for the number of AT pixels, and

    new genetic operator for template optimization.

Using these enhancements, DRC with 16 AT pixels is expected to provide better compression ratios than with 12 AT pixels.

5   Hardware Implementation

This section explains a prototype version of the EHW implementing DRC, which consists of a data compression/decompression FPGA (DC-FPGA) and a PCI board. This is a prototype for an LSI version and was developed to test the performance of the DRC. Due of capacity shortages, this FPGA version has some limitations in its functionality;

(1)   GA for optimizing templates is executed on a host PC

(2)   the number of reference pixels is 10, and

(3)   the field of AT pixel locations is restricted to 8 lines and 32 columns.

The rest of this section explains the PCI board and the DC-FPGA

5.1   PCI Board

Fig. 10 is a block diagram of the PCI board.

The *PCI connector* is connected to the PCI slot of the host PC, and the *PCI controller* controls access from the host PC.

The *FPGA control signal generator* generates signals to control the memories and registers in the DC-FPGA. It has control the registers and status registers, and resets the other components according to commands from the PC, and conveys the status of the DC-FPGA to the PC.

5.2   Data Compression/Decompressions FPGA

Fig. 11 is a block diagram of the DC-FPGA. This consists of several components, of which the major ones are described in this section:

The *control register* stores commands from the host PC, and status information of the DC-FPGA.

The *image data memory* is composed of a 2-port RAM (320 x 32 bit x 10 lines). This receives image data from the PCI board at compression, and temporally stores the expanded images at decompression. The *reference buffer* has two buffers of 8 lines x 64 bits, and they mutually work to compensate the data loading time. They receive part of image data from the line memory, and send it to the context circuit. The *context circuit* generates the 10 bits data from the 256 bits data sent by the reference buffer according to the template data. It has 10 multiplexers (256 to 1) . This corresponds to the reconfigurable hardware in the EHW.

The template data, consisting of 8 bits x 10, is sent from the host PC and stored in the *template memory*. It is used to choose 10 bits from the 256 bits in the context circuit.

The DC-FPGA has two MQ-Coders for compression and decompression. The *MQ-Coder (C)* receives the 11 bits data from the context circuit and executes the compression procedure. The *MQ-Coder (D)* receives the compressed data from the PCI board and 10 bits data from the context circuit, and executes the decompression procedure.

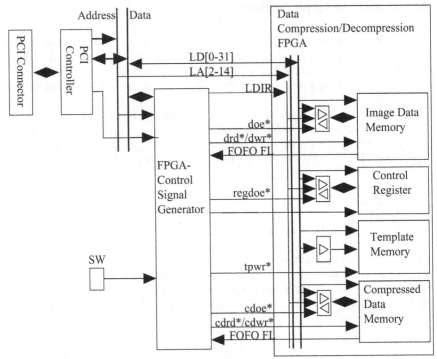

Fig. 10. Block diagram of PCI Board

6   Promotion Activity toward International Standardization

Currently, we are working as part of ISO/TC130/WG2 and ISO/IEC JTC1/SC29/WG1 toward the standardization of DRC.

The scope of ISO/TC130/WG2 is prepress data exchange, and our final goal is that DRC will be included within the international standards (e.g. TIFF/IT [12]) as a data compression method for halftone images. We have already submitted technical reports at three meetings [13].

Within ISO/IEC JTC1/SC29/WG1, whose scope is still image coding, we are promoting the JBIG2 Amendment 2 project for the extension of adaptive template for halftone coding. We have submitted for the first time a technical report at the Arles meeting in July 2000, and the project has been approved at the New Orleans meeting, Dec. 2000 [7]. At the Stockholm meeting, July 2001, a draft proposal based on DRC was approved as the working draft of JBIG2 Amendment 2 [14][15]. The promotion schedule is as follows:

Oct. 2001:   Proposed Draft Amendment 2 (PDAM2)
Feb. 2002:   Final Proposed Draft Amendment 2 (FPDAM2)
July 2002:   Final Draft Amendment 2 (FDAM2)
Nov. 2002:   Amendment 2 (AMD2)

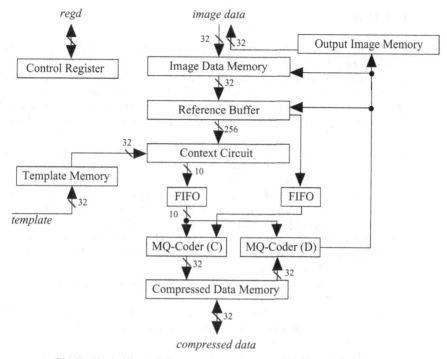

Fig.11. Block diagram of data compression/decompression FPGA

If there are no problems or objections, DRC will be included in the international standard during 2002.

7   Conclusions

This paper has described the Dispersed Reference Compression (DRC) method, which has been developed to compress binary halftone images. Using a Genetic Algorithm, this can change its compression method according to the characteristics of the target image to be compressed. The results of experiments showed that DRC provides compression ratio about 30% better than JBIG2, which is the current international standard for binary image coding. A prototype version of EHW implementing DRC, consisting of a data compression/decompression FPGA and a PCI board was also presented in this paper.

Currently, standardization projects are in progress within ISO/TC130/WG2 and ISO/IEC JTC1/SC29/WG1. In particular, a working draft for the JBIG2 Amendment 2 for the extension of an adaptive template for halftone coding was approved at the 24th ISO/IEC JTC1/SC29/WG1 meeting held in New Orleans, Dec. 2000. DRC may be included in JBIG during 2002.

References

[1] H. M. Fenton, et al.: *On-Demand Printing Ð The Revolution in Design in Customized Printing*, p.320, Graphic Arts Technical Foundation, 1997.

[2] International Telecommunication Union: *Standardization of Group 3 facsimile terminals for document transmission*, ITU-T Recommendation T.4, 1995.

[3] International Telecommunication Union: *Facsimile coding schemes and coding control functions for Group 4 facsimile apparatus*, ITU-T Recommendation T.6, 1995.

[4] JBIG Committee: *Information Technology Ð Coded Representation of Picture and Audio Information Ð Lossy/Lossless Coding of Bi-level Images*, ISO/IEC JTC1/SC29/WG1 N1359, 1999.

[5] M. Salami, et al.: *On-Line Compression of High Precision Printer Images by Evolvable Hardware*, Proc. of the 1998 Data Compression Conference (DCCÕ98), pp.219-228, IEEE Computer, Society Press, 1998.

[6] H. Sakanashi, et al.: *Evolvable Hardware Chip for High Precision Printer Image Compression*, Proc. of the 15th National Conf. on Artificial Intelligence, pp. 486-491, AAAI Press (1998).

[7] J. H. Holland: *Adaptation in Natural and Artificial Systems*, University of Michigan Press, 1975.

[8] H. Sakanashi, et al.: *Evolvable Hardware Chips and their Applications*, Proc. of the 1999 IEEE Systems, Man, and Cybernetics Conference (SMC'99), pp. V559-V564, 1999.

[9] B. Martins, et al.: Bi-level Image Compression with Tree Coding, Proc. of 1996 Data Compression Conference, 1996.

[10] International Telegraph and Telephone Consultative Committee (CCITT): *Progressive Bi-level Image Compression,* Recommendation T.82, 1993.

[12] TIFF/IT: International Organization for Standardization (ISO): Graphic technology -- Prepress digital data exchange -- Tag image file format for image technology (TIFF/IT), ISO 12639, 1998.

[13] H. Sakanashi: *Status Report: Dispersed Reference (DR) Compression Method DR compression discussed in SC29/WG1*, ISO/TC130/WG2 N930, 2001.

[14] WG1/JBIG: *AMD2 of JBIG2 (JTC 1.29.10) for the Extension of Adaptive Template for halftone coding*, ISO/IEC JTC1/SC29/WG1 N1993, 2000.

[15] F. Ono: *Report of SG on JBIG meeting held in Stockholm, Sweden*, ISO/IEC JTC1/SC29/WG1 N2226, 2001.

Evolvable Optical Systems
and Their Applications

Hirokazu Nosato[1], Yuji Kasai[2], Taro Itatani[2], Masahiro Murakawa[2],
Tatsumi Furuya[1], and Tetsuya Higuchi[2]

[1] Dept. of Information Sciences, Toho University,
2-2-1 Miyama, Funabashi, Chiba, 274-8510 Japan,
[2] National Institute of Advanced Industrial Science and Technology,
AIST Central 2, 1-1-1 Umezono, Tsukuba, Ibaraki, 305-8568 Japan

Abstract. This paper describes evolvable optical systems and their applications developed at the National Institute of Advanced Industrial Science and Technology (AIST) in Japan. Three evolvable optical systems are described: (1) an evolvable fiber alignment system, (2) an evolvable interferometer system, and (3) an evolvable femtosecond laser system. As the micron-meter resolution alignment of optical components usually takes a long time, to overcome this time problem, we propose 3 systems that can automatically align the positioning of optical components by genetic algorithms in very short times compared to conventional systems. In the evolvable fiber alignment system, the positioning of a fiber can be aligned automatically according to 5 degrees of freedom (DOF) in three minutes. In the evolvable interferometer system, the optimal positioning of the plane mirrors is determined automatically. And, in the evolvable femto-second laser system, the positioning of laser components can be aligned automatically within thirty minutes; something which in conventional system would take technicians more than five days to achieve. Moreover, the automatic alignment system makes possible the compact implementation of optical systems.

1 Introduction

In recent years, the market for optical technology has been growing rapidly, particularly due to the advances in optical communication systems. In general, optical systems consist of many components, such as light sources, mirrors, lens, prisms, semiconductor elements, and optical fibers. In order to obtain the optimal performance from an optical system, it is necessary to align these optical components to the optimal position with micron-meter precision. Moreover, as shifts in the positioning of each component impacts on the alignment of the whole system, each component must be repeatedly adjusted until the optimal alignment is achieved. As the number of experienced technicians is extremely limited, these difficulties increase the manufacturing times and the costs of optical systems.

Although algorithms for automatic alignment have been devised, they are often not suitable for systems where the parameters to be set optimally are

Y. Liu et al. (Eds.): ICES 2001, LNCS 2210, pp. 327–339, 2001.

numerous, where there are local optimum, and where the parameters are interdependent. In this paper, we propose methods of automatic alignment using genetic algorithms. These methods have the following four advantages.

1. Automatic adjustment for maintenance-free systems
   The performance of an optical system is affected by the temperature and the stability of the environment in which it is used. However, variations in performance can be adjusted for automatically by our algorithms on-line. This means that the optical systems are maintenance-free, making them easy for non-experts to use.
2. Cost-reduction
   Using less expensive mechanical parts in optical systems, unfortunately, often leads to longer adjustment times, because the precision of these components is usually inferior. However, as our algorithms provide a quick and flexible way of adjusting performance, it is possible to use less expensive mechanical parts in order to reduce the costs of the system.
3. Compact implementation
   With automatic adjustment methods, it is possible to reduce the spaces between the components in an optical system, which are necessary when adjustments are made by technicians. Thus, it is possible to downsize optical systems.
4. Realization of on-site optical systems
   Advantages 1. and 3. mentioned above also make the realization of on-site optical systems possible. Compact and maintenance-free systems open up new application areas, such as portable on-site environment measurement systems, which is something that has been impossible to date due to the difficulties of using high-performance lasers and spectrums outdoors.

This paper describes three optical systems using GA-based automatic alignment algorithms.

1. Evolvable Fiber Alignment System
   The alignment of optical fibers according to five degrees of freedom can be completed within a few minutes, whereas it would take a human technician about half an hour.
2. Evolvable Interferometer
   The adjustment of plane mirrors in an interferometer is difficult for non-experts. However, the automatic adjustment method eliminates this problem making it possible to use interferometers outdoors.
3. Evolvable Femto-Second Laser
   It often takes five days for experienced technicians to adjust the physical positioning of femto-second laser components in order to achieve optimal performance. The GA-based adjustment can align the laser components automatically within 30 minutes.

2   Evolvable Fiber Alignment System

2.1   Background

With the growth in optical fiber communications, fiber alignment has become
the focus of much industrial attention. This is a key production process because
its efficiency greatly influences the overall production rates for the opto-electric
products used in optical fiber communications. Fiber alignment is necessary
when two optical fibers are connected, when an optical fiber is connected to a
photo diode or a LED, and when an optical fiber is connected to an optical wave
guide.

Metallic wire connection is relatively easy because an electric current will flow
as long as the two wires are in contact. The connection between two optical fibers,
however, requires much greater precision, in the order of sub-micron-meters.
Therefore, experienced technicians are needed for fiber alignment, but as such
technicians are in limited supply, this causes a bottleneck in the mass production
of opto-electric components for optical fiber communication. To overcome this,
various algorithms for automatic alignment have been devised. However, existing
automatic-alignment algorithms are only capable of aligning fibers according to
three degree of freedom (DOF). Thus, the present authors have devised a new
algorithm based on GAs that is capable of aligning fibers according to five or
more DOF.

The advantages of the algorithm are summarized below:

1. Improvement in transmission efficiency
   Unless two optical fibers are connected in an optimal way, there will be a
   reduction in transmission efficiency. However, as alignment is possible quickly
   and according to a greater DOF, there are no reductions in transmission
   efficiency.
2. Improvement in reliability
   Automatic alignment reduces the numbers of tasks to be conducted by technicians, which leads to the improved reliability of products.
3. Improvement in production efficiency
   The fiber alignment process is often a critical part in the entire production
   process, and so by reducing fiber-alignment times, production efficiency can
   be improved considerably.
4. Cost reductions
   Less expensive opto-electric components are often inferior in terms of precision. However, variations from the desired specifications for each component
   can be compensated by using our alignment algorithm for greater DOF connections. This makes it possible to reduce the overall costs of a system.

The present authors have devised a fiber alignment algorithm for 5 DOF
connections. Conventional alignment systems cannot achieve connections with
5 DOF because such precise fiber alignment is impossible given practical time
constraints. However, our algorithm successfully carried out alignments for 4
different initial conditions in a few minutes. In the following subsections, the
details of the alignment system and the algorithm are described.

2.2   Optical Fiber Alignment

Transmission loss due to non-optimal connections. Although metallic wire connections are very easy, optical fiber connections require extreme precision. This is because optical signals can only pass along the core of a fiber, which is only a few micron meters in diameter. Therefore, when connecting optical fibers, the cores of the two fibers must be aligned to face each other exactly with sub-micron-meter precision. In making such alignments, there are a number of factors that must be considered which can reduce transmission efficiency, such as shifts in the light axis, bending, and when core surfaces are not flush. To deal with these factors, alignment algorithms capable of greater DOF connections are required.

Automatic alignment system. Automatic alignment systems can shift slightly the light axes of the two optical fibers to minimize transmission loss. Once alignment is complete, the light axis is fixed by laser processing and a setting resin. Figure 1 shows the organization of the automatic alignment system. The system consists of a light source, alignment stages, a stage controller, a power meter to measure the light intensity, and a controlling PC. An alignment stage moves the tip of one optical fiber with micron-meter precision using step motors. The PC collects information from the power meter and feeds this back to the stage controller through a GP-IB interface in order to control the alignment stage. The control signals are generated by the PC where the automatic-alignment algorithm is executed.

Conventional automatic alignment algorithms. Alignment time depends completely on the performance of the automatic alignment system. Various algorithms are available on the market. For example, the alignment software sold by Suruga Seiki provides two search programs; a field search and a peak search program. The field search program is used when the two light axes are shifted greatly. Once this program has been executed, the peak search program is initiated. However, these programs have the following weakness; (1) local optimum,

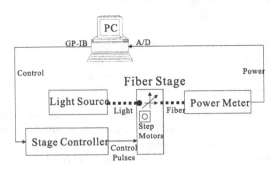

Fig. 1. An Automatic Fiber Alignment System

(2) problems with angle adjustments, and (3) noise-sensitivity. Due to these weaknesses, these software programs are not capable of alignments according to 5 DOF, have long alignment times, and result in large transmission losses.

Automatic alignment with GA. Our alignment algorithm uses a GA called the Minimal Generation Gap (MGG) model [1]. The MGG model has two advantages; (1) it can maintain the diversity within a population, and (2) the convergence speed is fast. We also employ a local learning method in order to improve search efficiency.

Alignment is conducted in the following way. First, the positions of the light axes are coded as chromosomes, and then an initial population is generated. GA operations, such as selection, crossover, and mutations, are then carried out on the population and local learning is conducted. The fitness function is light intensity. With the GA-based alignment algorithm, the optimal positions for the light axes are obtained to minimize transmission loss.

2.3 GA-Based Automatic Alignment Experiments

Figure 2 shows an experimental system for automatic alignment. It consists of a light source, a power meter, a control PC, and the stage controller. Four experiments were conducted in which we prepared 4 different initial conditions by altering the light axes from the positions set by an experienced engineer. In each experiment, alignment was successfully completed in three minutes. The details of the experiments are described below.

Experimental conditions. The objective of the experiments was to align the tips of two optical fibers according to 5 DOF. The 5 DOF in fiber alignment correspond to the following 5 parameters; the coordinates in the X, Y, and Z planes, as well as the two angles of inclination in the X-Z and Y-Z planes for optical fiber which is moved, as shown in Figure 3. In the first three experiments,

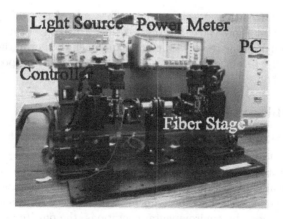

Fig. 2. The Fiber Alignment System Used for Development

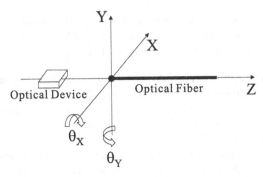

Fig. 3. Optical Axis

Table 1. Alignment Results for the Experiment 1

Power(dBm)	Avg.	Max.	Min.	Std.
Conventional	2.58	3.94	1.23	0.73
GA without LL	3.26	4.31	2.44	0.66
GA	4.41	4.73	3.85	0.27
HC	3.83	5.00	1.52	1.08

three of these parameters were altered, while two were changed in the fourth experiment.

In these experiments, the performance of our proposed method was compared to three other alignment algorithms. The first was a conventional alignment algorithm that is only capable of alignments according to 3 DOF. The second was a genetic algorithm that does not employ local learning, which is incorporated within our proposed method. The third was our GA method with local learning. The forth method was a hill-climbing alignment algorithm. There were 10 trials in each experiment.

Experimental Results. We show only the result of the first experiment in Table 1. Figure 4 shows how the optical fiber tips converged in the experiment. Although only the result of the first experiment is shown, the best average performances in all four experiments were attained by the proposed automatic alignment method.

3   Evolvable Interferometer

3.1   Background

Advances in spectrum technology are making important contributions to the analysis of substances that cause environmental pollution. There is an increasing demand to be able to carry out spectrum analysis rapidly *on-site*. However, because the instruments used for spectrum analysis are very large, and performance is greatly influenced by environmental conditions, the on-site use of spectrum-analysis instruments has been virtually impossible. If portable spectrum-analysis

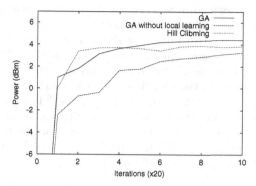

Fig. 4. Power versus Iterations for Experiment 1

instruments were developed, they would greatly accelerate the development of surveillance and analysis technology for environmental pollutants.

The challenge in developing portable instruments is to realize the automatic adjustment of the optical components of spectrum analysis instruments. Most analysis instruments consist of optical components that must be physically positioned with micron-meter precision for optimal performance. If such instruments are used on-site outdoors, then the optical components have to be re-adjusted every time an instrument is moved, but this kind of adjustment usually requires experienced technicians. Therefore, automatic adjustment methods for optical components are essential in order to realize portable spectrum-analysis instruments. We have developed an evolvable interferometer, which is used in Fourier-transform Infrared Spectroscopy (FTIR). FTIR performance is heavily dependent on the performance of the internal interferometer. The evolvable interferometer is a version of a Michelson interferometer, involving mirrors, that requires automatic adjustment in order to realize portable FTIRs.

The advantages of the evolvable interferometer are as follows:

1. Portable FTIR
   When the FTIR is used outdoors, variations from the optimal positions for the optical components can be adjusted on the spot.
2. Compact FTIR
   The automatic adjustment of the interferometer does not require the large spaces needed by engineers when manually adjusting interferometers. This makes it possible to downsizing the FTIR considerably.
3. Greater versatility for environmental measurements
   With an automatic adjustment method, it is possible to employ optical components that are more suitable to the environment. For example, users can exchange prisms according to the target.

After outlining the Michelson interferometer, the evolvable interferometer is described in detail.

The Michelson Interferometer. The Michelson interferometer consists of a light source(S), a beam splitter (BS), a fixed mirror (M1), and a movable mirror(M2), and a testing chamber (P), as shown in Figure 5. The BS splits the light from S into two beams sending them towards M1 and M2. These beams are reflected back by these mirrors towards BS that then synthsizes them to send out a single beam in the direction of P. The synthesized beam, called an inter-ferogram, is used to analyze substances by the FTIR. In order to obtain an optimal inter-ferogram, the movable mirror, M2, has to be adjusted to the right position with micron-meter precision. As interferometer performance is influenced by the temperature and transportation, it is very difficult to implement portable FTIR without automatic adjustment.

Evolvable Interferometer. We have realized an automatic adjustment system for interferometers. Figure 6 shows the overall organization of the system. The system automatically adjusts the position and the inclination of the movable mirror in the interferometer. This adjustment can be conducted according to 3 DOF. The 3 DOF here are the position of the mirror, its angle of inclination, and its angle of rotation. In Figure 6, the PC reads the light intensity from a detection board and the genetic algorithm on the PC determines the three mirror parameters. The genetic algorithm employs the MGG model.

4   Evolvable Femtosecond Laser system

4.1   Background

Laser systems, invented in the 1960s [2] [3], have been applied to various fields, such as semiconductor lithography [4], optical communications [5], and sampling

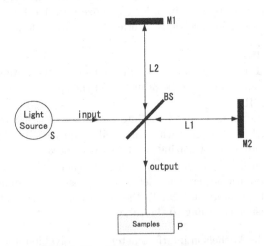

Fig. 5. The Michelson Interferometer

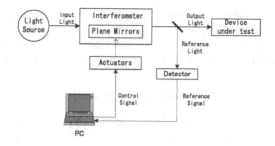

Fig. 6. An Automatic Adjustment System for the InterFerometer

measurement methods [6]. During the last decade, the development of femtosecond lasers, based on Kerr-lens mode-locking techniques, has progressed rapidly and many commercial products are now available.

However, as the Kerr-lens mode-locking technique used in femtosecond pulse lasers requires precise positioning of the focusing mirrors within the laser cavity, femtosecond lasers are difficult to align, which makes them less attractive for industrial use. For example, even a 10μm discrepancy in a focusing mirror will prevent an optical resonation with 1.0W pumping power from forming. It, therefore, typically takes about a week to manually align a femtosecond laser.

In order to overcome these difficulties, we propose an Evolable Laser System (ELS) for femtosecond lasers, which can adjust the positioning of the laser cavity components (e.g. the mirrors and prisms) by genetic algorithms (GAs). This laser system has three advantages:

1. Higher Power
   When the ELS starts up or the pumping power is changed, the GA can be executed to optimize the laser power.
2. Automatic and Reliable Adjustment
   As the GA is capable of adjusting the system automatically and reliably, manual adjustment is not required with the ELS. Moreover, because the need for manual adjustment is eliminated, the risk of human eyes being damaged while making such alignments is also effectively removed.
3. Compactness
   Conventional laser systems requiring manual adjustment tend to be rather large, as there must be sufficient space to allow human hands to make the physical adjustments. However, by virtue of the incorporated automatic adjustment mechanisms, the ELS can be made much smaller, even making the development of portable femtosecond laser systems feasible in the future.

As this approach can be applied to a wide variety of laser systems, femtosecond laser technology can become a more widely-used industrial technology.

The ELS developed in this study has 10 picomotors that determine the positioning of the mirrors genetically. Each picomotor can adjust the x-position, yaw and pitch of a mirror with a resolution of μm. Manual adjustment with picomotors is very difficult because the light has to travel many times within a laser cavity before returning to the focus point with μm resolution. In an initial experiment, an output power was obtained that is 2.3 times higher than the power achieved with a manual alignment made in our laboratory.

4.2 Overview of the Evolvable Laser System

This section provides an overview of the proposed system. Figure 7 is a schematic representation of the evolvable laser system, and Figure 8 is a photograph of the developed system. YVO$_4$ green laser is used in the system as the pumping beam (input to the laser system), which enters at the point marked IN in Figure 7. Four focusing mirrors, M3–M6, which form an optical resonator, output the laser light from the point marked OUT. The positioning of these four mirrors can be adjusted subtly to correct for positioning errors, which can greatly reduce output performance. This adjustment involves ten picomotors in total (the picomotor is a piezo that turns a screw [7]).

The screw positions of the picomotors are determined by the motor controller in Figure 7. Their absolute positions can be set by downloading a binary string to the controller. We call this string the control bits. In the genetic learning, the control bits are regarded as GA chromosomes. GAs are robust search algorithms which use multiple chromosomes and apply natural selection-like operations in seeking improved solutions [8]. The GA identifies the optimal control bits for the evolvable laser system. The goal of this evolution is to obtain the maximum power level for the laser pulse from the laser cavity.

Every chromosome is downloaded into the controller. A fitness value is calculated by observing the laser outputs. The positioning of the mirrors is varied to improve the performance of the laser system.

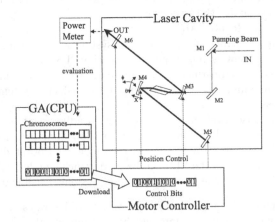

Fig. 7. Schematic Representation of the Evolvable Laser System

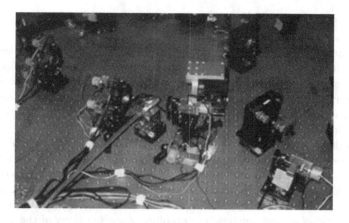

Fig. 8. Photograph of the Developed Evolvable Laser System

4.3   Configurations of the Evolvable Laser System

This section describes the evolvable laser system in detail.

The layout of the cavity is shown in Figure 9. The cavity is a standard Ti:sapphire laser [9]. The main cavity consists of four mirrors, M3 to M6, and the 20-mm-long Brewster-angled Ti:sapphire gain medium placed at the center of the cavity. This z-cavity configuration is essential for femtosecond laser systems.

The pumping laser is a diode-pumped frequency-doubled YVO$_4$ laser. The wavelength is 530 nm and noise is less than 0.5% of the output power. The pumping beam is focused into the Ti:sapphire gain medium by the plane mirrors M1, M2, M5, M6, and the concave mirrors M3 and M4. The output power of

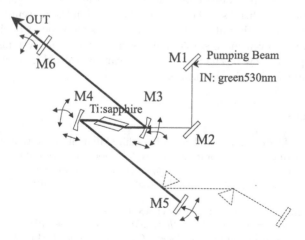

Fig. 9. Cavity Configuration for the Ti:sapphire Laser

the laser from mirror M6 is measured by a power meter. A plane coupler before the power meter has a transmission of approximately 2.0%.

The output power is greatly affected by discrepancies between the actual and ideal positioning of the mirrors M3–M6. For example, even a 10μm discrepancy in the focusing mirror M4 can prevent an optical resonation with 1.0W pumping power from forming. In order to correct for such discrepancies, the mirrors in conventional femtosecond laser systems can be aligned manually by subtly moving adjustment screws.

In order to overcome the problems of positioning and non-linear lens effects, we propose the evolvable laser system, which has ten picomotors for automatic adjustment, which is determined by genetic learning. The yaws and pitches of mirrors M5 and M6 can be adjusted with two picomotors, and the x-positions, yaws and pitches of mirrors M3 and M4 can be adjusted with three picomotors. With the aid of potentiometers, the absolute screw positions of the picomotors can be specified by downloading a bit string to the controller. By treating the bit strings as GA chromosomes, the proposed system can adjust the positioning of the mirrors. In tests, automatic adjustment was successfully completed within 30 minutes. Currently, we are also developing a very compact ELS for industrial usage.

5 Conclusion

This paper has described three evolvable optical systems;(1) an evolavable fiber alignment system, (2) an evolvable interferometer system, and (3) an evolvable femtosecond laser system. The adavantage common to all these systems is the inclusion of GA in order to automatically execute the micron-meter precision alignment of optical components, such as mirrors and prisms, very quickly and very efficiently. This opens up the possibility of developing new optical instruments that can be used on-site outside for the analysis of environmental pollutants. Although the fitness function for the GA used in the present three systems was light intensity, other fitness functions are possible which would make automatic alignment even more stable and faster. The present systems are simply early examples of how mechanical evolution can be utilized in a wide variety of industrial applications in the future.

References

1. H. Satoh, M. Yamamura, and S. Kobayashi. Minimal generation gap model for gas considering both exploration and exploitation. In *Proceedings of the Fourth International Conference on Soft Computing (IIZUKA 96)*, pages 494–497, 1996.
2. T. H. Maiman. Stimulated optical radiation in ruby. *Nature*, 187(493):493–494, 1960.
3. A. E. Siegman. *Lasers*. University Science Books, 1986.
4. W. M. Moreau. *Semiconductor Lithography*. Plenum Press, 1988.
5. A. Yariv. *Optical Electronics in Modern Communications*. Oxford University Press, 1996.

6. J. A. Valdmanis. *Semiconductor and Semimetals*, chapter 4. Academic Press, 1990.

7. http://www.newfocus.com/. U.s. patent no. 5,410,206.

8. J. H. Holland. *Adaptation in Natural and Artificial Systems*. The University of Michigan Press, 1975.

9. D. E. Spence, P. N. Kean, and W. Sibbett. 60-fsec pulse generation from a self-mode-locked ti:sapphire laser. *Optics Letters*, 16(1):42–44, 1991.

Author Index

Lecture Notes in Computer Science

For information about Vols. 1–2371
please contact your bookseller or Springer-Verlag